# INFRARED THERMOGRAPHY
# RECENT ADVANCES AND FUTURE TRENDS

## Edited By

## Carosena Meola

*Department of Aerospace Engineering*
*University of Naples Federico II*
*Italy*

# eBooks End User License Agreement

# CONTENTS

# FOREWORD

Infrared (IR) radiations, discovered in 1800 by Herschel waited only 30 years to be measured by the first detectors (thermopiles). Fifty years later, in 1880, the first IR photographic image was obtained by William de Wiveleslie Abney. Finally, after eighty-three years, in 1963, the first mono-detector IR camera, produced by AGA, appeared. This event was a real revolution for science, industry and medicine. However, after a period of enthusiasm, some disappointment occurred due to the rather qualitative performances and exploitations of the cameras.

The second revolution occurred in the last decade of the XX[th] century, due to the conjunction of an important effort in the understanding and modelling of the thermal phenomena involved in the observation of the scenes by passive and active thermographic techniques and of the appearance of focal plane array cameras which allowed higher thermal sensitivities and more precise time and space analysis of thermal fields. To conclude this ultra short historical overview of the IR thermography, let me recall that in 2010 an IR camera with a thermal sensitivity of 1mK was presented. It should be clear that such a sensitivity opens up a harvest of new perspectives.

The consequence of these technical progresses is the booming development of thermographic techniques in more and more diversified application fields. This situation requires that seminars, conferences, journals and books allow the practitioners to confront their respective approaches, since such a comparison is the source of mutual enrichment and innovation. The present ebook participates to this multidisciplinary dissemination of information.

In the first part of the ebook, the roots of thermography are described. A chapter details the origin of the technique and its theoretical basis, another one recalls the history and the new trends of the infrared detectors which constitute the "heart" of the IR cameras.

The second part, much more extended, provides a wide overview of thermographic applications in many fields like medicine and veterinary, foodstuff conservation and livestock health, thermo-fluid-dynamics and combustion, non destructive evaluation, civil engineering including preventive maintenance, energy saving, repair follow-up…

These practical applications, based on theoretical approaches that consider both direct and inverse problems, demonstrate that thermal phenomena can be related to many physical phenomena, which explains the rich variety of applications and the success of infrared thermography. Furthermore, a chapter is specially devoted to stimulated – or active – thermography, with its two main variants using pulsed and modulated sources.

Each chapter written by an expert in the related field, gives an updated vision of the techniques, and makes this eBook a useful tool for thermography practitioners active in research laboratories and industries.

*Daniel L. Balageas*
ONERA
The French Aerospace Laboratory
Châtillon, France

# PREFACE

Over two centuries passed since the first stone posed by Herschel in the *foundation* of the infrared radiation world. For many years, infrared thermography IRT had been a subject of dispute and investigation with great enthusiasm and wide scepticism. Today, the usefulness of infrared thermography has been amply demonstrated leading to a proliferation of infrared devices of different sizes and different performances to fulfil the requirements of the multitude of users in the vast variety of applications.

The basic role of an infrared imaging system is to supply information about the temperature map over a body surface, but it is the successive processing of such a map that transforms the infrared system into a multifarious tool. It becomes an hypertermia detector for a doctor, a hot spot indicator for an electrician, a thermal junctions detector for an architect or a civil engineer, a nondestructive evaluation (NDE) technique for a product's quality inspector, and a variety of other instruments depending on who uses it and for what it is used. Indeed, IRT can be advantageously used in many and many fields. Of course, complete exploitation of an infrared device requires the understanding of basic theory and of application standards. This ebook attempts to comply with both requirements. In fact, it includes information about: basic theory, infrared detectors, signal digitalization, the working concepts of an infrared device as well as a panorama of today's infrared cameras, and examples of applications in many fields such as medicine, foodstuff conservation, fluid-dynamics, architecture, non destructive testing and evaluation of materials and structures.

The eBook is organised in two main parts: Part I and Part II and further into several chapters.

The Part I includes two chapters. The first one, by the Editor, deals with basic theory, which is described following the historical steps by eminent scientists, from Herschel, to Nobili, Melloni, Stefan, Boltzmann to Planck and others. The radiation mechanisms with the most important parameters, which play a key role in the acquisition and interpretation of thermal images, are recalled and discussed. A section is devoted to detectors used for infrared technology. The main steps in detectors' development following the technological progress are also drawn. The second chapter is by Roberto Rinaldi of the Infrared Training Centre (ITC) by Flir Systems in Milan (Italy). This chapter is concerned with an overview of infrared imaging devices from the first prototype developed in 1958 to the multitude of models, which are available today. The historical evolution of the infrared technology is traced within the key features of each model. In particular, some basic characteristics and performance are described, which may help the reader in the choice of the most appropriate device for the specific application.

Part II is subdivided into four sections and many chapters, which are numbered following part I.

The first section includes applications to medicine (Chapter 3) and veterinary (Chapter 4). The study of the temperature of the human body has been associated with health as far back as the 1st century BC, when Hypocrites (the father of medicine) used *the sense of touch* for skin surface temperature anomalies to determine the *health* of his patients. Even today, monitoring the body temperature variation aides in both diagnosis and treatment planning. Chapter three was prepared by Boris G. Vainer of the Institute of Semiconductor Physics of the Russian Academy of Sciences. This chapter reports on the IRT's state of the art in medicine with methodological approaches and a variety of applications such as in the diagnosis of breast cancer, in ophthalmologic surgery, in cardiovascular surgery, in the visualization of ischemic tissues and in many others. Chapter four presents the application and use of infrared thermography in farm animals and veterinary medicine. This chapter was supplied by Petr Kunc and Ivana Knizkova of the Institute of Animal Science-University of Prague (The Czech Republic). The addressed areas include reproduction, thermoregulation, animal welfare and the milking process. The application of IRT to veterinary medicine is particularly useful to predict inflammation since, contrary to human beings, animals cannot reveal any symptom before the illness has become important.

Section two includes a chapter (5) on the use of Infrared thermography in foodstuff conservation by Klaus Gottschalk of the Leibniz-Institut für Agrartechnik Potsdam (Germany). It shows the usefulness of IRT to control the conservation conditions of fruits and vegetables. The main advantage of using an infrared device lies in the possibility to control and improve the climate, which is essential in prolonging the shelf life of crops.

Section three explains the applications of IRT to industrial engineering. The chapter six, prepared by Giovanni M. Carlomagno of the Department of Aerospace Engineering-University of Naples Federico II (Italy), is an overview on IRT to thermo-fluid-dynamics. After recalling the first historical attempts in measuring heat transfer coefficients, this chapter describes the most useful heat flux sensors, supplies information about thermal restoration of data and shows several examples of convective heat transfer measurements in complex fluid flows, ranging from natural convection to hypersonic regime. The attention of chapter seven is focused on the application of IRT to combustion. This contribution is by Christophe Allouis and Rocco Pagliara of the Combustion Institute CNR in Naples (Italy). It demonstrates the usefulness of an infrared imaging system for understanding the fluid-dynamics phenomena associated with the combustion processes in turbine burners. The chapter eight by Ralph A. Rotolante of Vicon Infrared in Boxborough, MA (USA) explains the use of IRT for nondestructive inspection purposes. The main pulse and lockin techniques are described with some application examples, including the inspection of real aircraft parts. Indeed, a remote imaging system offers many advantages over other methodologies since it is fast and two-dimensional and guarantees the safety of the part integrity.

Section four is concerned with the application of IRT in architecture and civil engineering. This is a relevant topic for infrared thermography applications after Building Regulation (2007) for Conservation of Fuel and Energy. Chapter nine by Ermanno Grinzato of CNR-ITC in Padua (Italy) reports some examples of structural analysis aided by IR thermography. In particular, it stresses the impressive help, which is given to the comfort monitoring by the distributed temperature map measured by an infrared device. The attention also goes to the possibility, using a novel method, to "see" the main environmental quantities, such as air temperature, relative humidity and velocity, obtained from thermographic readings.

Besides those herein described, an infrared imaging system can be advantageously used for many other applications. Infrared thermography is an excellent condition monitoring tool to assist in the reduction of maintenance costs on mechanical equipment. One of the biggest problems in mechanical systems is the heat generated by friction, cooling degradation, material loss or blockages. The infrared technique allows for the monitoring of temperatures and thermal patterns, on a wide variety of equipments including pumps, motors, bearings, pulleys, fans, drives, conveyors *etc.*, and also while the equipment is online and running under full load. Information acquired from thermographic images enable a company to predict equipment failure and to plan corrective actions before a costly shutdown, equipment damage, or personal injury occurs. What is more, the inspection can also be performed far away from any dangerous condition without additional costs in terms of workers' health care.

However, it has to be pointed out that infrared thermography is still not completely exploited. It could be employed in a lot of other novel applications; it is only a matter of fantasy and skill!

*Carosena Meola*

Department of Aerospace Engineering
University of Naples Federico II
Italy

# List of Contributors

**Allouis Christophe**
Istituto di Ricerche sulla combustione – CNR P.le V. Tecchio 80 – 80125 Napoli, Italy

**Carlomagno Giovanni M.**
Department of Aerospace Engineering, University of Naples Federico II, Piazzale Tecchio, 80 80125 Napoli, Italy

**Gottschalk Klaus**
Leibniz-Institut für Agrartechnik Potsdam-Bornim e.V. (ATB) Max-Eyth-Allee 100, D-14469 Potsdam, Germany

**Grinzato Ermanno**
Building Technology Institute (ITC)-National Research Council (CNR), Corso Stati Uniti, 4, 35127 Padova, Italy

**Knizkova Ivana**
Institute of Animal Science, Praha Uhrineves, The Czech Republic

**Kunc Petr**
Institute of Animal Science, Praha Uhrineves, The Czech Republic

**Meola Carosena**
Department of Aerospace Engineering, University of Naples Federico II, *Via* Claudio, 21 80125 Napoli, Italy

**Pagliara Rocco**
Istituto di Ricerche sulla combustione – CNR P.le V. Tecchio 80 – 80125 Napoli, Italy

**Rinaldi Roberto**
Infrared Training Center, FLIR Systems, *Via* L. Manara 2, 20051 Limbiate, Milano, Italy

**Rotolante Ralph A.**
Vicon Infrared 98 Baldwin Dr. C-8 Boxborough, MA 01719 USA

**Vainer Boris G.**
A.V.Razhanov Institute of Semiconductor Physics, Russian Academy of Sciences, Siberian Branch, Novosibirsk, Russia.

2

# CHAPTER 1

# Origin and Theory of Infrared Thermography

## Carosena Meola[*]

*Department of Aerospace Engineering, University of Naples Federico II, Via Claudio,21, 80125 Napoli, Italy*

**Abstract:** The intention of this chapter is to trace the origin of infrared thermography (IRT) and to supply its theoretical basis. The speech goes through the milestones by eminent scientists and is supported by the most important basic relationships. The reader is also plunged into radiation mechanisms which are fundamental as starting points for comprehension and then application of infrared thermography. A section is devoted to detector' technology with a description of the basic features, which should be taken into account for the choice of the most appropriate equipment for the specific application. The main historical steps in detector' development are also traced owing to the dynamic evolution of the infrared devices. Indeed, following the technological progress, the performance of infrared devices improves ever more, leading to novel applications.

**Keywords:** Infrared thermography, origin of infrared thermography, basics of infrared radiation, radiation mechanisms, technology of infrared detectors, infrared detectors performance, behavior of materials in the infrared spectrum, materials emittance, surface emissivity.

## 1. INTRODUCTION

Roughly speaking, an infrared camera can be assimilated to a video camera we use to take pictures since the result is a detailed image of the scene, which looks like that viewed by naked eyes. The substantial difference between visible and infrared (IR) images lies in the type of information supplied.

A visible image (Fig. **1a**) is made of different tonalities depending on how the incident light is reflected: the object is a reflector and is visible only in presence of light. Instead, the different tonalities in the infrared image (Fig. **1b**) are a measure of the energy emitted by the viewed object: the object is a source and it appears also in the dark in absence of light.

<table>
<tr><td align="center">a) Visible picture.</td><td align="center">b)Thermal image.</td></tr>
</table>

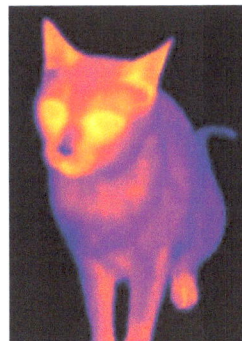

**Figure 1:** Comparison of images: a) Visible; b) Thermal.

The electromagnetic radiation emitted by an object is related to its surface temperature. The temperature is a measure of the object internal energy (the energy of motion of atoms and molecules the object is made of). Every object at a temperature $T$ above absolute zero (*i.e.*, $T > 0K$) emits electromagnetic radiation in the form of rays which fall into the infrared portion of the electromagnetic spectrum.

**\*Address correspondence to Carosena Meola:** Department of Aerospace Engineering, University of Naples Federico II, *Via Claudio*, 21, 80125 Napoli, Italy; E-mail: carmeola@unina.it

The electromagnetic spectrum, as shown in Fig. **2**, is roughly divided into a number of wavelength regions called bands. The infrared spectral band, which comes from 0.75 to 1000 microns, is located between the visible (after the red) and radio waves. The region that extends from 0.1 to 1000 microns is called heat region and includes all the infrared, all the visible and a portion of the ultraviolet. For this reason the IR radiation is also called thermal radiation.

## 2. HISTORICAL STEPS

The origin of infrared radiation dates back to the early 1800s when the German-born British astronomer Friedrich Wilhelm Herschel (better known as Sir William Herschel) discovered thermal radiation beyond the deep red in the visible spectrum. At first, there was widespread scepticism about such invisible light because of contradictory ideas about the nature of luminous, thermal and chemical radiations [1]. Of fundamental importance was the discovery of the thermopile in 1829 by the Italian Leopoldo Nobili and thanks to another Italian Macedonio Melloni that the analogy between light and heat was definitely ascertained in 1840.

Later in 1860, the German Gustav Kirchhoff introduced the concept of blackbody and provided standards for comparison of radiation sources. In 1879, the Slovak Joseph Stefan experimentally found that a blackbody emitted an amount of energy proportional to the fourth power of its absolute temperature [2]. Such conclusion was also reached *via* theoretical thermodynamic analysis by the Austrian Ludwig Boltzmann in 1884 [3]; this was named the Stefan-Boltzmann law.

In the meantime (1865) the Scottish James Clerk Maxwell predicted the theoretical existence of electromagnetic waves and proposed their similarity with light waves. In 1888 the German Heinrich Rudolf Hertz confirmed the Maxwell's hypothesis [4] and found that electromagnetic waves, like light waves, were reflected and refracted and, which is most important, that they travelled at the same speed of light but had a much longer wavelength.

At the end of the nineteenth century, it became easy to recognise the different kinds of radiations of which the electromagnetic spectrum was composed and many attempts were made to derive the basic laws of infrared radiation. The first law was proposed in the early 20th century by two British scientists, John William Strutt, 3rd Baron Rayleigh better known as Lord Rayleigh and Sir James Hopwood Jeans, and is known as the Rayleigh-Jeans Law:

$$E_{\lambda,b} = 8\pi k_b \frac{T}{\lambda^4}$$

(1)

where: $E_{\lambda b}$ is the blackbody monochromatic radiation intensity, $\lambda$ is the wavelength of the radiation being considered, $k_b$ is the Boltzmann's constant ($k_b = 1.38 \times 10^{-23}$ J/K) and $T$ is the absolute temperature of the blackbody. This law was derived from classical physics arguments. It agreed with experimental measurements for long wavelengths, but completely disagreed with experiments at short wavelengths where it diverged and predicted an unphysical infinite energy density. This failure is known as the *ultraviolet catastrophe*, which was one of the first clear indications of problems with classical physics. The solution to this problem led to the development of an early form of quantum mechanics.

It was the German Max Planck who postulated that electromagnetic energy did not follow the classical description, but could only oscillate or be emitted in discrete packets of energy proportional to the frequency. Then, he derived in 1900 the law of radiation, which precisely describes the spectral distribution of the radiation from a blackbody:

$$E_b = \frac{2\pi h c^2}{\lambda^5 \left( e^{hc/\lambda k_b T} - 1 \right)}$$

(2)

where: $h$ is the Planck's constant ($h = 6.6 \times 10^{-34}$ Js) $c$ is the speed of light ($c \cong 3 \times 10^8$ m/s). The Planck's formula is plotted in Fig. **3** for several absolute temperature values in the range 200-6000K. A family of curves is obtained; for each curve, $E_{\lambda b}$ goes to zero for $\lambda = 0$ then increases rapidly to a maximum and decreases towards zero again at very long wavelength values. As can be seen, the higher the temperature, the shorter the wavelength at which the maximum occurs.

In 1905 the great German Albert Einstein exploited the Planck's idea to show that an electromagnetic wave such as light could be described by a particle called the photon with a discrete energy dependent on its frequency. This led to the quantum description of electromagnetism [5-8].

By differentiating the Planck's law (Equation 2) with respect to $\lambda$ and by finding the maximum radiation intensity, the Wien's (Wilhelm Wien) displacement law is obtained:

$$\lambda_{max} = \frac{d_w}{T} \tag{3}$$

$d_w$ is called Wien's displacement constant and is approximately equal to 2898. Then, Equation 3 mathematically states that at ambient temperature (about 300K) the radiation peak lies in the far infrared at about 10µm. The sun radiation (about 6000K) peaks at about 0.5µm in the visible light spectrum (Fig. 3). Conversely, at the temperature of liquid nitrogen (77K) the maximum radiation intensity occurs at 38µm in the extreme infrared wavelength. Note that, in Equation 3, $\lambda_{max}$ and $T$ must be, respectively, given in micrometers and degrees kelvin. By integrating the Planck's law over the entire spectrum ($\lambda = 0 - \infty$), the total hemispherical radiation intensity (Stefan-Boltzmann's law) is obtained:

$$E_b = \sigma T^4 \tag{4}$$

with $\sigma$ the Stefan-Boltzmann's constant ($\sigma = 5.67 \times 10^{-8}$ W/m² K⁴).

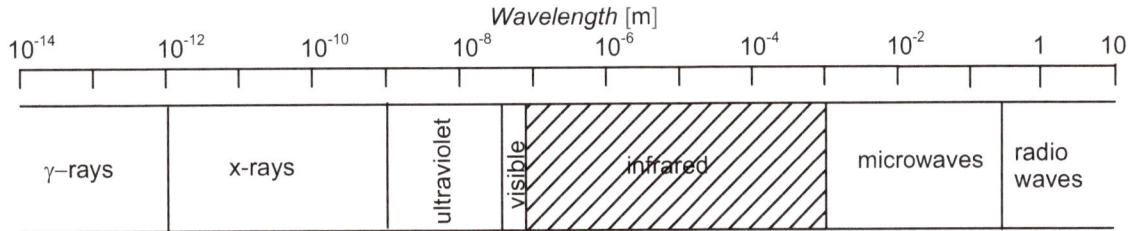

**Figure 2:** Electromagnetic spectrum.

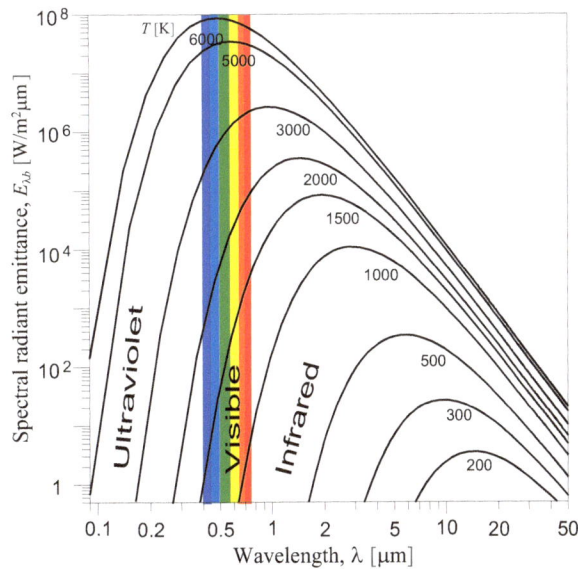

**Figure 3:** Distribution of emissive power with wavelength.

## 3. RADIATION MECHANISMS

Thermal radiation plays an important role in thermodynamics because it is through thermal radiation that two bodies, non in contact, can exchange heat [9-12]. More specifically, a body radiates an amount of energy due to the vibration of the electrons of the substance of which it is made of. Only some relationships which are required for the comprehension of infrared radiation mechanisms are recalled here; for more detailed discussion the reader is readdressed to specific literature on electromagnetisms and optics [13-15].

As anticipated in the previous section, the energy is emitted not as a continuous flow but in the form of discrete quanta, or photons. Each photon has energy $E_{ph}$ equal to its frequency $f_{ph}$ multiplied the Planck's constant $h$ and since, according to Einstein, photons move at the speed of light $c$ the following relationship can be written:

$$E_{ph} = hf_{ph} = h\frac{c}{\lambda} \qquad (5)$$

which states that the energy is inversely proportional to the wavelength, *i.e.,* the higher the energy, the shorter the wavelength. The magnitude of emitted radiation depends on the temperature and on the characteristics of the emitting surface.

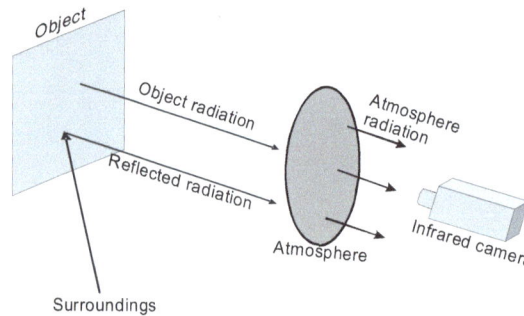

**Figure 4:** Influence of surroundings on the detected radiation.

In a real situation, the electromagnetic energy detected by the infrared camera $E_d$ includes the energy emitted by the object $E_{obj}$ and the reflected energy coming from the surroundings $E_{ref}$ and the energy emitted by the atmosphere $E_{atm}$

$$E_d = AE_{obj} + BE_{ref} + CE_{atm} \qquad (6)$$

A, B and C are constants that account for the characteristics of the object surface and of the atmosphere between the object and the camera and will be explicated later. A schematic representation is depicted in Fig. **4**. Therefore, apart from the infrared detector (or detectors) performance [16-18], which will be discussed in a successive chapter, the precision of temperature measurements depends on three main factors that are: the characteristics of the object surface, the interposed medium (atmosphere, or other) and the surroundings.

### 3.1. Surface Characteristics

Depending on the characteristics of the surface and of the bulk material beneath the surface, the energy striking an object can be absorbed, reflected, or transmitted. Therefore, applying the energy balance to a surface element, the following relationship is obtained:

$$\alpha + \rho + \tau = 1 \qquad (7)$$

which connects the total absorptance $\alpha$ to the total reflectance $\rho$ and to the total transmittance $\tau$. Equation 7 still applies for an elementary spectral interval:

$$\alpha_\lambda + \rho_\lambda + \tau_\lambda = 1 \tag{8}$$

with $\alpha_\lambda$ the spectral absorptance, $\rho_\lambda$ the spectral reflectance and $\tau_\lambda$ the spectral transmittance.

Of course, this is true to the extent that some phenomena (*e.g.*, Raman scattering, *etc.*) can be neglected. In a real object, the incident radiation $I$ is in part reflected $I_{refl}$, in part absorbed $I_{abs}$, in part transmitted $I_{trans}$ and in part scattered $I_{scat}$ as sketched in Fig. **5**.

A surface may exhibit selective behaviour not only with respect to wavelength but also with respect to the direction of the incident energy; more specifically, the energy, absorbed by a surface, comes not from the whole hemisphere but only from a certain direction. Assuming a direction $\varphi$, the Kirchoff's law states that, at a local thermodynamic equilibrium, the release of energy from a surface occurs through the same direction a surface is able to absorb energy:

$$\alpha_{\lambda,\varphi} = \varepsilon_{\lambda,\varphi} \tag{9}$$

The symbol $\varepsilon$ indicates the emissivity that will be discussed in a section to follow. This also means that the object remains at constant temperature with the rate at which it absorbs energy equal to the rate at which it emits energy; otherwise the object would warm, or cool.

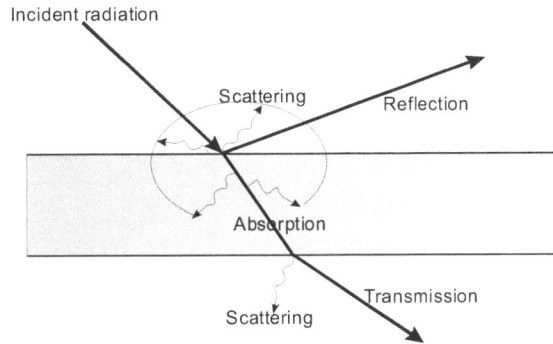

**Figure 5:** Sketch of surface behaviour with respect to the incident radiation.

### 3.1.1. Absorption

Absorption occurs in two fundamental processes, which are the electron and lattice absorption, with reduction of the number of photons in the forward direction. In particular, the intensity $I$ of the electromagnetic wave travelling through a homogeneous medium of thickness $x$ is attenuated due to absorption; the decay is described by the empirical relationship:

$$I(x) = I_0 e^{-\alpha x} \tag{10}$$

$I_0$ is the intensity of the incident beam, $\alpha$ is the absorption coefficient (or absorbance) that is equal to:

$$\alpha = \frac{4\pi\kappa}{\lambda} \tag{11}$$

with $\kappa$ the extinction index. Equation 10 may be found in literature as: Beer's law, or Beer-Lambert law, or Lambert-Beer law, or Beer-Lambert-Bouguer law because it was independently discovered (in various forms) by the French Pierre Bouger in 1729, by the German Johann Heinrich Lambert in 1760 and by the German August Beer in 1852.

***Electronic absorption-***characterises good electric and thermal conductors and is mainly observed in the high frequency region of the infrared spectrum. The absorption occurs through the interaction between incident radiation and the motion of free electrons (or holes) within the material. Only electromagnetic radiation with energy (Equation 5) sufficient to promote the electron movement between valence and conduction bands is absorbed by this mechanism. The transitions of electrons supply information about the short wavelength absorption edge.

In conductors the presence of a cloud of free electrons allows for a continuous absorption whose magnitude increases with the square of the incident wavelength over the whole infrared region. In semiconductors the absorption depends on the size of the energy band gap ($E_g$) at a given temperature and on the material homogeneity grade. In particular, three behaviours may be observed:

1.    $E_{ph} = E_g$ the excited electron just moves to the conduction band;

2.    $E_{ph} > E_g$ an electron-hole pair is generated and the excess energy $E_{ph} - E_g$ is dissipated as heat;

3.    $E_{ph} < E_g$ an electron-hole pair is generated only if there are available energy states due to impurity, or defects.

Generally, optical materials that are opaque in the visible region, because of small bandgaps energy ($E_g \leq 1.25\text{eV}$), are arbitrarily classified as infrared semiconductors whilst materials of larger bandgaps energy are named insulators.

***Lattice absorption***-is the absorption mechanism that occurs in insulators and is observed in the lower frequency regions, in the middle to far-infrared wavelength range. It is caused by a coupling effect between the motions of thermally induced vibrations of the constituent atoms of the substrate crystal lattice and the incident radiation. The absorption happens not in a continuous way but some regions of transparency are present. More specifically, at certain frequencies the incident radiation is allowed to propagate through the crystal lattice (transparency); whilst, at other frequencies, when the incident radiation is at resonance with any of the properties of the lattice material, propagation is forbidden and radiation causes vibrational motion of atoms with production of thermal energy (creation of phonons).

Hence, it is important to note that real materials are bounded by limiting regions of absorption caused by atomic vibrations in the far-infrared (>10μm), and motions of electrons and/or holes in the short-wave visible regions. In the interband there is insufficient energy to promote transition of electrons from the valence to the conduction band; in such a region the material is loss-free (transparent region). Depending on the presence of impurities, or defects, in the material, the transparent region may be affected by absorption from charge carriers.

***Dielectric dispersion***-the presence of spurious particles (imperfections) in the material causes scattering of the electromagnetic wave. For particles much smaller than the wavelength of the light (Rayleigh scattering) an attenuation effect similar to that induced by absorption (Equation 10) is observed, which can be described by the relationship:

$$I(x) = I_0 e^{-N\sigma_s z} \tag{12}$$

with $N$ the number of scattering centres and $\sigma_s = \frac{2\pi^5 d^6}{3\pi^4} \left(\frac{n^2-1}{n^2+2}\right)^2$ the scattering cross-section where $d$ is the diameter of the particle and $n$ is the refractive index. The refractive index (often called index of refraction) is a factor scale for the phase velocity $\upsilon_e$ of electromagnetic radiation in a material with respect to the speed of light in vacuum. In vacuum:

$$c = \frac{1}{\sqrt{\mu_0 \varepsilon_0}} \tag{13}$$

with $\mu_0$ magnetic permeability and $\varepsilon_0$ electric permittivity. In a material medium:

$$\upsilon_e = \frac{1}{\sqrt{\mu_e \varepsilon_e}} \tag{14}$$

and thus:

$$n = \frac{c}{n} = \sqrt{\frac{\mu_e \varepsilon_e}{\mu_0 \varepsilon_0}} = \sqrt{\mu_r \varepsilon_r} \tag{15}$$

For a non-magnetic material $\mu_r$ is very close to 1 and

$$n = \sqrt{\varepsilon_r} \tag{16}$$

$\varepsilon_r$ is also called dielectric constant. The refractive index (Equation 15) and the absorption coefficient (Equation 11) are important quantities which describe the dispersive and absorptive nature of a material. These two aspects are related by the complex refractive index $\tilde{n}$:

$$\tilde{n} = n + i\kappa \tag{17}$$

the imaginary part $\kappa$, which is the extinction index, represents the absorption loss when the electromagnetic wave propagates through the material ($\kappa$ is related to the absorption $\alpha$ in Equation 11. Both $n$ and $\kappa$ depend on the frequency (wavelength). The variation of $n$ with frequency (except in vacuum) is known as dispersion and causes many phenomena in optics. The real part of the refractive index tends to increase with frequency in non-absorbing regions. While, nearby absorption peaks, the curve of the refractive index takes a complex form given by the relation:

$$n(\omega) = 1 + \frac{2}{\pi} C \int_0^\infty \frac{\omega' \kappa(\omega')}{\omega'^2 - \omega^2} d\omega' \tag{18}$$

where $C$ is constant for the material (Cauchy principal value of the integral) and $\omega$ the angular frequency ($\omega = 2\pi/\lambda$), ($\omega'$) indicates all the frequencies for which there is no absorption. This relationship was introduced by the German-American Ralph Kronig and by the Dutch Hendrik Anthony Kramers in 1926 and is named the Kramers-Kronig dispersion relation.

***Materials classification***-the materials behaviour with respect to the propagation of electromagnetic waves depends on the electric permittivity $\varepsilon_e$ (already discussed) and on the electric conductivity $\sigma_e$. Dielectric materials (insulators) are characterised by $\sigma_e/\omega\varepsilon_e \ll 1$; at low frequencies they allow propagation of electromagnetic waves and are also called low-loss materials since absorption is practically null ($\tau \approx 0$). However, at higher frequencies, absorption may increase reducing the material transparency. Good conductors instead are characterised by $\sigma_e/\omega\varepsilon_e \gg 1$; the presence of large amount of loss inhibits the propagation of electromagnetic waves (absorption is good).

Another factor which affects the absorptive materials capability is temperature. As temperature increases, the refractive index declines towards long wavelengths because of the lattice absorption.

At a given temperature the thermal conductivity $k_T$ is proportional to the electrical conductivity; as the temperature increases $k_T$ increases too while $\sigma_e$ decreases. This behaviour, which is expressed by the Wiedemann-Franz Law:

$$k_T = N_L T \sigma_e \tag{19}$$

with $N_L$ the Lorenz number, is based on the fact that heat and electrical transport both involve the free electrons in the metal. A rise in temperature increases the average molecular velocity, which favours the transport of energy and in turn increases the thermal conductivity. On the contrary, the electrical conductivity decreases because particle collisions divert the electrons from carrying the charge.

### 3.1.2. Transmittance

Surfaces may be classified as transparent, or opaque, to infrared depending on their behaviour with respect to the infrared portion of the electromagnetic spectrum. This is of utmost importance in view of using infrared thermography as measurement technique.

However, both transparent and opaque to infrared materials are of interest for IR technology.

**Opaque materials**-for IRT measurements the surface must be opaque to infrared. For opaque materials no energy is transmitted ($\tau_\lambda = 0$) and Equation 8 simplifies to:

$$\alpha_\lambda + \rho_\lambda = 1 \tag{20}$$

**Transparent materials**-materials that are perfectly transparent do not absorb ($\alpha_\lambda = 0$) and Equation 8 reduces to:

$$\tau_\lambda + \rho_\lambda = 1 \tag{21}$$

It has to bear in mind that do not exist surfaces perfectly transparent, or completely opaque; but, some substances are more or less transparent to radiation at certain wavelengths. For example, glass is transparent to wavelengths within the visible range of the spectrum, but it generally absorbs radiation in the infrared region. However, glass displays changeable behaviour with wavelength. Most solids are opaque to thermal radiation, and emission and absorption of radiation are surface phenomena.

Materials, which are transparent in the infrared band, are used for the fabrication of optics (lenses and windows). However, as will be shown later, these materials are not transparent in the entire infrared region, but only in confined bands and are used for optics of systems working in such bands. Infrared windows are used in certain applications (*e.g.*, monitoring of a model inside a wind tunnel) which require an interface between the infrared camera and the object under measurement (this will be discussed later in the applications section).

For this reason, users of infrared thermography should be familiar with the material behaviour in the infrared portion of the electromagnetic spectrum.

The spectral transmittance can be simply calculated through the spectral refractive index $n_\lambda$:

$$\tau_\lambda = \frac{2n_\lambda}{n_\lambda^2 + 1} \tag{22}$$

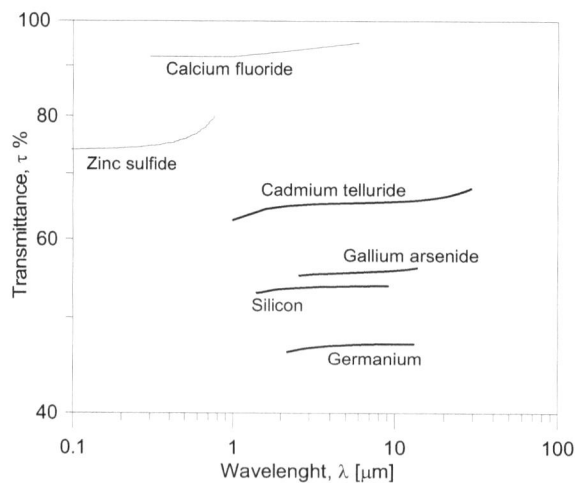

**Figure 6:** Transmittance of several materials.

In Fig. **6** are reported the peaks transmittance value (%) with the wavelength of several materials which are generally employed for the fabrication of optics in infrared systems. The transmittance of a material can be raised towards 100% by anti-reflection coatings which eliminate the reflective losses; of course, depending on the coating thickness also the wavelength is shifted towards a wider band.

The selection of materials for lens and window it is a compromise between optical and physical properties.

### 3.1.3. Reflectivity

According to Equation 8 all real surfaces (except the blackbody for which $\rho = 0$) reflect part of the incident radiation. Such a phenomenon more specifically consists of radiation that bounces off the target and is redirected. It is indifferently termed: reflectivity, or reflectance, or reflection. In this paper the term reflectance is used. Two types of reflectance may occur.

***Specular reflectance***-is observed when all (or almost all) the incident energy is thrown out the surface in a single direction. This happens when a surface is very smooth and highly polished (*i.e.*, a mirror-like surface). For a perfect mirror $\rho = 1$.

***Diffuse reflectance***-is observed when the incident energy leaves the surface almost uniformly in all directions. This is the case of rough surfaces.

These are the extreme ends of the way a surface reflects energy, the behaviour of common objects lies between perfectly specular and perfectly diffuse reflectors; the reflectance of a surface depends on the surface roughness and the wavelength of the incident radiation. More specifically, diffuse reflectance will dominate if the wavelength of the incident radiation is much smaller than the surface waviness.

The reflectance should be small ($\rho \rightarrow 0$) for accurate temperature measurements with an infrared system. Otherwise, radiation emitted by the surroundings may strike the surface of the object of measure and may be reflected being added to the radiation really emitted.

### 3.1.4. Emissivity

The ability of a surface to emit energy is of utmost importance in view of being monitored with a non contact infrared imaging system. First of all, to avoid confusion, it is worthwhile to clarify that the terms emittance and emissivity are often used indifferently to mean the same concept. To be more accurate the emittance is the radiation mismatch between an object and a blackbody whereas the emissivity is the emittance of a particular material under certain conditions (*i.e.*, emissivity is a material property). Herein, the term $E$ is used to indicate the emittance (*i.e.*, the energy radiated by a surface) and $\varepsilon$ to indicate the emissivity (the ability of a surface to emit energy).

***Blackbody definition***-The concept of blackbody was introduced in 1860 by Kirchhoff who defined a blackbody as the surface that neither reflects ($\rho = 0$) nor transmits ($\tau = 0$), but absorbs all incident radiation, independent of direction and wavelength. A blackbody, as schematically shown in Fig. **7**, can be represented by a cavity with an interior surface at uniform temperature in communication with the surroundings through a small hole; the energy entering the cavity is in part absorbed and in part reflected within the interior surface to ultimately be absorbed. In addition to absorbing all incident radiation, a blackbody is a perfect radiating body:

$$\alpha = \varepsilon = 1 \tag{23}$$

Real objects almost never comply with this law, although they may approach the behaviour of a blackbody in certain spectral intervals. A real object generally emits only a part $E_\lambda$ of the radiation emitted by a blackbody $E_{\lambda b}$ at the same temperature and at the same wavelength:

$$\varepsilon_\lambda = \frac{E_\lambda}{E_{\lambda,b}}$$

(24)

$\varepsilon_\lambda$ is called spectral emissivity coefficient, Equations 2 and 4 can be rewritten for real bodies by simply multiplying their second term by $\varepsilon_\lambda$.

For some non-blackbody objects the emissivity does not vary with the wavelength; these objects are called *greybodies*. A comparison between the spectral radiant emittance of a blackbody, a greybody and a real body, at the same temperature, is shown in Fig. **8**. From this plot, it is clear that the radiation curves of a greybody are identical to those of a blackbody except that they are dropped down on the radiated power density scale. Conversely, the radiation distribution of a real object varies with the wavelength.

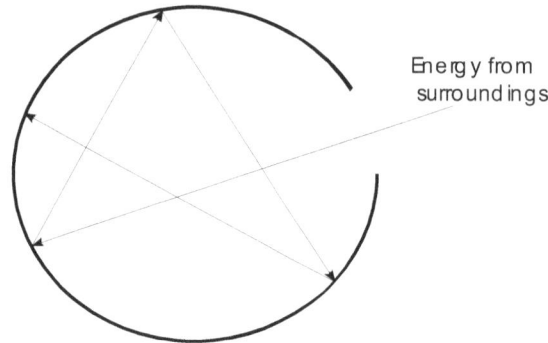

**Figure 7:** Sketch of the blackbody principle.

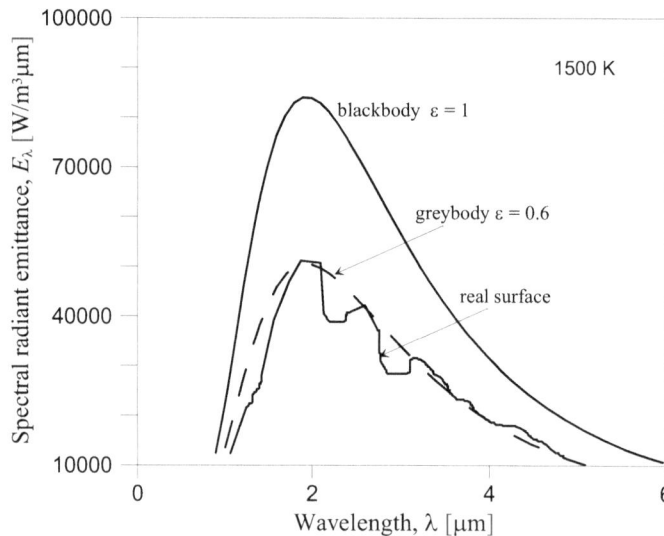

**Figure 8:** Comparison of emissive power for black, grey and real surfaces.

***Directional emittance*-**as previously said, a surface does not emit the same in all directions. As sketched in Fig. **9**, the maximum emission occurs in the direction normal to the radiating surface and decrease becoming null for a direction parallel to that surface. Only a blackbody emits radiation uniformly in all directions and the distribution takes the appearance of semi-circumferences.

A comparison between the directional emissivity of a blackbody and that of a non black one is shown in Fig. **10**. For the real body  the value of 0.7 has been assumed in normal direction (angle of observation $\theta = 0°$); this value remains constant until $\theta$ about 50° and then decreases first slowly and after sharply towards zero for $\theta$ approaching 90° (direction of observation parallel to the surface). This is an important aspect to be taken into account when using an infrared imaging system.

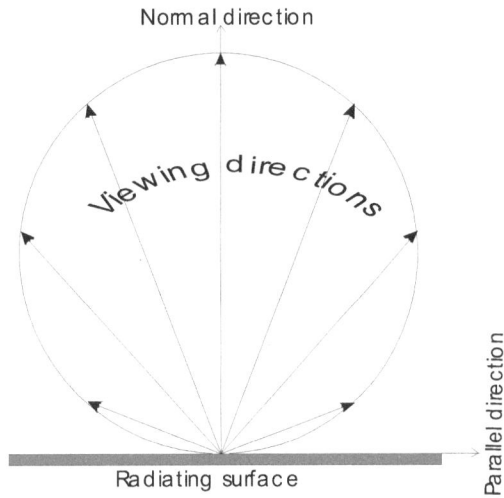

**Figure 9:** Angular variation of radiation.

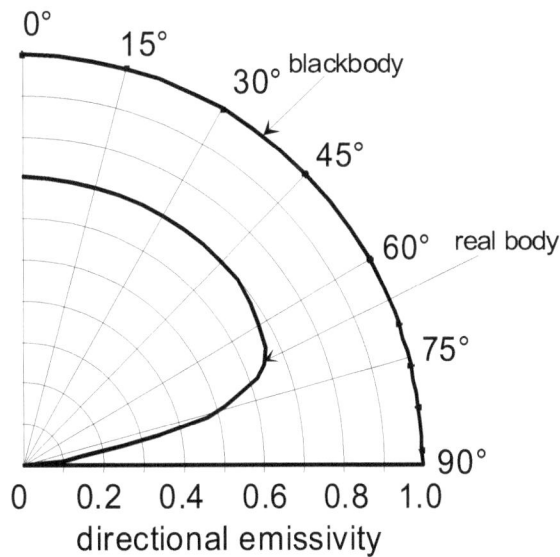

**Figure 10:** Directional emissivity for real and black bodies.

The radiation upon an element of surface area may consist of contributions which come from different directions. It is necessary to identify the radiation intensity $I$ which is incoming inside an infinitesimal arc of solid angle $d\Omega$; such a solid angle is sketched in Fig. **11** and is equal to:

$$d\Omega = \frac{dA_e}{r^2} \tag{25}$$

But $dA_e$ is the projection of $dA$ normal to the direction of radiation:

$$dA_e = dA\cos\theta \tag{26}$$

then:

$$d\Omega = \frac{rd\theta\, r\sin\theta\, d\phi}{r^2} = \sin\theta\, d\theta\, d\phi \tag{27}$$

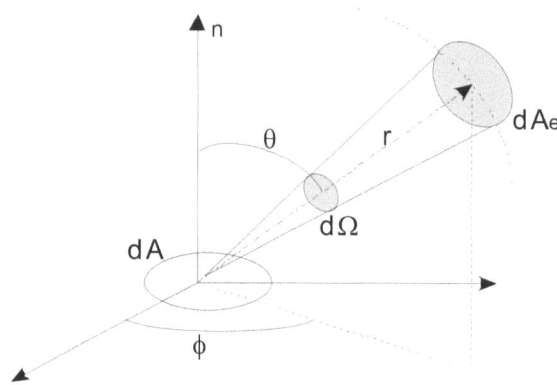

**Figure 11:** Angle of radiation emission.

The intensity per unit area and unit angle is:

$$I = \frac{dq}{d\Omega \, dA \cos \theta} \tag{28}$$

where $dq$ is the energy emitted per unit area and unit time, $I$ is the total intensity:

$$I = \int_0^\infty I_\lambda \, d\lambda \tag{29}$$

$I_\lambda$ being the monochromatic intensity. The total heat flux is obtained by integrating over angles $\theta$ and $\phi$:

$$q = \int_0^{2\pi} \int_0^{\pi/2} I(\theta, \phi) \cos \theta \, \sin\theta \, d\theta \, d\phi \tag{30}$$

and the total (hemispherical) emissive power $E_t$ of a surface $A$ is then:

$$E_t = \frac{q}{A} \tag{31}$$

***Diffuse emitter***-the radiation whose intensity is independent of the direction (the intensity is the same in all directions) is called isotropic radiation and the surface is a diffuse emitter. Such a surface is also called Lambertian surface because the total radiant power is described by the Lambert's cosine law for which the spectral directional emissive power is equal to the normal component multiplied the cosine of the angle $\theta$:

$$E_{\lambda b}(\lambda, \theta) = E_{\lambda b,n}(\lambda)\cos\theta \tag{32}$$

and being the integration over $\Omega$ equal to $\pi$ the total emissive power is:

$$E_t = \pi I \tag{33}$$

This is true for a blackbody for which the emissivity is equal to unity at every directions; in a polar plot the emissivity can be represented by a circumference of radius equal to unity.

***Emissivity of non-black surfaces***-the deviation of a non-black surface with respect to a black one is dictated by its geometrical structure and chemical composition. It is generally known that the emissivity of a real body (non-black) is always lower than 1; this is a consequence of the Kirchhoff's law of thermal radiation. Moreover, there are two caveats and one may be cautious when dealing with the thermal/optical behaviour of materials. First, the surface may be diffractive *i.e.*, the energy incident at one angle is partially reflected to another angle, and then the emissivity at one angle can unity. Second, the surface may be nonlinear; *i.e.*, the power incident power at one wavelength is re-emitted at another wavelength, and then the emissivity at some wavelengths can exceed unity. However, in both cases the emissivity will be lower than unity by integrating over all angles, or all wavelengths.

Roughly speaking the emissivity of a body might vary between 1 and 0. The first value applies to the blackbody behaviour that is a diffuse emitter. The second value instead would mean that according to the Kirchhoff's law:

$$\alpha = \varepsilon = 0 \tag{34}$$

A surface that do not absorb energy cannot transmit energy and then:

$$\tau = 0 \tag{35}$$

As a consequence, from Equation 7 it follows that:

$$\rho = 1 \tag{36}$$

all the incident energy is reflected and the surface is a specular reflector. In general, no surface is perfectly diffuse or specular, but similar behaviours can be approximated by either rough surfaces, or by polished or mirror-like surfaces.

The emissivity value for a rough surface is generally >0.1 and <1 in the direction normal to the radiating surface and it remains constant until the viewing angle reaches a value of about 50° and then decreases first slowly and after very rapidly becoming null when the viewing direction is parallel to the surface. For mirror-like surfaces $\varepsilon$<0.1, this value remains constant until an angle of 50° and then it increases.

### 3.1.4.1. Practical Considerations for Emissivity Evaluation

The knowledge of the thermal emissivity is essential for accurate temperature measurements with an infrared imaging system. The values of many most common materials are listed in literature; however, some difficulties arise when interpreting data present in literature because of two main reasons. Sometimes, data available in literature appear very accurate since four types of emissivity are reported: the total hemispherical emissivity $\varepsilon_T$, the directional total emissivity in normal direction $\varepsilon_{Tn}$, the spectral directional emissivity in normal direction $\varepsilon_{\lambda n}$ and the spectral directional emissivity in the normal direction for a given wavelength. These values are not at all useful when dealing with an infrared system which, as will be seen later, operates in a specific range of wavelengths, and viewing the real surface under a certain angle of observation. Another misleading source is that often we find tables with the parameter $\varepsilon$ simply termed emissivity without any additional explanation about the direction and the wavelength.

***Theoretical approaches***-The emissivity can be calculated relating to the optical and electrical materials properties and to relationships developed for absorption of electromagnetic waves since the validity of the Kirkoff's law (Equation  34).

*Free Carrier Absorption theory (FCA)* the spectral normal emissivity is predicted in terms of the Kirkoff's law for which:

$$\varepsilon_\lambda = 1 - \rho_\lambda \tag{37}$$

and the Fresnel's formula which relates the reflectivity to the refractive index $n_\lambda$ and to the extinction index $\kappa_\lambda$:

$$\rho_\lambda = \frac{(n_\lambda - 1)^2 + \kappa_\lambda^2}{(n_\lambda + 1)^2 + \kappa_\lambda^2} \tag{38}$$

and then:

$$\varepsilon_\lambda = \frac{4n_\lambda}{(n_\lambda+1)^2+\kappa_\lambda^2} \tag{39}$$

Sometimes the simplified expression:

$$\varepsilon_\lambda = \frac{4n_\lambda}{(n_\lambda+1)^2} \tag{40}$$

is encountered in literature. This is valid for a perfect dielectric ($\kappa = 0$); in this case the emissivity depends only on the refractive index, or better on optical characteristics.

*Hagen-Rubens theory* for metals and for relatively longer wavelengths (longer than 1 micron) the normal spectral emissivity may be simply extracted (with some simplifications) from the square root of the electrical resistivity according to the theory developed by Hagen and Rubens in 1900:

$$\varepsilon_\lambda \cong \sqrt{\frac{2\omega}{\pi\sigma_e}} \tag{41}$$

where $\sigma_e$ is the electrical conductivity that is equal to the inverse of the electrical resistivity and $\omega$ is the frequency of the radiation at wavelength $\lambda$. On the other hands, this formula represents the result of the more general Drude's model for intraband absorption with the assumption of very small frequency values (far infrared, non-relaxation region) for which $n_\lambda \approx \kappa_\lambda$.

### 3.1.4.2. Experimental Methods

It has to be taken into account that the emissivity is mainly a surface property (rather than a material property) and thus the real surface characteristics of the object under test must be considered. Many methods have been developed to experimentally measure the emissivity of the surface of interest [19, 20].

One method suitable for metals is the direct-heating with a dc current passing through the sample; the emissivity is evaluated by considering the increase of surface temperature and the electrical power lost by radiation from the surface. For non metals a Fourier Transform Spectrometer (FTR) is generally used.

The general criterion is to relate the energy emitted by a body to its effective temperature; the inherent difficulty lies in the surface temperature measurement. A convenient way is to measure the thermal emissivity with the infrared system itself. The procedure consists simply in comparing the radiation emitted by the material sample and that emitted by a blackbody at the same temperature. For directional emissivity measurements, the sample should be positioned on a rotating traversing system. Of course, care must be put to thermally insulate the pipes for the flowing fluid to have both blackbody and sample at the same thermostatic bath temperature. Almost every infrared thermographic system is equipped with software for emissivity calculation.

***How to deal with low emissivity surfaces***-For thermographic measurements, it is preferable to work with high-emissivity surfaces. It is possible to increase the surface emissivity of highly polished metals, or reflectors, with deposition of thin films of paint, or grease. This of course can be done in laboratory measurements and when the coating does not affect the surface performance. Almost every final part includes a protective coating of the surface; in most cases (except when specularity is a requisite), the choice of an opaque paint may fulfil both requirements: assure surface protection and allow for thermographic monitoring.

If a coating is not practicable, others expedients should be adopted; one may be to prevent undesired reflected radiation from the surroundings with screens.

## 3.2. Interposed Medium

Infrared systems are non-contact and thus, the radiation, emitted by a body, passes through the atmosphere (Fig. **4**), before entering the detector, and it may be affected by absorption, scattering, emission and turbulence. Therefore, the distortion induced by the atmosphere must be taken into account.

Normal atmosphere (*e.g.* the air we breathe) is composed of a mixture of gases (Table **1**). Generally, carbon dioxide absorbs infrared radiation significantly in the 15 micron band; this is the broad region of maximum intensity for the Planck function for emitters at characteristic atmospheric temperatures (about 180K to 300K). Water vapour absorbs thermal infrared in the 6.3 micron band and in several regions between 3 and 0.7 microns. Except for ozone, which has an absorption band in the 9.6 micron region, the atmosphere is relatively transparent from about 8 to 13 microns. The atmospheric transmittance is represented in Fig. **12**; the coloured areas indicate the atmosphere transparency.

The presence of aerosol (salt particles, water droplets, dust, pollution haze) also contributes to absorption. However, this contribution is generally negligible with respect to the molecular absorption. More than absorption, the aerosol causes scattering (Mie scattering); *i.e.,* a redistribution of the radiation into all directions with loss in the travelling direction from source to detector. One may remark that also molecules cause scattering (Rayleigh scattering). Generally, molecular scattering is negligible with respect to aerosol scattering. In any case, the scattering phenomenon is stronger in the visible region and in IR region close to the visible one; as the wavelength increases the transmittance improves. In fact, IR cameras are used to see through the fog, or in presence of sand storm.

**Table 1:** Atmosphere composition.

| Gas | Chemical Formula | Amount % |
| --- | --- | --- |
| Nitrogen | $N_2$ | 78 |
| Oxigen | $O_2$ | 21 |
| Water vapor | $H_2O$ | 1-4 |
| Argon | Ar | 0.93 |
| Carbon dioxide | $CO_2$ | 0.35 |
| Neon | Ne | 0.0018 |
| Helium | He | 0.0005 |
| Methane | $CH_4$ | 0.00017 |
| Krypton | Kr | 0.00011 |
| Hydrogen | $H_2$ | 0.000055 |

Since Equation 9 the atmosphere emits its own radiation. However, this contribution is important only in specific test conditions, *i.e.,* for large distances between the object and the IR camera and for measurements of low-temperature object (close to the ambient temperature). The last phenomenon of turbulence occurs in presence of wind, or convection transport effects. Turbulence induces a random fluctuation of the refractive index and this results in smearing of the generated image. This generally occurs for large distances object-detector (order of hundreds meters) and with high sensitivity IR cameras.

The attenuation induced by atmosphere could sometimes be a heavy burden for the operator since there are normally no easy ways to find accurate values of emittance and atmospheric transmittance for the actual case. An easy and convenient way is to account for the atmospheric interference during calibration of the infrared system. In some specific applications the infrared camera views the object through a solid medium (*e.g.*, a window of a closed wind tunnel) and this poses the problem of design and material choice.

### 3.2.1. Practical Suggestions

The type of material depends on both optical and mechanical considerations. For optical requirements, the material must be chosen according to its transmittance in the wavelength region of the employed infrared detector (Fig. **6**). In addition, the IR window must resist the stresses caused by temperature and pressure differences between inside and outside. Considering that the in-use IR systems work in the middle-far infrared (generally from 3 to 12μm), from Fig. **6**, a material suitable for window purposes is germanium. In particular, with specific coating the transmittance of germanium can be raised up to 95-99%.

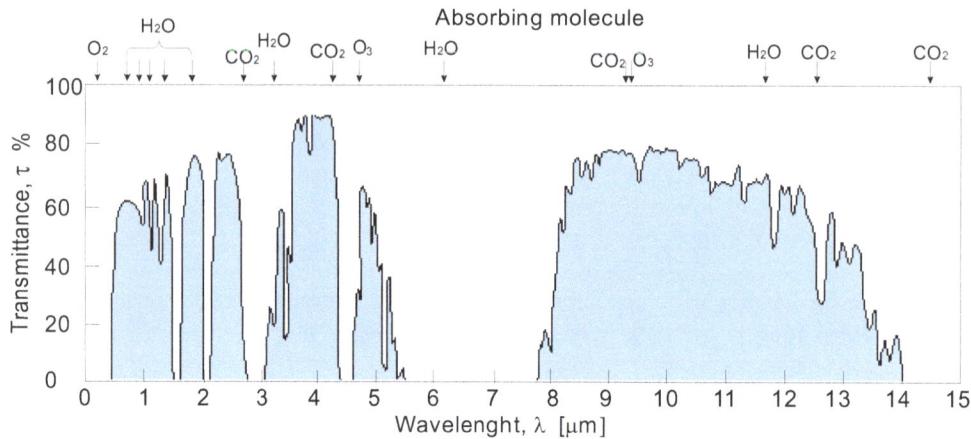

**Figure 12:** Atmosphere transmittance.

In absence of pressure differences between the test room, the environment and of heavy fragments splattering, the IR window may be replaced by a (more economical) thin polyethylene foil.

Sometimes, an IR camera is employed to monitor materials removal/shaping processes; also in these cases it is necessary to protect lenses from splattering of substances such as material fragments, or lubricating liquids. Therefore, a protective medium must be interposed; again, the choice of the type of protection involves considerations about the type of substance to protect against.

### 3.3. The Surroundings

One factor that strongly affects measurements is the radiation coming from the surroundings. As depicted in Fig. **4**, this radiation impinges on the surface of the object under measurement and, depending on the object surface characteristics, two main cases may arise:

a)   Highly polished surface – the incident radiation is almost totally reflected.

b)   Opaque surface ($\varepsilon \gg 0$, $\varepsilon < 1$) – a part of the incident radiation is reflected and a part is absorbed. The second part may increase the surface temperature and, as a consequence, also the emitted radiation increases.

In any case, there is an incorrect estimation of the effective object temperature. Of course, this is a problem if the surroundings contain large and intense radiation sources. Coming back to Equation 6, the three constants A, B and C can now be explicated:

$$\begin{cases} A = \varepsilon_{obj}\tau_{atm} \\ B = \left(1 - \varepsilon_{obj}\right)\tau_{atm} \\ C = 1 - \tau_{atm} \end{cases} \tag{42}$$

where $\varepsilon_{obj}$ is the emissivity of the object surface, $\tau_{atm}$ is the atmosphere transmissivity.

A solution to the problem is to put the object under measure, whenever possible, in a dark room, or to eliminate reflections by adequate screens. The atmosphere absorption is a difficult task; but it can be easily taken into account by, as already suggested, perform calibration of the infrared system in the same test conditions.

## 4. INFRARED DETECTORS

Infrared instruments are often classed as total-radiation radiometers and are considered to be based on the Stefan-Boltzmann's law even if their detectors sense radiation in only a limited bandwidth of the IR region. Actually, infrared devices perform measurements mainly in two main IR bands: middle wave infrared MWIR ($2-5\mu m$) and long wave LWIR ($8-15\mu m$). Systems which work in the NIR window are tailored for specific applications. As shown in Fig. **13** the atmosphere is opaque to infrared in the region between 5.5 and $7.8\mu m$ and thus no measurements are possible in that region. Therefore, measurements obtained with infrared radiometers are generally based on Planck's law.

The ultimate result of an infrared system is a surface temperature map. However, up to a few years ago, the sensing element was zero-dimensional (practically a point). Two-dimensionality was achieved by a scanning mechanism which consisted of oscillating mirrors, or rotating refractive elements (such as prisms), and allowed object scanning in both vertical and horizontal directions. Nowadays, the new systems are based on the staring focal plane array (FPA) technology; but, often they are still referred to as infrared scanning radiometers (IRSRs) since their output comes from an electronic scanning.

The overall performance of an IR imaging system is conventionally evaluated in terms of useful and accurate information that can be acquired per unit of time. This can be expressed through several parameters such as thermal sensitivity, or equivalent random noise level; scan speed, or update rate of the scanning mechanism; image resolution, or number of independent measurement data points of which the image is composed; and intensity resolution, or number of intensity levels, which allows for fine temperature differences, *i.e.,* dynamic range.

The core of the IR thermographic system is the detector (or array of detectors) [21-41]. Basically, it is a transducer that absorbs the IR energy emitted by the object (being measured) and converts it into a signal, usually an electrical voltage or current, and then, it can be transformed into a temperature map as already said.

Detectors used for infrared technology can be grouped into two main categories: thermal and quantum (or photon) detectors.

### 4.1. Thermal Detectors

These detectors are also called energy detectors, or photon absorber because they absorb the incident energy and warm up; the temperature changes are measured through the variation of a temperature-dependent property of the material such as the electrical resistance. The main advantage is their response at room temperature and over a large range of the electromagnetic spectrum. The main disadvantage is their slow response time, which makes them not suitable for high frequency events.

Examples of detectors that belong to this family are: thermopiles, bolometers, pyroelectric detectors, microcantilevers.

Thermopiles are based on the thermoelectric effect. A thermopile is composed of a number of thermocouples connected in series. As well known a thermocouple is obtained by joining two dissimilar metals. Heating the junction causes a voltage (Seebeck effect) that is proportional to the temperature change; thus, the temperature is computed from the measured voltage.

The first thermopile was simply built using fine wires thermocouples (*e.g.*, copper-constantan). The temperature of the hot junctions increases until an equilibrium is reached between the rate of incident energy and the loss by conduction, convection and radiation to the surroundings. The thermoelectric output is measured by a suitable device and transformed into temperature (*i.e.*, the temperature of the target). The modern technology relays to thin films deposited on a substrate; this allows for reduced dimensions. The sensitivity can also be enhanced by encapsulating the thermopile in a low thermal conductivity medium (*i.e.*, avoid losses to the surroundings).

Bolometers are based on the resistance change with temperature of a resistor element. Basically, it consists of two platinum strips, covered with lampblack, one strip is shielded from the radiation and the other one exposed to it. The strips form two branches of a Wheatstone bridge; the resistance in the circuit varies when the strip, which is exposed to electromagnetic radiation, heats up and changes its electrical resistance.

Nowadays, advances in silicon micromachining have lead to the microbolometer technology which includes a grid of vanadium oxide, or amorphous silicon heat sensors, atop a porous silicon bridge as thermal isolating and mechanical supporting structure. The microbolometer grid is commonly found in different sizes: 160×120, 320×240, 640×512, or more arrays.

Pyroelectric detectors are based on the change of surface charge with temperature. In fact, the pyroelectric effect consists of a change in the crystal surface charge as a consequence of the change in the dipole moment when the crystal temperature is raised. For continuous operation, the common technique is to incorporate a chopping device inside the optical system and to create an AC output signal; the spurious signals caused by low-frequency ambient temperature can be filtered. Triglycinesulfate (TGS), lithium tantalate ($LiTaO_3$) and polyvinyl fluoride (PVF2) are materials that exhibit the pyroelectric effects.

Microcantilevers are based on the dissimilar thermal expansion of bimetals (*e.g.*, silicon nitride and gold film). More specifically, a cantilever forms a capacitance in combination with a reference plate. The exposition to infrared radiation causes temperature increase with consequent bending of the cantilever which, in turn, alters the capacitance of the structure. The intensity of IR radiation is measured as being proportional to the bending extension.

### 4.2. Photon Detectors

Photon detectors generate free electrical carriers in response to photon absorption. The main advantage is a very short response time (of the order of microseconds). A disadvantage is the need of cooling down to cryogenic temperature to get rid of excessive dark current. In fact, electrons can be excited from the valence to the conduction band by photons having an energy $E_{ph}$ that is larger than the energy $E_g$ between the bands; there exists a wavelength $\lambda_c$ named cut-off wavelength which corresponds to the energy $E_g$:

$$\lambda_c = \frac{hc}{E_g} \tag{43}$$

beyond which no emission occurs. The value of $E_g$ varies from semiconductor to semiconductor; it tends to increase as temperature increases (wavelength decrease). Therefore, the detector must be maintained at low temperature.

In the past, the detector was often located in the wall of a dewar flask which was filled with liquid nitrogen, $LN_2$ (77K). The new generation systems include miniature coolers based on the thermoelectric Peltier effect or the Stirling cycle and do not require any external cooling source. There are four main types of photon detectors: photovoltaic, photoconductive, photo-emissive and Quantum Well Infrared Photodetector (QWIP). In particular, the photovoltaic and photoconductive types may be intrinsic, or extrinsic.

Photovoltaic intrinsic (electromotive force generation). The structure of a photovoltaic (PV) detector is based on a p-n junction device (two dissimilar materials) which, under IR radiation, generates photocurrents (current flows across the junction of the two materials). More specifically, when the detector is knocked by photons of energy greater than or equal to the energy band-gap, electrical carriers are swept across the photodiode and become separated by the potential barrier of the p-n junctions to create a current. PV devices operate in the diode's reverse bias region; this minimizes the current flow through the device which in turn minimizes power dissipation. These detectors are generally fabricated from silicium (Si), germanium (Ge), gallium arsenide (GaAs), indium antimonide (InSb), indium gallium arsenide (InGaAs), and mercury cadmium telluride (HgCdTe) also called (MCT).

Photoconductive intrinsic (conductance change). In photoconductive (PC) detectors, the incident radiation, with energy greater than or equal to the energy band-gap of the semiconductor, generates majority electrical carriers. The electric conductivity of the material is improved; there is an internal photoelectric effect. Common materials are lead sulfide (PbS), lead selenide (PbSe) and mercury cadmium telluride (HgCdTe).

Extrinsic detectors are similar to intrinsic detectors. The difference lies in the fact that carriers are excited from the impurity levels and not over the band-gap of the basic material; this is achieved by doping the semiconductor material.The mostly used materials are silicium and germanium doped with impurities such as boron, arsenic and gallium. The spectral response of these detectors can be controlled by the doping level.

Photo-emissive detectors are bi-materials involving a metal layer superimposed to a semiconductor layer; a typical example is platinum silicide (PtSi) on silicon (Si). The process, which is also called external photoelectric effect, consists in the emission of carriers from the metal into the semiconductor under photon absorption. The main advantage is a more uniform response since the response depends on the characteristics of the metal. However, being the photon absorption proportional to the square of the wavelength, this type of detector is more indicated for long wavelengths.

Quantum Well Infrared Photodetector (QWIP). The basic principle is similar to that of extrinsic detectors with the peculiarity that the dopants are concentrated in microscopic regions and creates the quantum wells. The radiation is absorbed by the entire quantum well, not only by a single doping atom, and thus the absorption is increased (and also the response) with respect to extrinsic detectors. QWIPs generally consists of layers of thin gallium arsenide (GaAs) alternated to layers of aluminium gallium arsenide (AlGaAs).

## 4.3. Detectors Performance

The performance of a detector is evaluated through three parameters that are: Responsivity $R_v$, Noise Equivalent Power (NEP) and Detectivity $D^*$.

The responsivity $R_v$ is a measure of the signal output $S_{out}$ (voltage, or current) per incident radiation $E_{mc}$ over the active area of the detector $A_{det}$:

$$R_v = \frac{S_{out}}{E_{mc} A_{det}} \tag{44}$$

The active area is determined as the ratio $A_{det} = A_0/A_e$ between the optical area $A_o$ and the actual *electrical* area $A_e$. The responsivity can also be written as:

$$R_v = \frac{\lambda \eta_q Q_e p_g}{hc} \tag{45}$$

with $\eta_q$ the quantum efficiency, $Q_e$ the electron charge, $p_g$ the photoelectric gain.

The Noise Equivalent Power (NEP) defines the intrinsic noise level of the detector, or better the detection limit of the detector. It can be expressed by:

$$NEP = \frac{E_{mc} A_{det}}{\frac{S_{out}}{N_{out}}\sqrt{\Delta f}} = \frac{N_{out}}{R_v \sqrt{\Delta f}} \tag{46}$$

with $N_{out}$ the noise output and $\Delta f$ the noise bandwidth.

The detectivity $D^*$ defines the resolving power of the detector and is expressed in terms of the signal-to-noise ratio $(S_{out}/N_{out})$ with respect to the incident power:

$$D^* = \frac{S_{out}}{N_{out}} \frac{1}{E_{mc}} \frac{\sqrt{\Delta f}}{\sqrt{A_{det}}} = \frac{\sqrt{A_{det}}}{NEP} \tag{47}$$

The common units of $D*$ are cmHz$^{1/2}$/W; often, in literature, this quantity is measured in Jones in honour to D. Clark Jones who defined this magnitude in 1959.

The noise that affects detection of IR radiation comes from two main sources: the IR detector itself with its circuits and the background fluctuations. The first contribute is generally negligible with respect to the second one and then the detection limit is evaluated by accounting only for the background radiation. This contribution is called Background Limited Infrared Photodetection (BLIP).

Photovoltaic detectors are characterised by a $\sqrt{2}$ higher BLIP; in fact, the relationship for a photovoltaic detector is:

$$D^*_{BLIP} = \frac{\lambda}{hc} \frac{\sqrt{\eta_q}}{\sqrt{2\Phi_b}}$$

(48)

while for a photoconductive is:

$$D^*_{BLIP} = \frac{\lambda}{2hc} \frac{\sqrt{\eta_q}}{\sqrt{\Phi_b}}$$

(49)

$\Phi_b$ being the background photon flux density (also called dark current). The value of $\Phi_b$ received by the detector depends on its responsivity to the wavelengths contained in the radiant source and on its Field Of View (FOV) of the background.

A relevant parameter for infrared systems is the Noise Equivalent Temperature Difference (NETD); it can be defined as the temperature change, for incident radiation, that gives an output signal equal to the r.m.s. noise level $N_L$:

$$NETD = N_L \frac{\Delta T}{\Delta S_m}$$

(50)

$\Delta S_m$ is the signal measured for the temperature difference $\Delta T$.

## 4.4. Historical Steps in IR Detectors Technology

The first infrared detector may be considered the thermometer used by Herschel; *i.e.*, the simple mercury thermometer that responds with a temperature variation to the incident radiation. The history of infrared detectors for remote sensing really originates with the thermoelectric effect at the junction of two dissimilar metals, discovered by the Estonian Thomas Johann Seebeck in 1821. This effect, which is also called Seebeck effect, consists in a direct conversion of temperature differentials into electric voltage and allowed the invention of the thermopile by Nobili in 1829. Later, in 1880, the property of metals to change electrical resistivity with temperature variation lead to the invention of the bolometer by the American Samuel Pierpont Langley.

A milestone in detectors development is certainly the photoelectric effect, which was firstly observed by Hertz in 1887, and explained and formulated by Einstein in the early 1900s. Closely related to the photoelectric effect is the photoconductive effect that is the increase of electrical conductivity of non metals when exposed to electromagnetic radiation.

The first phoconductive-based detector, a thallous sulphide, was developed by Theodore Willard Case in 1917. Such a detector suffered from problems of noise and stability; it was subject of study during World War II and later dropped in favour of lead sulphide (PbS) which had higher performance. Indeed, from the late 1940s and continuing in 1950s, new materials were developed such as lead selenide (PbSe) and lead telluride (PbTe) which allowed a shift of spectral sensitivity from the (1-3μm of PbS) to (3-5μm) infrared wavelength.

The end of the 1950s and the beginning of the 1960s saw the introduction of semiconductor alloys, such as the indium antimonide (InSb) and the mercury cadmium telluride (HgCdTe) or (MCT), which allowed the spectral response to be custom tailored from mid (MWIR) to long (LWIR) wavelength infrared. MCT was at once considered the most important and versatile material for IR detector applications; indeed, this material has been and is still now under study. MCT allows for both photoconductive and photovoltaic detection and has inspired the development of the three generations of detector devices. Initially, some difficulties arose in producing high-uniformity detectors. One main negative aspect was the weak Hg-Te bond which entailed instability. Therefore, initially, single element detectors were produced and the scenery image was obtained through a mechanical scanning mechanism. These detectors belong to the so-called pre-first generation.

The first generation involved linear arrays. However, the fabrication of arrays of sensors started with the development of photolithography in the early 1960s; PbS, PbSe and InSb detectors were first used in linear array technology. In the late 1960s and beginning of 1970s linear arrays of intrinsic MCT photoconductive detectors were developed. As technology further advanced, linear arrays of 60, 120, or 240 pixels were produced with acceptable uniformity and called linear focal plane arrays (FPA). The video image was obtained by scanning, through a rotating mirror, the linear arrays along with a line-filling scheme in the vertical axis. The resulting image had a quite good resolution (comparable to TV standards) in the horizontal direction and a low one in the vertical direction. IR systems with detectors of the first generation are still in use.

The second generation consists of two dimensional (2D) arrays of detectors that are now in wide production. The basic idea of this type of detector originated from the invention of the charge coupled device (CCD) in 1969 by Willard Boyle and George Smith. It was possible to have detection and readout implemented on one silicon chip. The detectors are configured in a 2D array generally referred to as staring FPA. These arrays are electronically scanned by readout integrated circuits (ROICs). The second generation is a full-framing system and has a great number of pixels (about three orders of magnitude) than the first generation. There are also intermediary systems which include linear arrays scanned with multiplexed devices; of course, these systems are characterized by a time delay integration (TDI).

FPAs can be developed in different architecture and generally grouped into: hybrid FPAs and monolithic FPAs. The monolithics, as the term lets to suppose, are compact with both the IR sensitive material and the signal transmission paths on the same layer. The basic element in a monolithic approach is a metal insulator semiconductor (MIS) with the main function of charge transfer device. In other words, a MIS detects and integrates the IR generated photocurrent (multiplexing function). This results in a more simple and low-cost manufacture, but with the penalty of a low fill factor (*i.e.*, the ratio of active IR sensing material to inactive row and column borders) and consequently low sensitivity with respect to their hybrid counterpart. In hybrid FPAs, multiplexing can be achieved with two different circuits the CCD and the complementary metal oxide semiconductor (CMOS).

The third generation essentially includes large format multicolour and avalanche photodiodes (APD). A significant advantage of multicolour (or multi spectral) technology is their capability of sensing in two or more spectral bands; this can provide added information that is not available with traditional single-colour systems such as, with the assumption of gray emitting body, to avoid to know the emissivity coefficient to determine its temperature. A primary requisite of third generation detectors (TGD) is to provide high spatial resolution (of the order of $1 \times 10^7$ pixels against the maximum of $3 \times 10^6$ of the second generation staring FPAs), which also means reduction of pixel pitch. Indeed, the goal of TDG is to achieve high frame rates (simultaneous or sequential reading) and better thermal resolution with very low NETD of about 1mK against the 20-30mK of FPAs. As demonstrated by Rogalski [22] the NETD is inversely proportional to the square root of the integrated charge and then a low NETD value requires a high charge storage capacity. This is not possible with standard CMOS capacitors [21].

The $Hg_{1-x}Cd_xTe$ ternary alloy [22, 23] is considered as the near ideal material since it can be tuned from SWIR to LWIR by adjusting the x composition of the alloy. The HgCdTe for third generation detectors involves a novel technology based on molecular beam epitaxy (MBE), which may be germanium (mainly used by the French SOFRADIR), or silicon (mainly used in USA by Raytheon Vision Systems (RVS)). The connection with indium bumps to a ROIC chip as a hybrid structure, also called sensor chip assembly (SCA), allows for very large array manufacturing.

Research organizations are working for the development of large multi-band IR detectors. Among them, in USA the RVS has developed a 1280×720 pixels dual-band MW/LWIR FPA under the U.S. Army's Dual-Band FPA Manufacturing (DBFM) program. Such dual-band FPA architecture was fabricated from MBE-grown HgCdTe triple-layer heterojunction (TLHJ) wafers and is sponsored to provide highly simultaneous temporal detection in the MWIR and LWIR bands using time-division multiplexed integration (TDMI) incorporated into the ROIC. In France, the SOFRADIR is involved with the CEA-LETI/LIR within a Design of Excellence for the Future of Infrared (DEFIR). In the LWIR region, the HgCdTe material is limited by metallurgical problems of the epitaxial layers such as uniformity and number of defective elements [24].

A low-cost alternative technology is the QWIP. A 1024×1024 pixels QWIP FPA working in MWIR and LWIR has recently been developed and tested by a NASA lead-team. This is a multi-quantum-well structure which consists of coupled quantum wells of GaAs, InGaAs thin (10, 20Å) doped layers and un-doped AlGaAs barriers [25, 26]. QWIP dual band detectors provide good resolution but involves integration times of the order of 5-10ms while, for fast changing scenes, a lower integration time (typically 1ms) is required.

In this context, antimonide based type II superlattices have emerged as third candidate for third generation infrared detectors [21, 27]. Type II InAs/GaInSb superlattice structure has great potential for LWIR and very LWIR spectral ranges it is a mechanically robust III–V material and has a direct band gap with wavelength tunability from 3 to 25μm [24]. A major advantage is the higher potential performance compared to HgCdTe with the same cut-off wavelength. The lower production costs are a good premise for commercial market applications in the future. However, they are an early stage of development and problems still exist in material growth, processing, substrate preparation, and device passivation.

## 5. APPLICATIONS

By now, infrared thermography has amply proved its usefulness leading to a proliferation of infrared devices to fulfil desires of the multitude of users in the vast variety of applications. In fact, an infrared imaging system can be tailored for specific requirements. And thus, we have the availability of a wide selection of infrared devices, which differentiate for weight, dimensions, shape, performance and of course costs.

Upon a correct choice of: infrared system, test procedure and successive data analysis, it is possible to use infrared thermography in a lot of application fields and for many different purposes. For example, IRT may be used for diagnosis (in medicine, architecture, maintenance), or for understanding of complex fluid dynamics phenomena (flow instability, flow separation and reattachment), or for material characterization and procedures assessment which can help improving design and fabrication of products. However, an infrared device can be advantageously exploited for process control and maintenance planning with reduction of undesired production stops and with consequent money saving.

Below there is a list of some of the most important application fields.

### Aerial IR Imaging

➢ **Agriculture** (management of irrigation devices, acquire information about the freezing process in plants, *etc.*).

➢ **Asphalt** control.

> ➢ **Electrical** (Urban and rural distribution lines).

> ➢ **Environmental** impact surveys.

> ➢ **Marine surveyor**.

> ➢ **Roofing** inspection.

> ➢ **Animals** finding and counting (warm-blooded animals).

## Building Surveys

> ➢ **Envelope** (air tightness testing, visualization of energy losses, thermal bridges, *etc.*).

> ➢ **Masonry structures** (detection of construction failures, material degradation, presence of cracks and voids, detection of moisture, *etc.*).

> ➢ **Buried structures** (check underfloor heating, pipes, *etc.*).

<u>**Cultural Heritage**</u> (control the conservation state of artworks, frescoes, paintings).

<u>**Medical**</u> (diagnosis of melanoma, management of neuropathic pain, hypertermia, *etc.*).

<u>**Veterinarian**</u> (control of animals welfare, *etc.*).

## Predictive Maintenance

> ➢ **Electrical** inspection (Power generation: exciters, connections, motor control centers; Substation: switchgear, breakers, transformers, capacitor banks, *etc.*).

> ➢ **Electronic** inspection (individuate: improper soldering of circuitry, broken traces between components, heat up of circuitry, *etc.*).

> ➢ **Facility** inspection.

> ➢ **Manufacturing** (control of machine, engine,valve, *etc.*).

## Research & Development

> ➢ **Fluidynamics** (boundary layer development over model wings, separation and reattachment lines over a wing at angle of attack, convective heat transfer in static and rotating channels, heat transfer from a plate to impinging jets, *etc.*).

> ➢ **Material science** (Control in-process: milling, drilling, cutting; nondestructive testing, fatigue tests, *etc.*).

This list is by no means exhaustive since, as technology advances, infrared devices evolve too and offer ever more new opportunities for application. Indeed, any process which is temperature-dependent may benefit from the use of an infrared device. What it is more, an infrared imaging system may be also fruitfully exploited in teaching for the comprehension of many physical phenomena like energy transformation and heat transfer mechanisms. At last, when approaching to hang a painting, or something else,to a wall, often a doubt rises that drilling a hole, a buried structure such as a feeding, ora sewage pipe, or an electric conduit may be intercepted causing a serious inconvenient. In such cases, it suffices a small cheap infrared camera to prevent this problem.

Some applications of infrared thermography are fully described in the successive chapters of this eBook; for a more exhaustive panorama the reader is addressed to specific literature [42-64].

## REFERENCES

[1]   Dulong PL, Petit AT. Des recherches sur la mesure des températures et sur les lois de la communication de la chaleur (Seconde partie, des lois du refroidissement). Annales de Chimie et de Physique 1817; 7: 247.

[2]   Stefan J. Über die Beziehung zwischen der Wärmstrahlung und der Temperatur, Sitzungberichte, Akademie der Wissenschaften Wien 1879; 79 (2): 391–428.

[3]   Boltzmann L. Ableitung des Stefan'schen Gesetzes, betreffend die Abhängigkeit der Wärmestrahlung von der Temperatur aus der electromagnetischen Lichttheorie. Annalen der Physik und Chemie 1884; 22: 291–94.

[4]   Maxwell JC. A dynamical theory of the electromagnetic field. Philosophical Transactions of the Royal Society of London1865; 459–512.

[5]   de Broglie, Prince Louis Victor Pierre Raymond, Recherches sur la théorie des quanta, Thesis, University of Paris (Sorbonne), Paris, 1924 (published in Annales de Physique, 10ᵉ série, Tome III, Masson et Cie Editeurs Paris 1924, pp. 22–128).

[6]   Singh V. Einstein and the quantum Special section: the legacy of Albert Einstein. Current Science 2005; 89 (12): 2101-12.

[7]   Smith GS. An introduction to classical electromagnetic radiation Cambridge University Press 1997.

[8]   Smith RA, Jones FE, Chasmar RP. The detection and measurement of infrared radiation. Clarendon Oxford 1958.

[9]   Edwards DK, Denny VE, Mills AF. Transfer Processes: An Introduction to Diffusion, Convection, and Radiation (2nd edition), Hemisphere Publishing Corporation, Washington, 1979.

[10]  Siegel R, Howell JR. Thermal Radiation Heat Transfer, Hemisphere Publishing Corporation, Washington 1992.

[11]  Brewster MQ. Thermal Radiative Transfer & Properties, John Wiley & Sons, New York, 1992.

[12]  Modest MF. Radiative Heat Transfer, McGraw-Hill, New York, 1993.

[13]  Barr ES. Historical survey of the early development of the infrared spectral region. Am J Phys 1960; 28: 42-54.

[14]  Kruse PW, McGlauchlin LD, Mcquistan RB. Elements of infrared technology. Wiley, New York, 1962.

[15]  Handbook of Infra-red detection technologies edited by M. Henini and M. Razeghi, Elsevier 2002.

[16]  Levinstein H, Mudar J. Infrared detectors in remote sensing. Proc of the IEEE 1975; 63(1): 6-14.

[17]  Soloman S. Sensors Handbook McGraw Hill Professionals 1998.

[18]  Rogalski A. Infrared detectors: status and trends. Progress in quantum electronics 2003; 27: 59-210.

[19]  Giulietti D, Lucchesi M. Emissivity and absorptivity measurements on some high-purity metals at low temperature. J Phys D Appl Phys 1981; 14: 877-81.

[20]  Hersketh PJ, Zemel JN, Gebhart B. Polarized spectral emittance from periodic micromachined surfaces. II Doped Silicon: angular variation. Physical review B 1988; 37(18): 10803-816.

[21]  Soloman S. Sensors Handbook McGraw Hill Professionals 1998.

[22]  Rogalski A. HgCdTe. Infrared detector material: history, status and outlook Report. Prog Phys 2005; 68: 2267-36.

[23]  Wachter EA, Thundat T, Oden PI, Warmack RJ, Datskos PG, Sharp SL. Sharp Remote optical detection using microcantilever. Rev Sci Instrument 1996, 67: 3434-39.

[24]  Levinstein H, Mudar J. Infrared detectors in remote sensing. Proc IEEE 1975, 63: 6-14.

[25]  Kreisler AJ, Gaugue A. Recent progress in high-temperature superconductor bolometric detectors: from the mid-infrared to the far-infrared (THz) range, Supercond Sci Technol 2000; 13: 1235-45.

[26]  Rogalski A. Infrared detectors: status and trends. Prog Quant Elect 2003, 27: 59-210.

[27]  Konstantatos G, Howard I, Fischer A, Hoogland S, Clifford J, Klem E. Ultrasensitive solution quantum dot photo detectors. Nature Letters 2006; 13: 442.

[28]  Armani N, *et al.* Defect-induced luminescence in high-resistivity high-purity undoped CdTe Crystals. J Phys Cond Matt 2002; 14: 13203-209.

[29]  Tribolet P, Vuillermet M, Destefanis G. The third generation cooled Ir detector approach in France, Advaancesin Proc. SPIE Advances in Optical Thin Films II, C. Amra, N. Kaiser and A. H. Macleod Ed. , 2005, 5964: 49-60.

[30]  Cabanski W, *et al.* Third generation focal plane array IR detection modules and applications in Proc. Infrared Technology and Applications XXXI B. F. Andresen, G. F. Fulop Ed. 2005, 5783: 340-49.

[31]  Sizov FF. Infrared detectors: outlook and means Semiconductor Physics. Quantum Electronics and Optoelectronics 2000; 3: 52-58.

[32]    Janesick J, Putman G. Developments and applications of high-performance CCD and CMOS imaging arrays. Annu Rev Nucl Sci Part Sci 2003; 53: 263-300.

[33]    Rose A. Concepts in photoconductivity and allied problems. Interscience New York 1963.

[34]    Rogalski A. Competitive technologies of third generation infrared photon detectors. Opto-Electron Rev 2006; 14: 87-101.

[35]    Norton P, Campbell J, Horn S, Reago D. Third generation infrared imagers. Proc SPIE 2000; 4130: 226-36.

[36]    Tribolet P, Destefanis G. Third generation and multi-color IRFPA developments: a unique approach based on DEFIR. Proc SPIE 2005; 5783: 37-53.

[37]    Rogalski A, Martyniuk P. InAs/GaInSbsuperlattices as a promising material system for third generation infrared detectors. Infrared Physics and Technology 2006; 48: 39-52.

[38]    Gunapala SD, *et al.* 1024×1024 pixel mid-wavelength and long-wavelength infrared QWIP focal plane arrays for imaging applications. Semicond Sci Technol 2005; 20: 473-80.

[39]    Nesher O, Klipstein PC. High-performance IR detectors at SCD present and future. Opto-Electron Rev 2006; 14: 61-70.

[40]    Rehm R, *et al.* InAs/GaS bsuperlattice focal plane arrays for high-resolution thermal imaging. Opto-Electron Rev 2006; 14: 19-24.

[41]    Handbook of Infra-red detection technologies edited by M. Henini and M. Razeghi, Elsevier 2002.

[42]    Beaudoin J-L, Merienne E, Danjoux R, Egee M. Numerical system for infrared scanners and application to the subsurface control of materials by photo thermal radiometry. Infrared Technology and Applications Proc SPIE 1985; 590:287–92

[43]    Balageas DL, Boscher DM, Deom AA, Fournier J, Gardette G. Measurement of convective heat transfer coefficients in wind tunnels using passive and stimulated infrared thermography. Rech Aerosp 1991; 4:51–72.

[44]    Andreopoulos J. Heat transfer measurements in a heated jet pipe flow issuing into a cold cross-stream. Phys Fluids 1983; 26: 3201–10.

[45]    Balageas DL, Krapez JC, Cielo P. Pulsed photothermal modeling of layered materials J Appl Phys 1986; 59:348–57.

[46]    De Luca L, Carlomagno GM, Buresti G. Boundary layer diagnostics by means of an infrared scanning radiometer. Exp Fluids 1990; 9: 121–28.

[47]    Baughn TV, Johnson DB. A method for quantitative characterization of flaws in sheets by use of thermal-response data. Mater Eval 1986; 44: 850–58.

[48]    Meola C, de Luca L, Carlomagno GM. Azimuthal instability in an impinging jet: adiabatic wall temperature distribution. Exp Fluids 1995; 18: 303–10.

[49]    Meola C, de Luca L, Carlomagno GM. Influence of shear layer dynamics on impingement heat transfer. Exp Therm Fluid Sci 1996; 13:29–37.

[50]    Vicinanza D, Meola C, Carlomagno GM, Di Natale M. Temperature distribution of a hot water discharge in a wave environment. Proc Medcoast 01 Int Conference on the Mediterranean Coastal Environment (Hamamet) 2001; 3:1177–88.

[51]    Grinzato E, Bison PG, Marinetti S. Monitoring of ancient buildings by the thermal method. J Cult Herit 2002; 3: 21–9.

[52]    Meola C, Carlomagno GM. Recent advances in the use of infrared thermography. Meas Sci Technol 2004; 15: 27-58.

[53]    Carlomagno GM, de Luca L. Infrared thermography in convective heat transfer Handbook of Flow Visualization ed W J Yang (London: Taylor and Francis) 2001, ch 34, 547–75.

[54]    Wisniewski M, Lindow S, Ashworth E. Observations of ice nucleation and propagation in plants using infrared video thermography. Plant Physiol 1997; 113: 327–34.

[55]    Hooshmand H, Hashmi M, Phillips EM. Infrared thermal imaging as a tool in pain management an 11 year study: II Clinical applications. Thermol Int 2001; 11: 119–29.

[56]    Meola C, Carlomagno GM. Application of infrared thermography to adhesion science. J Adhes Sci Technol 2006; 20 (7): 589-632.

[57]    Wu D. Lockin Thermography for Defect Characterization in Veneered Wood, Qirt 94, Eurotherm Series 42, D. Balageas, G. Busse, G. M. Carlomagno Eds. EETI editions, 1994; 298-302.

[58]    Meola C. Infrared thermography of masonry structures. Infrared Physics and Technology 2007; 49 (3): 228-33.

[59]    Meola C, Carlomagno GM, Di Foggia M, Natale O. Infrared thermography to detect residual ceramic in gas turbine blades. Appl Phys A 2008; 91: 685-91.

[60]    Meola C, Carlomagno GM. Impact damage in GFRP: new insights with Infrared Thermography. Composites Part A 2010; 41: 1839-47.

[61]    Proc. Annual SPIE Thermosense Conferences, Bellingham, WA 1978.

[62]    Proc. Advanced Infrared Technology and Applications (AITA) 1991.

[63]    Proc. Biennial Quantitative Infrared Thermography (QIRT) Conferences, 1992.

[64]    Proc. Annual Inframation Conference 2000.

# CHAPTER 2

## Infrared Devices: Short History and New Trends

## Roberto Rinaldi[*]

*Infrared Training Center, FLIR Systems, Via L. Manara 2, I-20051 Limbiate, Milano, Italy*

**Abstract:** This chapter deals with the historical evolution of infrared technology from the first prototype developed in 1958 to the multitude of models, which are available today. Particular attention is devoted to the detectors' working principles, their characteristics and performance as well their main fields of application. A section is devoted to radiometric systems involving: image generation and uniformity correction.

**Keywords:** Infrared device, historical evolution of infrared devices, performance of infrared cameras, IR image generation, cooling systems of IR detectors, IR detector materials, uncooled detectors, radiometric systems.

## 1. INTRODUCTION

It is obvious that the development of Infrared (IR) Technology started for military purposes in the early 1950s. Sharing or not the idea about the origin of the investigation, as many other electronic technologies like this, we are now getting great benefit on using it for many civil applications. The fact that we can make the heat *visible* can lead us to corrective actions and conclusions much faster than before with more advantages in term of result efficiency, personnel safety and economical decisions. The applications can span in many areas of interest and they are just limited by the fantasy of the users.

Below (Fig. **1**) there is a short panorama of examples, just to introduce the benefits deriving from the use of an infrared imaging device. In particular, Fig. **1a** shows an injured limb. As known, the lower legs are prone to contusions as a consequence of running into or tripping over a piece of furniture or other obstacles. An infrared camera may help to visualize the affected area and locate eventual presence of hematoma within the muscle especially for deep contusions before any inflammatory response starts resulting in swelling and further tissue injury.

Fig. **1b** represents coupling between an electrical motor and a mechanical device. One of the two ball-bearings appears hotter accounting for local extra friction there. This clearly indicates that maintenance needs to be scheduled at any convenience. In this case, IR thermography is not the only diagnostic method; vibration analysis is required too. The advantage of using an IR camera resides in the possibility to locate the problem very quickly anyway.

Fig. **1c** displays a high temperature pressurized vessel. The hot zone, which is clearly visible on the wall (inside the rectangle) is showing us that there is a local weakness in the insulation material. The consequence of this may be a very serious problem since the external metal sheet can reach its critical temperature and compromise the whole process affecting also the safety of the personnel working around such a equipment. It is once again a preventive maintenance action easy and quick to perform with the aid of a remote infrared imaging device.

Fig. **1d** shows the result of a routine inspection on aircraft where is clearly visible the presence of water intrusion in the wall structure. The presence of water inside aircraft structures leads to ice formation at the cruise altitude and to liquid water again after landing. This also involves variation of volume with consequent mechanical stresses, which affect the structure life plus additional weight to carry on.

***Address correspoedence to Roberto Rinaldi:** Infrared Training Center, FLIR Systems, *Via* L. Manara 2, I-20051 Limbiate, Milano, Italy; E-mail:roberto.rinaldi@flir.it

Fig. **1e** is an example of a typical building inspection targeting heat losses on the wall surfaces. The cost of the energy today is very high, and it will be higher for sure in the future. Building classification is now mandatory in Europe and in many other countries in the world; in this context, an IR camera represents an helpful and perhaps unique tool for the classification procedures as well for indicating the due corrective actions to take.

In Fig. **1f** is shown a Photovoltaic solar panel (PV) picture with superimposed an IR image; the inspection was performed during standard working conditions. Some of the cells, which in figure are either enclosed in a rectangle, or indicated with a spot, show higher temperatures which are not in the normal range. In these conditions the yield of the module will be dramatically lower and since the PV panels are usually connected in series, the efficiency of the system will be lower. According to the selling price of the electricity in the installation place, a rapid economical calculation will make clear the convenience to replace the single module or not.

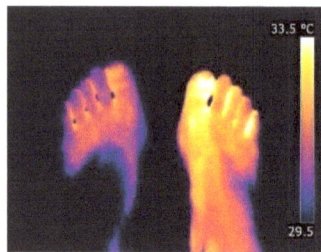

a) Medical field: contusion on the right lower leg.

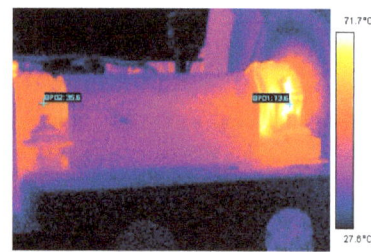

b) Preventive and predictive maintenance: hot ball bearing

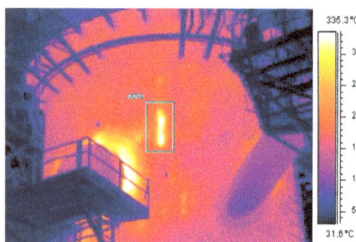

c) On line safety process control: worn refractory thickness

d) Non Destructive Testing: Water intrusion in aircraft composite structures

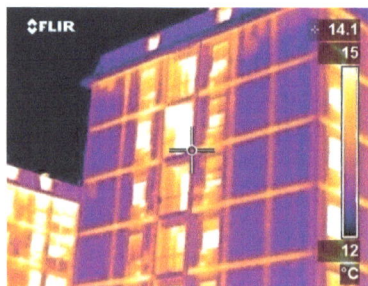

e) Energy efficiency control on buildings: Thermal bridges and insulation faults

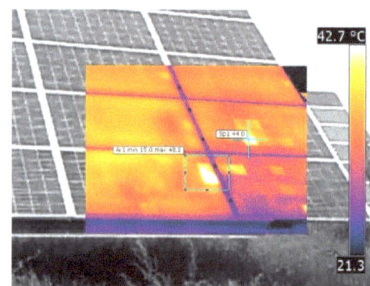

f) Research and development: defective cells

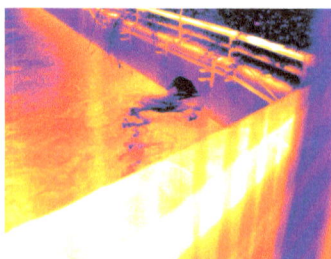

g) Environmental pollution control: Oil Spill

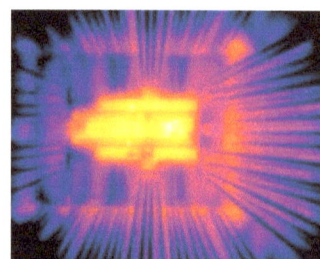

h) Electronic chip: image taken with macro lens

**Figure 1:** Panorama of thermal images © 2010 Flir systems.

Fig. **1g** is showing the possible use of the IR technology to detect oil and, more generally, other fluid leakages, which causes environmental pollution around industrial facilities.

At last, Fig. **1h** is describing how the electronic industry can benefit about the use of the IR technology to improve the reliability of the chip cores by analyzing the working temperature of it and prove the spec limits at certain working conditions.

## 2. THE EVOLUTION OF INFRARED DEVICES

IR Camera, Control Unit, tripod plus ancillary items constituted the equipment for the thermographer in the 1960s. The total weight for such a configuration was about 70kg and a van was part of the system configuration. Fig. **2a** is showing the typical work situation during an outdoor IR inspection.

Fig. **2b** is showing one of the smallest hand held infrared camera today's available, the weight is slightly above 300 grams. The performance of this camera model is much higher compared to the ancient predecessor. Table **1** below gives a forced comparison between the two camera models.

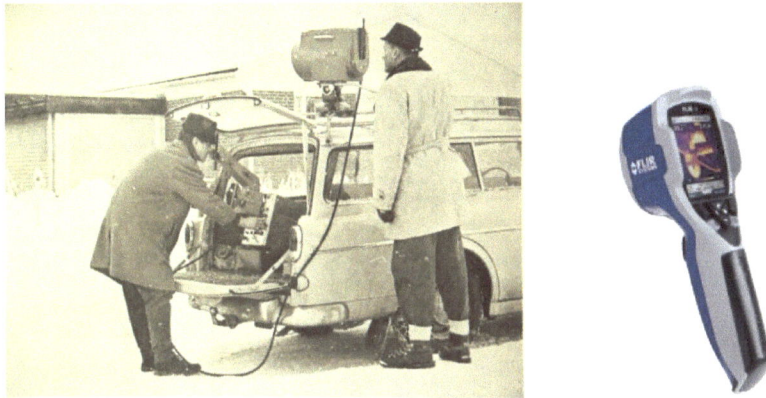

a) IR inspection in 1960s with the AGA Thermovision 661.    b) IR inspection in 2010 with the FLIR i7.

**Figure 2:** Past and present IR arrangements for outside inspection.

**Table 1:** Characteristics of past and today cameras.

|                     | **AGA Thermovision 661**                  | **FLIR i7**                |
|---------------------|-------------------------------------------|----------------------------|
| Weight              | 25kg (only camera)                        | 340 grams                  |
| Size                | Estimation: 450mm x 300mm x 400mm         | 223mm x 79mm x 83mm        |
| Sensitivity         | 0.2°C                                     | 0.1°C                      |
| IR Image size pixels| 80x80                                     | 120x120                    |
| Detector            | Single element InSb LN2 cooled            | Uncooled Micro bolometer   |
| Field of View       | 5°x 5°                                     | 25°x 25°                   |
| Display             | Standard Fairchild oscilloscope           | Color LCD 2.8"             |

The systems technology evolution is running today quite fast, while the products price level is becoming ever more affordable. The consequence of this will be the spread of the technology on the market with a tremendous increase of applications.

### 2.1. Some History of IR Systems

The idea to produce Infrared equipments started at the Swedish Defence Department in cooperation with AGA in the early 1950s. The very first prototype was based on a thermistor bolometer detector type, which is almost the technology we are using nowadays. The principle of operation was pretty simple and smart, but the time required to obtain the thermogram was very long, fifteen minutes.

In Fig. **3** it is depicted the scanning principle. The incoming heat radiation was deflected by the tilting plane mirror through other optical mirrors and then to the IR sensor. A chopper placed in the internal optical path was able to create the differential signal from the outside scene. This signal was used to modulate the intensity of a glow lamp placed on the back side of the flat scanning mirror and the generated light from the lamp impressed a standard photographic film. The time needed to create the final picture was 15 minutes, but then there was the need to develop the film and only after it was possible to recognize if the picture was good or not.

The system evolution went on and at the end of the 1950s a second prototype named TS1 (Thermal Sight) was presented. To make the image presentation faster, a higher number of sensors were used and a different scanning system was implemented. Fig. **4** is showing a photo of this prototype; below are listed the technical features of it.

Technical features of TS1:

- Vertical scanning: Lens controlled by camcurve for sawtooth scanning, 2.4Hz.
- Horizontal scanning by turning the camera body back and forth, 6.1s.
- Field of View 2.5 x 10 degrees.
- Thermal sensitivity for the first prototype in 1959 was 20°C.

The demand for higher sensitivity and faster frame rate, made the researcher to move further on and, in the beginning of 1960s, the Swedish Defence Department (FOA) started to use a new scanning principle together with a fast InSb LN2 cooled detector. Two prototypes were made available, one with a typical reflecting optic, vertical tilting mirror and a four sides reflecting drum. The other one was based on the same concept, but it included a six sides rotating drum. In Fig. **5** it is shown the camera prototype that was mainly developed to target tanks. Below are listed its technical features.

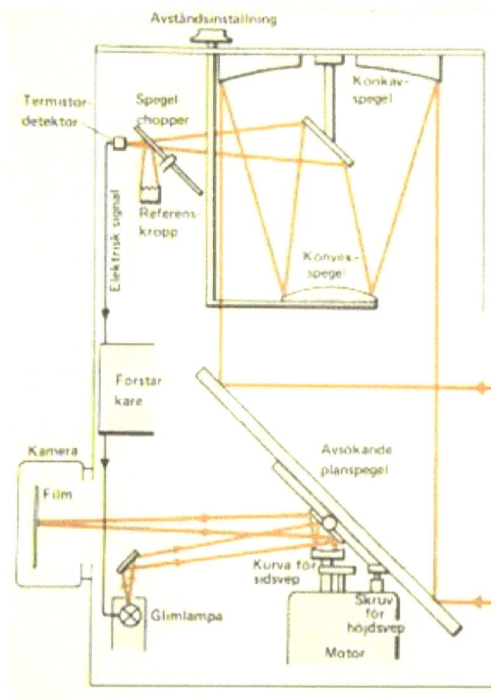

**Figure 3:** First IR camera prototype principle © 2010 Flir systems.

Technical features of AGA infrared Sight:

- Horizontal: Rotating 4-sided reflecting prism.

- Vertical: Tilting Mirror.

- Detector: Single element InSb LN2 cooled.

The military version available one year later:

- 6-sided reflecting prism.

- Tilting mirror.

- 10 fields per second.

- 83 lines per field.

- FOV 5x10 degrees.

- Single element detector InSb or MCT LN2 cooled.

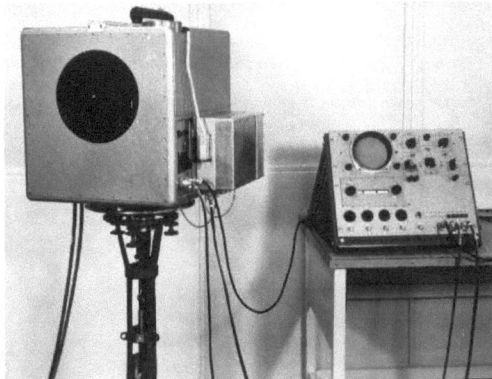

**Figure 4:** Photo of the Thermal Sight 1.

**Figure 5:** AGA Infrared Sight (1963).

In the meantime a new scanning concept was designed and patented by FOA and then tested by AGA. The novelty about such a scanning system was a fast refractive horizontal prism.

Between 1964 and 1965 AGA presented the very first commercial infrared system named AGA Thermovision. Three models were made available to the market, which were named 661, 665 and 669; they included the same camera body 1 but three different lenses. Fig. **6** is showing the camera Thermovision 661 at work. This system was mainly developed for predictive maintenance on electrical power lines and after adapted, using different lenses types, for other purposes.

Technical features of AGA Thermovision 661, 665, 669:

- Horizontal: Rotating refractive 8 sided Silicon prism 200 rps,1600 lines per second.

- Vertical: Tilting mirror 16Hz, 16 pictures per second.

- Detector: Single element InSb LN2 cooled with Dewar allowing 4 hours holding time.

- Sensitivity: 0.2°C.

- Lenses: 5x5 degrees (661), 11x11 degrees (665) and close up 5x5 degrees (669).

- Image presentation: Standard Fairchild oscilloscope.

**Figure 6:** AGA Thermovision 661 used for electric power line inspection. Picture taken in 1969.

The only way to document the thermographic inspection was to take pictures of the oscilloscope screen and, in the most sophisticated case, the picture could also be a color picture by using an eight colors filter wheel with an isotherm function on the live screen image. Each isotherm step was using an exposure time of 15 seconds. Fig. **7** is showing the photo adaptor device with a 16 mm cine camera mounted.

In the 1968 a new infrared camera was released. The optical scanning system was totally redesigned using the refractive principle for the horizontal and the vertical scan. The advantage of such improvement was to have a more compact camera body with straight optical path. This opened the way to have many interchangeable lenses using the same camera body. Fig. **8** is showing this camera model; below are listed its main features.

Technical features of AGA Thermovision 680:

- Scanning: 2 refractive prisms Hor.+ Vert.
- Detectors:

- SW-Single element InSb, LN2 cooled Dewar with 4 hours holding time.

- LW-Single HgCdTe (MCT) cooled in the same way.

- Lenses FOV: 8°x8°, 15°x15°, 25°x25°, 45°x45° and Microscope lens.

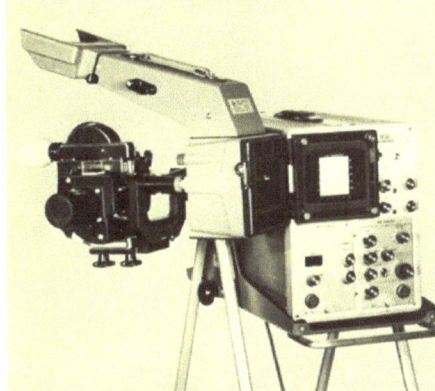

**Figure 7:** Photo recording attachment for AGA Thermovision 661/665/669.

The more compact optical path design allowed the possibility to internally add two mechanical wheels, one for the spectral filters and the other one for the apertures. This made possible the camera calibration on a wide temperature range. Fig. **9** is showing the set of calibration curves supplied with the camera.

An important set of accessories became quickly available. In Table **2** are mentioned the most important.

Based on the Thermovision 680 experience, in the 1973 a more compact IR camera model was released *i.e.*, the, Thermovision 750 (Fig. **10**). The main features of this camera were: a smaller size (downscaled the Thermovision 680 twice), small 5" display, battery operation for a real portable IR system. Fig. **10a** is showing this device that is carried by only one person, while Fig. **10b** shows a cut view.

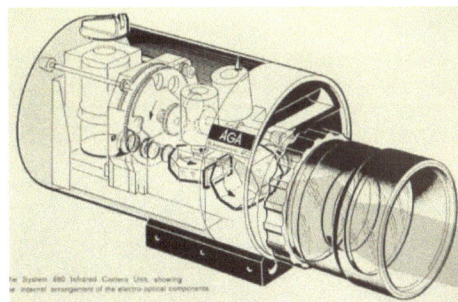

**Figure 8:** AGA Thermovision 680 – 1968 (SW version).

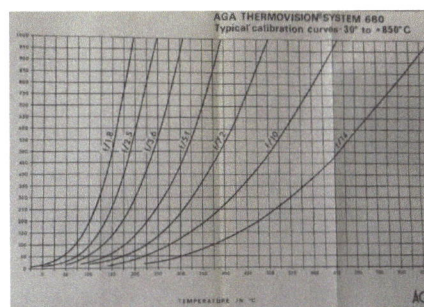

**Figure 9:** Typical calibration curve sheet for Thermovision 680, the range of calibration was from-30°C up to 850°C.

Since this system was really portable, it was especially dedicated for predictive maintenance check in an extended variety of industrial applications. A special version for the petro-chemical industry (furnace inspections) was also available.

Going further on in the camera evolution, 1978 was the year of presentation for the Thermovision 780. This camera represents the first most complete equipment solution for thermography investigations especially in R&D applications. Fig. **11** is showing the complete set up available with this system. A dual wavelength integrated camera system was a possible choice.

**Table 2:** Some of the most important accessories.

| | |
|---|---|
| | Color Monitor<br><br>Possibility to display up to 10 colors and gray scale on large screen. |
| | OSCAR<br><br>Off line System for Computer Access and Recording.<br><br>A/D converter for video signal and storage on digital tape. First post analysis system. |
| | Photo attachment for different camera models. Color wheel version available. |
| | Microscope lens.<br><br>Magnification factor: 15x, 50x and 125x |

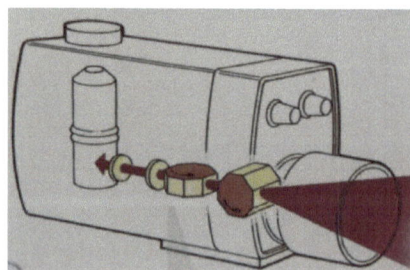

b

a

**Figure 10:** AGA Thermovision 750 – 1973 AGA Thermovision 750 – optical design.

**Figure 11:** Thermovision 780 configuration (1978).

The development of the Thermovision 780 was the 782 model, presented in 1982. This camera was suitable for field applications as well as for Laboratory applications. The configurations were including a computer for R&D analysis or a pocket calculator for direct temperature conversion on the field. Fig. **12** is showing these options.

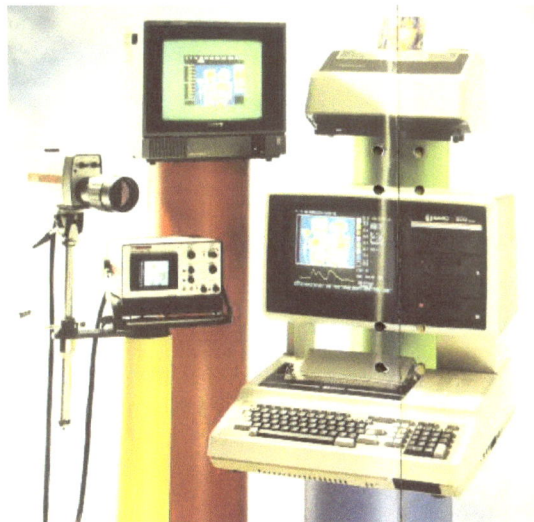

a) Thermovision 782 and BMC PC

b) Pocket calculator

**Figure 12:** Thermovision 782 configuration (1982).

1984 was the year of presentation of the Thermovision 110, which was the smallest IR imager without need of LN2 cooling and having good performance. Fully derived from military equipment, it was made available to the civil market by only doing some minor adaptations (connectors, battery packs and chargers). Fig. **13** is showing the very compact Thermovision 110. Below are listed the Thermovision 110 main specifications:

- Detector: 48 elements array PbSe detector vertically oriented, thermoelectrically cooled 2 stage.

- Scanning: Horizontal scanning by flat mirror.

- FOV 6x12 degrees.

Because this instrument was just an imaging system mainly used for maintenance applications, it was usually sold together with a standard optical pyrometer. The camera was used to locate the fault, the pyrometer was used to measure the temperature. This solution was considered the cheaper alternative to the more expensive Thermovision 782. Fig. **14** is showing the radiometric part for the Thermovision 110, the well known AGA TPT 80. This instrument was a result from a cooperation between AGA and Raytek Inc.

**Figure 13:** Photo of the Thermovision 110.

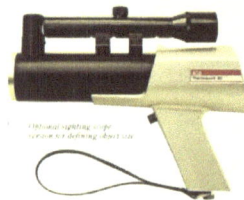

Detector: single element Thermopile

Measurement range: from ambient up to 3000°C

**Figure 14:** Photo of AGA TPT 80–1979.

Going further on, in the 1986, Thermovision 870 was introduced on the market and this was really a milestone event. The design of this product was a great success mainly for two reasons:

- Compact optical path back to the original reflecting principle.

- No liquid Nitrogen required.

Fig. **15** is showing the Thermovision 870 camera optical path cross-section where is visible the radiation path on the multiple mirrors available inside. Below are listed the Thermovision 870 Specifications:

- Scanning concept.

- Horizontal: 10 facetted polygon – 2500 lines, long life brushless motor.

- Vertical: Tilting mirror – 25 Hz – Fiber optics position controlled.

- 100 lines per field.

- Detector-Thermoelectrically cooled 4 stages Peltier cell, 1 element MCT SPRITE SW 2 to 5μm.

- Aperture and Filter turrets.

- All internal electronics controlled by μ processor.

Two internal micro reference sources were included to increase the temperature readout accuracy and avoid video drift.

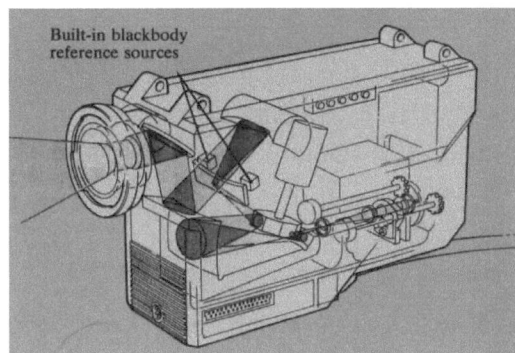

**Figure 15:** Thermovision 870 cross-section – 1986.

Thermovision 870 was the first of a camera generation that was capable to satisfy the most demanding IR applications for a long period of time. The reason of this success was the full range of detectors, lenses and accessories available. Figs. **16-19** show the possible system configurations available with this camera series.

Three detector options were possible:

- SWTE-Thermoelectrically cooled one element MCT SPRITE– 2-5.5μm.

- LW-LN2 cooled single element MCT – 8-12μm.

- SW-LN2 cooled single element InSb – 2-5.5μm.

Presented two years later, the further development of the Thermovision 870 concept was the Thermovision 470. The detector cooler free, the integration of more digital components on board, made possible to have a single piece portable camera with digital image recording on floppy disk. This represented a great improvement step for people working on maintenance facilities. Fig. **20** is showing this camera model appearing like a professional TV camera. Below are listed some specifications:

- Scanning concept LK4.

- Horizontal: 10 facetted polygon – 4000 lines.

- Vertical: Tilting mirror – 20Hz.

- 140x140 pixels per field.

- Detector: Thermoelectrically cooled,-60°C, SW MCT SPRITE – Sensitivity 0.1°C.

- 70 images stored on standard 3.5" floppy disk.

- Market introduction 1988.

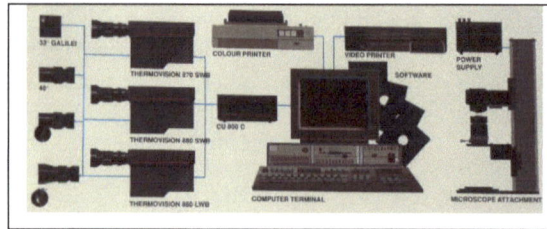

**Figure 16:** Thermovision 800 configurations.

**Figure 17:** Thermovision 800 Dual Version with 5° Fov motorized lenses.

**Figure 18:** Thermovision 800 scanner with 2.5° Fov catadioptric lens.

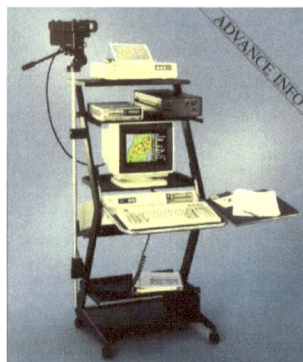

**Figure 19:** Thermovision 800 complete system setup for laboratory use with dedicated TIC 8000 computer interface.

**Figure 20:** Thermovision 470 with integrated 3.5" floppy disk.

The optical scanning module LK4 was further developed and designed to be a separate part of the camera body and this step made possible to release in 1991 two successful products such as Thermovision 900 and Thermovision 1000. The first one was dedicated to the research and development segment with a wide range of scanner versions, lenses and accessories. The second one was more intended to be used for surveillance applications. Fig. **21** is the drawing showing the compactness of the new LK4 module. The size of this module is about 10cm x 10cm x 6cm, all mirrors inside are made in aluminium polished with high precision diamond turning machine.

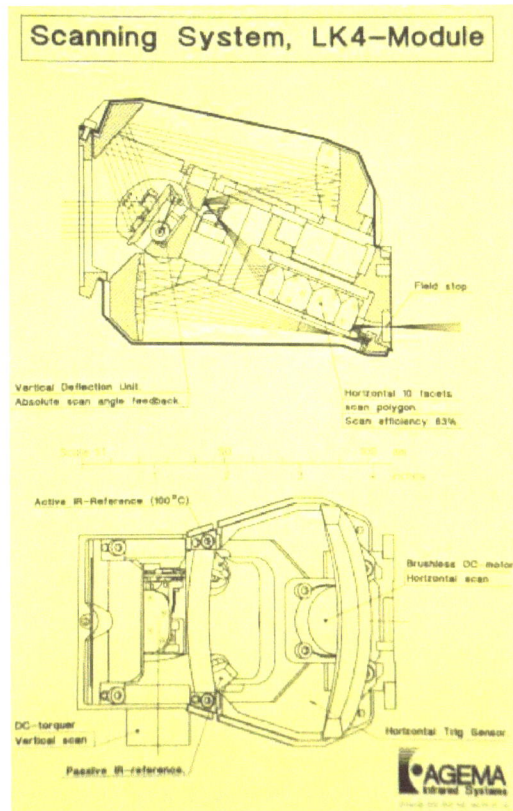

**Figure 21:** LK4 scanning module used on Thermovision 900 and 1000.

Fig. **22** shows the Thermovision 900 configuration. It was the first window based 12 bits digital IR camera. The dedicated Control Unit was capable to drive two scanners in real time with display and recording. Below are listed the main specifications:

- Scanner Detector type: LW: MCT single element.

- SW: InSb two element serial averaging.

- SW: MCT SPRITE.

- Cooling: LN2 or Stirling cycle micro 4 stages TE cooled (SPRITE version).

- Spectral Range: LW: 8 – 12µm.

- SW: 2 – 5.6µm.

- Frame frequency: 15Hz or 30Hz.

- 20Hz or 40Hz (SPRITE version).

- Spatial resolution: 230 pixels/line @ 50% modulation.

**Figure 22:** Thermovision 900 possible configuration.

- Fig. **23** is showing the Thermovision 1000; below are listed the technical specifications.

- Detector: Type 5 elements MCT SPRITE.

- Wavelength LW, 8-12μm.

- Cooling system: Integral Stirling cooler.

- Scanning system: opto-mechanical reflecting mirrors.

- Scanning direction Horizontal 18000RPM; Vertical 25 or 30Hz (Pal/Ntsc).

- IFOV/Line 565.

- Scanned image: NTSC/VGA PAL.

- Frame frequency: 30/25Hz (Ntsc/Pal).

- Line frequency: 3000Hz.

- Image Presentation: NTSC/VGA PAL.

- Sampling raster H/V 798/400 798/445.

- Electronic Zoom: Continuous 8x.

- Lenses integrated dual FoV: Wide Narrow.

- IFoV (HxV) 0.6x0.6 mRad 0.15 x 0.15mRad.

- FoV (HxV): wide 20° x 13°; narrow 5° x 3.3°.

- Near focus: 1m to 10m.

- Lens switch time: less than 1s.

- Physical dimensions: LxWxH = 310mm x 164mm x 221mm.

- Weight: 8kg.

Sophisticated high resolution radiometric thermal imager, thanks to the multi elements detector, it was able to deliver high geometrical resolution 12 bits thermal images. The stability and the good temperature readout accuracy made this system the right solution for thermographic inspections of electrical power lines on board of helicopters. Many other different variants of this model were produced to fulfill the majority of the surveillance applications.

The image quality of this camera was superb as it is possible to see in Fig. **24**.

**Figure 23:** Thermovision 1000 first version – 1991.

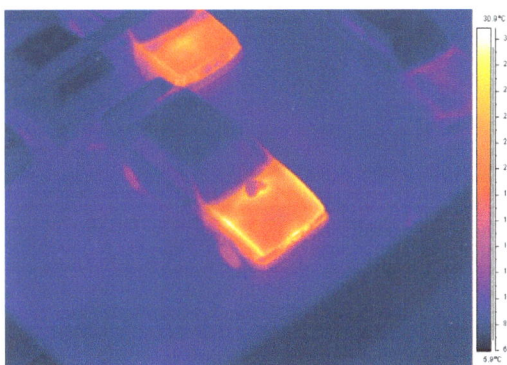

a) Image taken with 20° Lens.

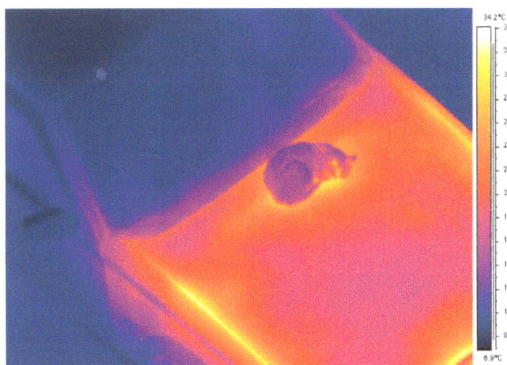

b) Image taken with 20° Lens.

**Figure 24:** Thermovision 1000 images taken with a) 20° Lens b) and 5° Lens, © 2010 Flir systems.

Even though Thermovision 110 and Thermovision 1000 were using a multi elements detector technology, the era for the detector array cameras (FPA detectors) started for the civil thermography market in the 1995. The first radiometric camera presented by Agema, at that time, was the THV 550. Fig. **25** is showing this product with a thermal image underneath; below there are the most important specifications:

- Detector FPA PtSi 320 x240 – Stirling cooler.
- Spectral range 3.6 to 5µm.
- Sensitivity 0.1°C @ 30°C.
- Integrated lens FOV 20°Hx15°V.

- Voice comment.

- Temperature range:-20°C to 250°C (+1500°C optional).

- Video NTSC/PAL output.

- 12 bit digital images.

- Image recorded on PCMCIA type III card.

- Battery belt with 4 batteries.

- Startup time <6 minutes.

a) THV 550 – 1995.

b) THV 550 image quality © 2010 Flir systems.

**Figure 25:** THV 550, a) photo and b) thermogram.

The FPA, or starring array, technology was mainly developed for the military. In 1997 Agema was able to present the first radiometric un-cooled camera based on a micro bolometer detector, (AGEMA 570). As Fig. **26** is showing, the camera was recalling the successful and ergonomic THV 550 design. Below are listed its specifications:

- Detector Focal Plane Array (FPA), Un-cooled microbolometer, 320 x 240 pixels 7.5-13 µm.

- Sensitivity 0.1°C @ 30°C.

- Integrated lens FOV 24°Hx18°V.

- Voice comment.

- Temperature range:-40°C to 500°C (+1500°C or 2000°C optional).

- Video NTSC/PAL output.

- 14 bit digital images.

- Image recorded on PCMCIA type III or type II card.

- Weight 2.3kg with battery.

- Start up time 45 seconds.

Fig. **26b** shows the image of a typical hand print on a wall taken with an AGEMA 570 un-cooled detector. It is visible the typical grained image available with the first generation of un-cooled detectors.

a) THV 570 – 1997.

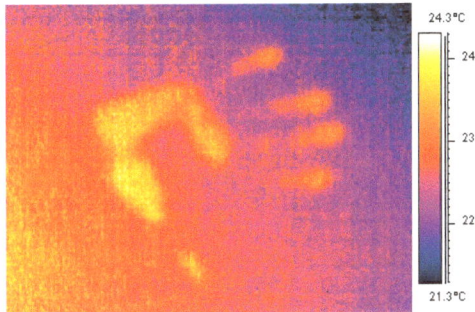

b) THV 570 image quality © 2010 Flir systems.

**Figure 26:** THV 570, a) photo and b) thermogram.

The good result obtained with this product opened the way to other important products. In the same year was presented the SC3000 model based on the same AGEMA 570 electronics. In Fig. **27a** is depicted the ThermaCAM SC3000 based on a new detector technology called QWIP (Quantum Well Infrared Photo detector) a cooled sensor with very high sensitivity. Fig. **27b** is showing the image quality of the ThermaCAM SC3000. In this case the camera was used to detect micro spy transceivers (bugs) hidden in the wall structure. The low noise detector allowed for a detailed and sharp image, considering the very narrow thermal span used, only 1°C.

a) SC3000 photos.

b) Image taken with the SC3000 © 2010 Flir systems.

**Figure 27:** ThermaCAM SC3000; a) photos; b) example of thermal image.

Below are listed the SC3000 specifications:

- Detector FPA QWIP 320 x240 – Stirling cooler.

- Spectral range 8 to 9µm.

- Sensitivity 0.03°C @ 30°C.

- Integrated lens FOV 20°Hx15°V.

- Temperature range:-40°C to 500°C (+1500°C optional).

- Video NTSC/PAL output.

- 14 bit digital images.

- Image recorded on PCMCIA type III or type II card.

- Startup time <6 minutes.

Due to the high image quality, Therma CAM SC3000 was mainly used for research and development, historical building restoration and medical applications.

Going further on with the detector evolution, the un-cooled micro bolometer detectors improved significantly their performance and, with a modified optical module, the sensitivity level went down to 0.08°C. Additional features such as Visual camera and text comment were added to the products to make reporting easier to the dedicated maintenance personnel. In the 2000 was presented the ThermaCAM 695 and the scientific version SC2000, a complete and versatile camera system. Fig. **28** is showing this camera model with a building façade image. Below are also listed its main specifications:

- Detector Focal Plane Array (FPA), Un-cooled micro-bolometer, 320 x 240 pixels, VOx – 7.5-13µm.

- Sensitivity 0.08°C @ 30°C.

- Integrated lens FOV 24°Hx18°V.

- Voice comment.

- Temperature range: -40°C to 500°C (+1500°C or 2000°C optional).

- Video NTSC/PAL output.

- 14 bit digital images.

- 50/60Hz image digital output to PC (only SC2000).

- Digital visual camera 640x480 pixels.

- Image recorded on PCMCIA type II card.

- Weight 2.2kg with battery.

- Start up time 45 seconds.

In the 2001 a new interesting project was presented to the market with a down sized 160x120 detector, ThermaCAM E2 (Fig. **29**). This model was the first compact and light weight camera that really started the low cost thermography market segment. With interchangeable lenses, simple user interface and very good ergonomics, this was the right product for a multitude of thermographic inspections and applications. In the beginning the camera could only measure temperatures and store ordinary.jpg images, but after a while also the radiometric capabilities were added to the system. *Via* a USB connection it was possible to download the images to a PC and prepare the inspection Report.

a) ThermaCAM 695.

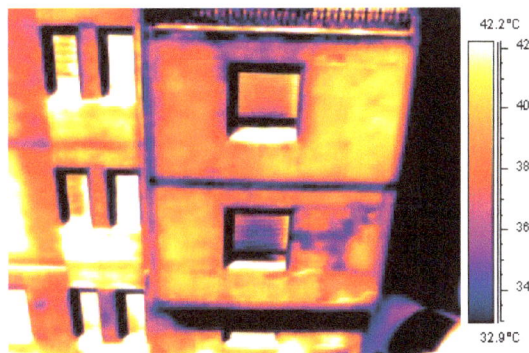

b) Thermogram, © 2010 Flir systems.

c) Visible picture

**Figure 28:** Photo of the camera 695 and an example of application to the building field.

**Figure 29:** ThermaCAM E2 – 2001.

Below are listed the main features of the ThermaCAM E2:

- Field of View: 25°x19° (interchangeable lens).

- Sensitivity: 0.12°C @ 25°C.

- Focus: Manual.

- Detector: Focal Plane Array (FPA), uncooled micro bolometer VOx 160 x 120 pixels – 7.5-13µm.

- Monitor 2.5" color LCD, 16K colors.

- Temperature range:-20°C to +250°C (+900°C optional).

- Accuracy ±2°C, ±2%.

- Measuring functions: Spot, area max, area min, area avg.

- Image storage: build in FLASH memory card (50 imgs).

- Laser pointer: class 2.

- Weight: 700g, battery enclosed.

- Size: 265mm x 80mm x 105mm.

- File format: Standard JPEG.

A further step of improvement for the professional Thermographer arrived in the 2002 with the presentation of the ThermaCAM P60. A 320x240 IR Camera with a very modern design and a large detachable display with build in remote control. Different versions of this model were developed to adapt this camera to the different market demands. It was also created a specific model for scientific applications named S60 with fast speed firewire video digital output and a unique model for petrochemical (furnace) applications, called P60F, with a special micro bolometer detector tuned at 3.9µm. Fig. **30** is showing the P/S 60 while below are listed the main specifications:

- Detector: uncooled IV generation microbolometer FPA, 320 x 240 pixel VOx – 7.5-13µm.

- Autofocus.

- Sensitivity < 0.08°C @ 30°C.

- Temperature range: -40°C to + 500°C (+1500°C or + 2000°C optional).

- Spatial resolution: 1.3 mrad (24°x18° integrated lens).

- Image frequency: 50/60Hz.

- Integrated 640x480 Visual camera.

- Integrated LCD high resolution color viewfinder (TFT).

- Detachable LCD display with remote control.

- Voice comment (30s).

- Text comment.

- Class 2 laser pointer.

- Measuring functions: combination of spots, areas, profiles and isotherms.

- USB and RS 232.

- Video NTSC/PAL and S-Video outputs.

- 14 bits image digitalization.

- IEEE 1394 radiometric Firewire interface (S60).

- Image storage on compact flash (128Mb).

- IRDA communication port.

- Weight: < 2kg, including LCD display.

**Figure 30:** ThermaCAM P60 – 2002.

The P Series established a new way to develop IR cameras with more communication features, by using standard interface protocols and standard storage devices. The development continued by increasing the image size and in the 2006 we could see the launch of a fast IR Camera using a large size cooled photon detector, the SC 6000 HS. Fig. **31** is showing this camera model and a typical thermal image; it was intended for R&D application and for use mainly in laboratory. This camera was made available with three detector types InSb, InGaAs and QWIP based on the same hardware platform.

Below are listed the main technical specification of the SC6000 camera:

- Detector 640 x 512 InSb, Spectral band 3-5μm (1.5-5μm optional), with linear Stirling cooler.

- Sensitivity 18mK.

- Integration type: Snapshot.

- Variable integration time from 10μs to 16ms.

- 14 bits Dinamics.

- Superframing mode for the dinamics extension to 22 bits using external software (Rtools).

- Image frequency: 125Hz @ full frame; up to 36kHz in windowing mode.

- User selectable detector window.

- Calibration range – 10°C a + 1500°C.

- Camera control *via* standard Gig-E.

- Image recording *via* Gig-E or Camera Link.

- PAL video output.

a) Photo.

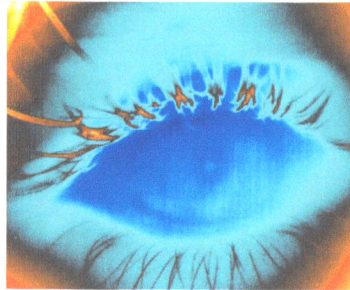

b) Example of 640 x 512 pixels image, © 2010 Flir systems.

**Figure 31:** SC 6000 HS – 2006.

In the 2007, based on the same camera design, was presented the SC 8000 with a square 1K detector size. The demand of larger detector size affected also the micro bolometer detectors and in the 2006 was presented a large format 640x480 radiometric hand held camera named ThermaCAM P640. In 2008 a better version of this detector type with higher sensitivity and innovative camera features such as GPS, high resolution Visual image, image fusion, made available the P660 (Fig. **32**). Below are reported some technical specifications:

- Sensitivity: <30mK.

- Accuracy: ±1% or ±1°C.

- Equipped with:

    - Alarm functions,

    - BURST rec. in camera,

    - MPEG 4 non-rad streaming to PC, Firewire/USB,

    - LaserLocatIR function,

    - Automatic GPS data,

Fig. **33** displays building details, which were taken with the ThermaCAM P640 (Fig. **33a**) and with the ThermaCAM P660 (Fig. **33b**). Owing to the same testing conditions and the same image setting of parameters (palette and temperature scale), it is clearly evident the superior resolution obtained with the P660. And then, it is obvious the choice of the latter when dealing with small and low thermal contrast buried anomalies.

**Figure 32:** P660 – 2008.

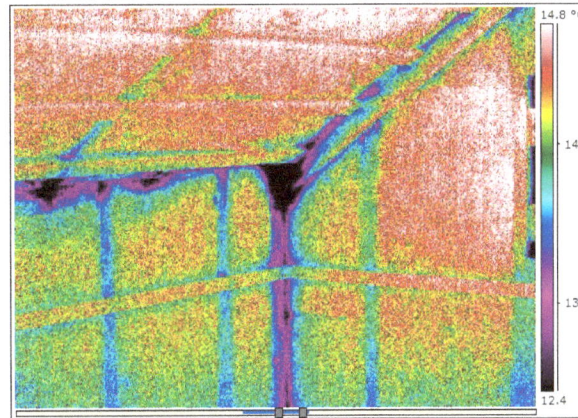

a) Image taken with ThermaCAM P640.

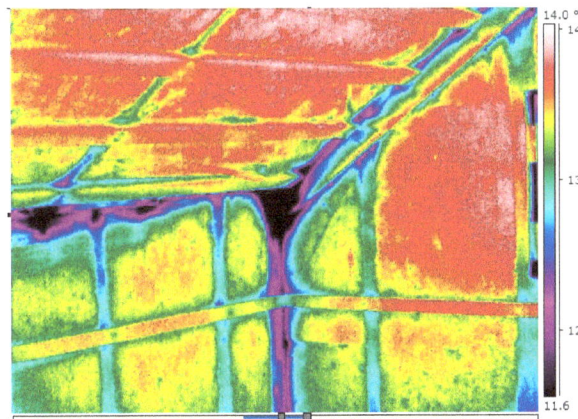

b) Image taken with ThermaCAM P660.

**Figure 33:** A comparison of thermal images taken with the ThermaCAM P640 (a) and the ThermaCAM P660 (b) © 2010 Flir systems.

## 3. DETECTORS

The detector is the core of an infrared device, it collects thermal radiation in some specific wavelengths. Since we are talking about radiative heat transfer in a means that is the atmosphere, the detector is designed to match the wavelengths where the radiative heat starts to occur and where the air is more transparent.

In principle it can be considered the right radio receiver tuned to see the heat waves naturally emitted by the objects having a certain temperature. Practically the scope of the infrared detectors is to generate an electrical signal when the incident IR radiation is hitting its surface.

### 3.1. Detector Families & Working Principles

Two detectors families are available and they are divided according to their physical working principle. In Table **3** below it is possible to find a summary for some detectors characteristics. In Table **4** are collected some examples of detectors encapsulated according to their working principle.

## 3.2. Typical Detector Materials

There is a large variety of detector materials available, which allows for production of different systems performances according to the final scope of the wanted equipment.

Thermal detectors can sense the infrared radiation in a quite large wavelength band, but as a drawback, because of their working principle, they are slow in terms of image refresh and changes of temperature.

Photon detectors are instead very fast in generating the signal and they are also able to capture fast temperature changes.

Detectors are also characterized for their Detectivity, generally called D*, which is indicating the level of sensitivity for a specific detector material. The detectivity is plotted against wavelength in Fig. **34** to respectively show the response for thermal (Fig. **34a**) and photon (Fig. **34b**) detectors. Table **5** reports a comparison between the main characteristics of the two detectors families.

**Table 3:** Characteristics of some detectors used for infrared technology.

| Thermal Detectors | | Quantum or Photon Detectors | |
|---|---|---|---|
| Working principle | Radiation hitting the detector surface changes the material temperature leading to the generation of a physical effect | Working principle | State change of electrons in a crystal as reaction from incident photons. |
| Detectors categories | - Bolometer<br>- Pyroelectric& Ferroelectric | | Radiation is directly converted in an electrical signal |
| Bolometer | Change in resistance; the generated signal is very stable in gain, is DC coupled, good to be radiometric | Advantages | - Very sensitive<br>- Very stable |
| Pyroelectric& Ferroelectric | Change in capacitor charge the generated signal is AC coupled, need chopper, and it is difficult to get radiometric | Disadvantage | Need cooling |
| Working temperature | Room temperature.<br>No cooling required | Working temperature | Need to be cooled at different temperatures according to the detector materials |

**Table 4:** Some examples of detectors.

| Single Element Detector | Array of Detectors |
|---|---|
| <br>MCT Sprite TE Cooled detector | <br>Uncooled Vox 320x240 array detector |
| <br>Dewar with LN2 cooled detector | <br>Stirling cooler (left) and typical FPA detector assembly (right) |

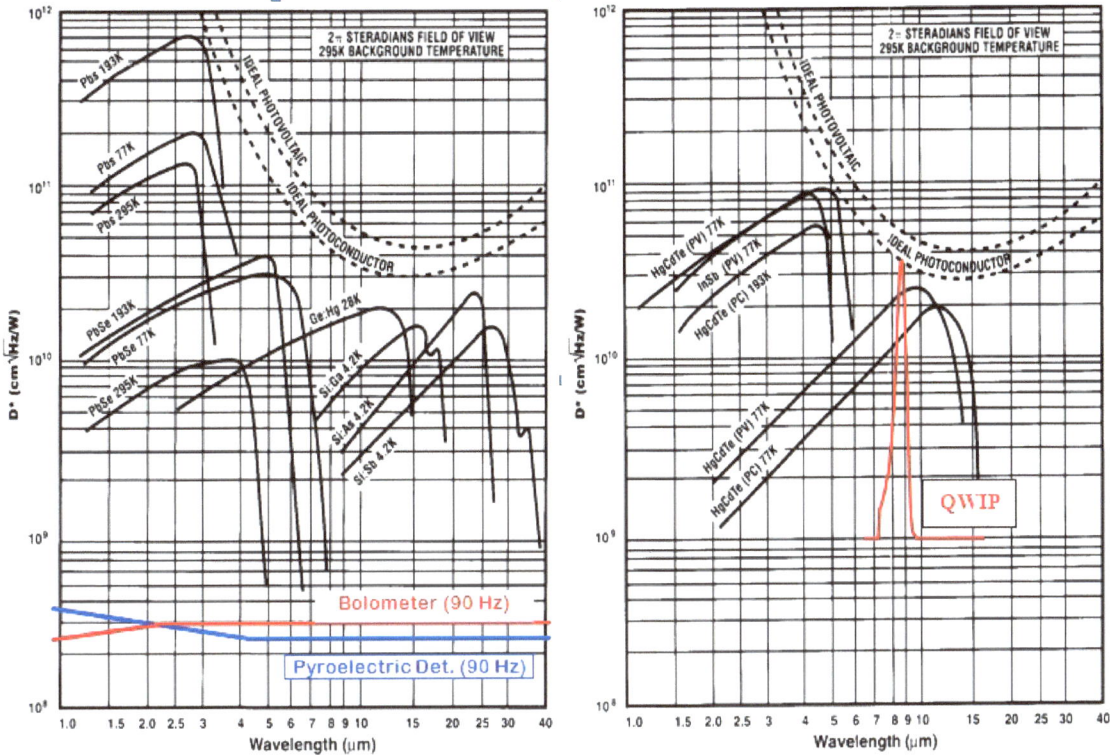

a) Thermal detector materials.

b) Photon detector materials.

**Figure 34:** Detectivity curves over wavelengths.

## 3.3. Cooling Systems

According to the detector material chosen to obtain the wanted system performance, it is necessary to control the substrate temperature of it.

For the Bolometer detector, it is extremely important to monitor the substrate temperature very well because the objects signal readout is just the difference from the detector signal. To do this there are two ways:

- Rely on an accurate and sophisticated substrate temperature measurement (pure un-cooled detectors).

- Stabilize the detector temperature, usually 30°-35°C, by using a Peltier element (stabilized detectors). A sketch is shown in Fig. **35**.

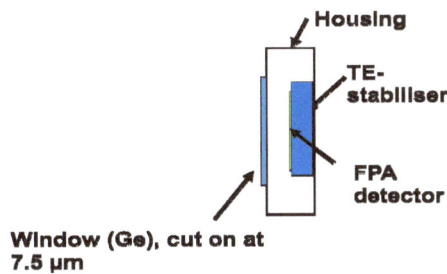

**Figure 35:** Sketch of the encapsulated bolometer detector.

For Quantum detectors where the need of very low substrate temperature is required, there are several ways to obtain the scope. A collection of cooled detector types is presented in Table **6**. The common technology on commercial infrared systems today is using un-cooled detectors and this is mainly for handheld systems. For laboratory systems with Quantum detectors, a rotary Stirling cooler is the preferred technology.

**Table 5:** A comparison between thermal and quantum detectors.

| Thermal Detectors (bolometers) | Quantum Detectors |
|---|---|
| Low sensitivity-$D*$» $10^{8}$-$10^{9}$ | High sensitivity-$D*$» $10^{11}$-$10^{12}$ |
| Slow response time-Time constant $t$ from 4 to 12ms | Fast response time –Time constant $t$ »µs (1µs ) |
| Sensitivity does not depend on the wavelength | Sensitivity strongly dependent on the wavelength |
| Uncooled (room temperature) | Cooled: (SPRITE (MCT –70ºC) (InSb:-100°, MCT:-196°C) (QWIP:-203°C) |

**Table 6:** Examples of cooled detector types.

| Cooler Type | Working Principle | |
|---|---|---|
| Multistage Peltier cells | Peltier effect: by applying voltage to a different material conjunction. Temperature max to 173K | Schema of 3 stage Peltier cooler |
| Open Cycle Joule/Thomson | Joule/Thomson effect: pressurized gas expansion causing the cooling effect. Temperature 77K | Schema of Joule/Thomson detector |
| Open cycle Dewar tank | Latent heat of boiling liquid, usually Liquid Nitrogen. Temperature 77K | Schema of Dewar detector |
| Closed cycle Stirling cooler | Closed cycle pressurized gas expansion. Temperature adjustable down to 70K | Schema of rotary Stirling cooler |

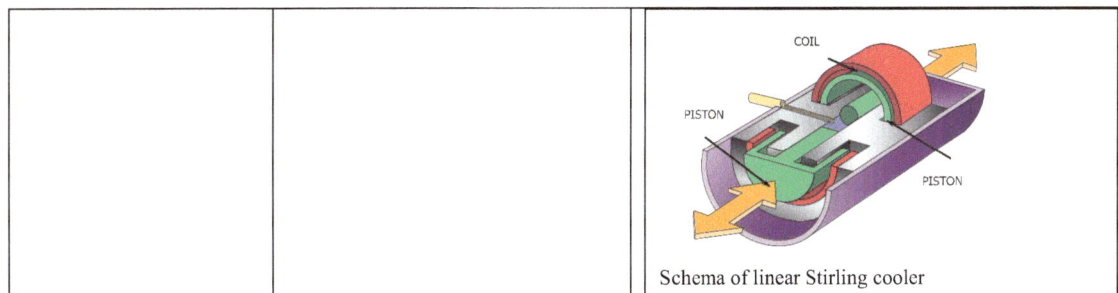

Schema of linear Stirling cooler

## 4. IR IMAGE GENERATION

### 4.1. Scanning Systems

Due to the fact that the first sensor technology available was made of a single element detector, the only way to obtain an IR image was to build up in front of the detector itself an opto-mechanical scanning system to generate the thermal pattern of the scene.

This technology went through several improvements in time since it was first introduced in the '50s. In particular, IR cameras with opto-mechanical scanning have been in production till the end of the '90s, after that the FPA technology took over.

So, having a closer look at the latest scanning methods available, we can find two working principles: refractive prisms path and reflective path.

#### *Refractive Prisms Path*

Two scanning prisms, one vertical and the other one horizontal, perfectly synchronized, allow the detector element to scan the viewing scene. The prism material is transparent and coated to maximize the heat radiation which is focused to the detector. An example of camera using the refractive prisms path is reported in Fig. **36**.

#### *Reflective Path*

One vertical scanning mirror and one horizontal drum plus a series of spherical optical mirrors duly shaped made of polished aluminum, allow the thermal radiation to hit the detector element. An example of camera using the reflective path is reported in Fig. **37**.

In both of the described systems, the prisms speed stability and synchronization are fundamental for the image quality. IR cameras adopting these types of scanning systems, due to the mechanical design, have some limitations in the image speed generation; in fact, the fastest camera can go up to 50/60Hz.

Hence, higher performance in terms of frame rate can be achieved by using more detector elements with an electronic scanning system instead of the mechanical one.

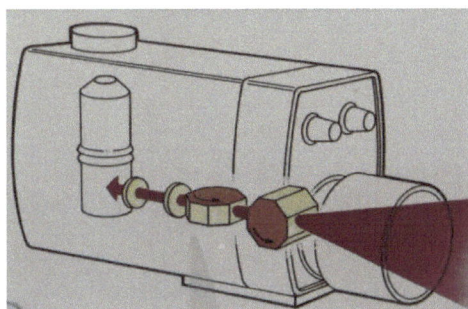

**Figure 36:** Example of IR camera using the refractive scanning principle. AGA Thermovision dated 1973.

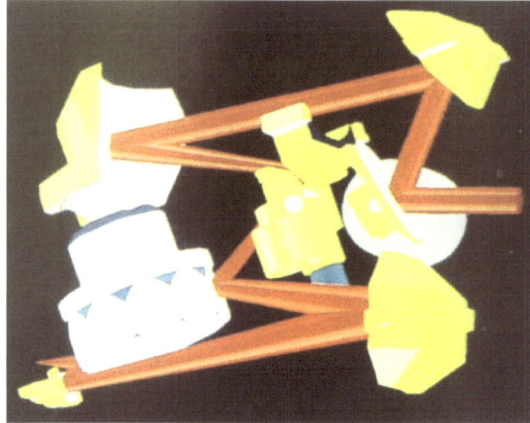

**Figure 37:** Example of IR camera using the reflective scanning principle. Agema THV 900 dated 1991.

## 4.2. Array Systems

### *4.2.1. Array Types*

As previously mentioned the most common detector array types used on commercial infrared cameras today are micro bolometers, for the handheld and automation models, and quantum sensors for high performance R&D applications.

As it is today (2010), the array sizes commercially available are generally of two types: microbolometer and quantum. Table **7** below is summarizing size and common sensor materials today in use for imaging and thermography purposes.

Higher resolution array are anyway available (1kx1k or 2kx2k) but not for civil applications. Due to military usage of this kind of sensors, there are severe export restrictions in place.

### *4.2.2. Image Uniformity*

To be able to generate a good image out of a Focal Plane Array detector, it is necessary to go through the characterization of each array pixels. Every pixel in the sensor array is acting like a single element detector with its own thermal response curve (offset & gain).

This means that if we put in front of the camera an object with a homogeneous temperature distribution, we would like to see each pixel of the array giving out the same signal, in terms of digital level, or grey level or color level. So, a raw image (Fig. **38**) is almost unusable without the due software corrections.

**Figure 38:** Detector Raw image example, © 2010 Flir systems.

### 4.2.3. Correction Maps

This process is the so called "Non Uniformity Correction" mapping (NUC).

In the Table **8** below, it is shown what a NUC process is practically doing. This procedure is part of the calibration procedure and in the Micro bolometer camera is factory prepared and stored in the camera memory.

For laboratory systems instead, this process can be either stored in the camera memory or performed by the user according to the working conditions.

The procedure consists in placing in front of the camera field of view two homogeneous black bodies, one at low temperature and another one at higher temperature. All the image pixels are scanned by the software and checked one by one for offset, gain and wrong response. Gain and offset values are identified and then stored in file maps corresponding to the image size. The non responding pixels are instead stored in another file map. The bad pixels will be replaced by the neighbours.

Once calculated, these maps will be applied on the fly to the raw image at camera scanning speed.

**Table 7:** Characteristics of the most commonly used sensors.

| Detector Type | Detector Material & Typical Spectral Band (μm) | Detector Size | Picture | Comments |
|---|---|---|---|---|
| Micro bolometer | - Vanadium Oxide (Vox) 7.5-14μm<br>- Amorphous Silicon 7.5-14μm | 320x240 or 640x480 | <br>Pixel construction detail for micro bolometer detector | The specified spectral bands are standard. Other wavelengths are available for special applications. Smaller sizes then 320x240 are available, but it is just an electronic downsizing of this array. It's used in low cost camera.<br>Less sophisticated Read Out Circuit, with functionality normally not accessible to the user. |
| Quantum | Indium Antimonide (InSb) 1.5-5μm<br>Mercury Cadmium Telluride (HgCdTe) or MCT 0.9-2.5μm; 3-5μm; 9-11μm<br>Indium Gallium Arsenide (InGaAs) 0.9-1.7μm | 320x256 or 640x512 | <br>Typical Quantum detector assembly with integrated Stirling cooler | The specified spectral bands are standard. Other wavelengths are anyway available for the same materials according to the final scope.<br>Highly sophisticated Read Out Circuit, with functionality normally accessible to the user, for advanced image analysis. Downsizing of the detector readout area, allow the user for very fast image recording. |

**Table 8:** Schematic representation of the NUC process © 2010 Flir systems.

| Image Type | Graphical Representation | Image Quality |
|---|---|---|
| Raw | | |
| Bad pixel replacement | | |
| Gain Corrected | | |
| Offset corrected | | |

## 5. LOOKING FORWARD IN THE FUTURE

Nowadays, the market of infrared technology is becoming a point of interest for company not being in this sector at all. The projection about the diffusion of this analysis technique seems to be very positive, according to the market experts, and this is creating a competing situation between the producers in advantage of the development and price of the equipment. The trend in term of units produced is clearly showing that IR technology is constantly expanding in the market of professional and semi professional users such as electricians, plumbers, home inspectors and this is creating the bases for a more capillary diffusion of the products. It is not uncommon that some users decided to buy the camera despite of the real need of it, but just because they were willing to investigate more the infrared science for the wide applications it can covers. This is of course happening because the combination of more affordable equipment cost and increased culture around this market segment.

A quick view for future development of the instruments in the short term will probably see the price trend to further go down slowly, but the equipment will benefit with higher performances in term of better hardware specifications and software features in all the instrument market segmentation. Higher resolution detectors, close to the High Definition video standard, more compact and refined design, will allow even more the market penetration of this technology. Looking in the medium term instead there is the possibility that IR technology will be only a part of a solution offered by integrators for an increasing number of more sophisticated applications.

Another orientation will be toward multifunctional instruments where the IR camera is only a small section of the instrument capability. A strong push will also come from the safety and security market where the infrared image associated to dedicated software will be able to create innumerable solutions to different problems.

## REFERENCES

[1]    AGA-Agema-FLIR Infrared Product History – Jan Dahlqvist November 2003.
[2]    Therma CAM P60 manual Publ. N° 1 557 534.
[3]    All images in this document are property of FLIR Systems Inc.

# Concluding Remarks to Part I

The aim of Part I was to introduce infrared thermography from the historical/theoretical perspective. Infrared thermography is becoming ever more popular embracing ever more numerous fields of applications. Now, there is availability of a vast multitude of infrared imaging devices to cover almost every type of applications. In the same time, infrared thermography is still not appropriately exploited mainly because of insufficient scientific and technical information.

Thus, Chapter 1 was conceived to serve as a starting point for the reader to be aware of the sophisticated technology which is behind an infrared instrument. For a correct use of the infrared imaging device and interpretation of its output, it is of utmost importance understanding:

- How an infrared detector works?

- How it is possible to transform the thermal radiation coming from an object into a temperature map?

- What are the factors affecting measurements and how to deal with?

And then, the radiation mechanisms were illustrated by recalling the basic relationships and by considering the object surface characteristics. In particular, a key parameter is the surface emissivity, to which thermographers must put attention to perform reliable measurements. A description of the different types of detectors was also supplied with the most relevant parameters that worth attention to assess the overall detector's performance.

Today, there is availability of a wide selection of infrared devices, which differentiate for weight, dimensions, shape, performance and of course costs; the infrared imaging system can be also tailored for the customer's requirements. This is thank to the technological progress since the first model, which appeared in the late fifties, was quite encumbering, difficult to use and of low performance. An excursion into the evolution of infrared imaging devices was provided in Chapter 2. Each model, was illustrated with attention to the incorporated detector characteristics and performance as well its main fields of application. It is possible to see, as otherwise for all electronic devices, the tendency towards miniaturization, concerning the physical appearance, and an ever-increasing performance in terms of resolution.

By now, infrared thermography has amply proved its usefulness leading to a proliferation of infrared devices to fulfil the requests of users in a vast variety of applications. In fact, upon a correct choice of infrared system, test procedure and successive data analysis, it is possible to use infrared thermography in many an many applications. Some of them will be analyzed in the following Part II. In particular, the use in:

➤ Medicine and veterinary in Section I (Chapter 3 and 4).

➤ Foodstuff conservation in Section II (Chapter 5).

➤ Industrial engineering in Section III (Chapter 6, 7 and 8).

➤ Architecture and civil engineering in Section IV (Chapter 9).

# CHAPTER 3

## Applications of Infrared Thermography to Medicine

## Boris G. Vainer[*]

*A.V.Rzhanov Institute of Semiconductor Physics, Siberian Branch of the Russian Academy of Sciences, 13, Lavrentyev av., Novosibirsk, 630090, Russia; Novosibirsk State University, 2, Pirogova str., Novosibirsk, 630090, Russia*

**Abstract:** Medicine is one of the most active fields in the use of infrared thermography (IRT). Medical diagnostics and monitoring today afford an opportunity to utilize the unique potentials of modern computer equipped FPA-based infrared cameras. At the same time, a large majority of IRT-assisted diagnostic methods and medical investigations are still modelled on the principles developed on the basis of instruments of previous generations. The aim of this chapter is to present the range of problems on up-to-date medical IRT in an orderly fashion and to unveil the unique capabilities of this method as applied to medicine.

**Keywords:** Infrared thermography (IRT), thermal imaging, focal plane array (FPA) based infrared camera, medical applications of IRT, topical problems in medical IRT, system approach to medical IRT, medical IRT methodology, surface thermal pattern, thermogram, mathematical treatments of medical thermograms, up-to-date definition of IRT.

## 1. INTRODUCTION

Otsukay K. and Togawa T. reminded us [1] that as far back as about 400 years before Christ, Hippocrates utilized wittily the basic features of temperature to diagnose diseases in humans. To achieve the expression of two-dimensional temperature distribution on the skin, the legendary healer recommended to enfold the patient's thorax with a piece of fine linen, which was soaked in warm moist finely triturated earth. Wherever it first dried, that was where the physician must cauterize or incise. A modification to this method was to apply the earth directly and looking for the place in the same way when the patient was wrapped in the linen. In the latter case, many people could be quickly diagnosed applying earth simultaneously to avoid that the first applied quantity became dry. In the stage of today's historical knowledge, it is unchallengeable, the Hippocrates's thermal measurement method can be acknowledged as the first thermographic (to be more precise, thermoscopic) technique in medicine.

It is agreed that medical infrared thermography (IRT) started from the middle of the fiftieths of the previous century, namely since publication of the scientific works [2, 3] by the Canadian surgeon-investigator Lawson R. The author of the above-mentioned works revealed the signs of breast cancer owing to visualization and estimation of the thermal radiation intensity emitted by women's breasts. Analysis of the diagram presented in [4] and plotted on the basis of systematization of scientific papers devoted to medical IRT shows that since the middle of the seventies of the $20^{th}$ century, a great interest initially taken in IRT was remarkably loosing. Such interest was revived in the middle of the nineties. This phenomenon can be mainly explained by the fact that the new generation focal plane array (FPA) based infrared cameras had become available to medicine. It was quickly recognized that FPA-based infrared cameras offer advantages in medical diagnostics over previous instruments [5-8].

There is no need to discuss how much the morphological methods and their medical images are popular in medicine today [9]. Among them, worth mention the traditional projection radiographs, with or without contrast and subtraction, nuclear medicine projection images, ultrasound images and the cross-sectional modalities of X-ray computed tomography (CT), magnetic resonance imaging (MRI), single photon

*Address correspondence to Boris G. Vainer: A.V.Rzhanov Institute of Semiconductor Physics, Siberian Branch of the Russian Academy of Sciences, 13, Lavrentyev av., Novosibirsk, 630090, Russia; E-mail: BGV@isp.nsc.ru

emission computed tomography and positron emission tomography [9]. Taking into account that in many cases not only morphological structure but also the local temperature of deep tissues is of great diagnostic consequence, medical practitioners and physicists endeavour to discover the physical factors which allow detection of local temperature change deep in the body with the help of "penetrating" ultrasonic and magnetic resonance imaging methods [10-12]. Nonetheless, despite the many attempts, no method can match IRT in the recognition of diseases or physiological and pathological deviations if diseases and deviations are accompanied by non-uniform body surface increase or decrease in temperature.

Long before the emergence of IRT, various contact methods and techniques intended for skin temperature measurement were devised. Thermocouples, thermistors and some other heat-sensitive elements were utilized as point thermometers. It was revealed in further investigations [13] that a correct measurement of temperature over a certain surface area requires use of several individual thermal sensors to be read simultaneously; the pressure which each sensor delivers to the skin must also be taken into account. Contact temperature measurements of open skin are still in use in medicine and physiology [14], although multipoint techniques are normally being replaced now by liquid crystal films whose colour depends on heating [15-17]. Infrared photography and distant point thermometry may be placed into the "intervening period" between contact thermography and IRT. Distant point thermometry was devised predominantly for non-contact measurement of tympanic temperature.

Sometimes, to verify the dynamic range of infrared imaging and the accuracy of infrared calibration, tympanic or contact thermometers are placed at several locations within the field of IRT measurements [18]. Usability of these thermometers is attractive to such an extent that investigators employ them not for intended application. For example, tympanic thermometer was arranged for detection of skin temperature differences in dorsal and palmar sides of hand [19]. In such kind of experiments, one should take into account that individual peculiarities of local surface vascularisation may introduce large errors into the measurements, as discussed in [20, 21].

The basics of the use of IRT in medical applications are well known. Relevant information can be found in a lot of articles and books [22-24, *etc.*]. For this reason, discussion on above-mentioned problem is omitted in this chapter. In brief, circulating blood is heated due to metabolism in the deepened human tissues. A part of warm blood travels to superficial tissues, including skin, and heats them in addition to their heating due to thermal conduction mechanisms. By that means, skin temperature distribution reflects the processes progressing in the body. Swain I. D. and Grant L. J. noted [25] that temperature is a surrogate marker of superficial blood flow. The last fact gives IRT a great advantage over many other medical diagnostic means and instrumentations because enables IRT to be a very sensitive method capable of detection of changes in the vascular system state, and therefore allows studying neurovascular reactions. The infrared camera captures the surface thermal patterns making them accessible to subsequent analysis.

Inexhaustibility of IRT for medical diagnostics is dictated by inexhaustibility of different ways of internal and external influence on the organism, because thermoregulatory processes of the organism normally respond keenly to any intervention. External and internal impacts on the organism provoke different neurovascular reactions which can be examined by IRT methods with a high degree of accuracy. Usually, healthy and affected organisms respond to stimuli or irritants in different ways that gives grounds to use the IRT results for biomedical diagnostic purposes.

The human organism is a self-regulated system, and its principle feature is the maintenance of the mean temperature within a very narrow interval constituting just 2% of the body thermodynamic temperature itself. Interest in thermoregulatory mechanisms responsible for this stability has continued till now [26, 27]. The most fantastic is that such a stability is maintained by a human organism under environmental conditions characterized by much wider than 2% diapason of surrounding temperatures, at the level of about 30-35%, from cold pole to equatorial heat. At the same time, even in the less aggressive conditions a human organism needs uninterrupted heat exchange with environment. It follows from the thermodynamics laws. Vital functions of the organism require uninterrupted consumption of energy. The last fact results in uninterrupted heat production, which is capable to bring the organism to boil if the released heat is not dissipated.

Let a certain mankind-average apparently healthy individual presents a typical result of human evolution. Compare this human with that affected by any lesion or physiological change. If the malfunction of the latter person is in active phase, the metabolic processes in his organism would be expected to be disturbed, and the altered heat production in the organism should lead to excess (or reduced) heat release into environment as against the amount of heat produced and dissipated by a healthy person, all other factors being the same (environmental conditions, behaviour, external loads, food intake, *etc.*). It will entail overheating (overcooling) of an affected person that cannot be allowed by a homoiothermal organism. While the reflectory conditioned struggle of the organism for recovery necessitates an increased (decreased) metabolism in affected part(s) of body, it might be supposed that, to keep a metabolic balance, the well organized systems of this organism would slacken (intensify) the function of intact parts of the body and reduce (increase) their heat production respectively. Thus, in a sense, the temperature contrast is changed (predominantly, increased) inside a body, and this change becomes apparent on the body surface as on outer boundary of the body volume. This model explains why the recording of surface thermal non-uniformity has a diagnostic significance in human medical examination.

It is clear that the model described in the previous paragraph is quite variable. In actual practice, outer temperature manifestations are influenced by a selection of particular factors. They imply the reduced food intake, changed intensity of physical loads, behavioural changes, *etc.*, that the previously healthy human could undertake in response to his ailment. By the way, the therapists often take an interest in these factors when interview a patient. If healthy areas of body remain, in outward appearance, indifferent to aberrations happened in the neighbouring tissues, mean temperature constancy is usually provided by the human behavioural changes or changes in human heat insulation of his body. Nevertheless, the physiologically conditioned and totally natural phenomenon involving both internal and outer thermal contrast alterations suggests itself as one of the most likely and significant signs, which are representative of the organism deviation from the norm. The proposed general model described above generates a need for development of methods and techniques meant for the investigation of spatial non-uniformity of the human body temperature and its transformation with time.

Unfortunately, a satisfactory apparatus making possible to display temperature tomograms of human viscera has not been invented yet. A possible technique approximating the above-mentioned apparatus is based on the measurement of thermal microwave radiation emitted from several centimetres depth of the body [28-30]. Though, low spatial resolution resulting from the long waves used, as well as low sensitivity of radiometers that are forced to detect electromagnetic radiation of extremely low intensity give no way of regarding this technique as a satisfactory solution to the problem in question. It is valid to say that the use of IRT is the optimal line of attack on the above-formulated problem to date.

Grand argument for objectivity of IRT-obtained medical data is that the human skin surface has been accepted, on the basis of experimental results, as a black body radiator characterized by emissivity value of about 1 measured within several bands lying in the range between 2 and 20μm [31-35]. The primary thermal pattern represented on a healthy human body surface is rather stable. It is well-known from a long-term practice, though the excessively chary investigators make the attempts to confirm it quantitatively [36]. Mainly due to the above-mentioned stability, IRT in humans is defensible.

The limiting factor of IRT is its capability to examine nothing but surface thermal patterns present on the human skin. Water, the basic part of the organism, does not allow infrared radiation to come out from deep layers of the body. In turn, the key advantage of the modern medical IRT is its extremely high sensitivity to surface temperature deviations (at a level of hundredths and thousandths of a degree) and exceedingly high spatial resolution attaining tens micrometers if necessary [21]. Both of these attributes enhanced by high-speed measurements (tens and hundreds frames per second) accessible to FPA-based infrared cameras open up intriguing opportunities for advancement of the methods for solving an inverse problem in the near future to obtain information from infrared data about real thermal state of deep tissues and organs in humans.

Forty years have elapsed since the epoch of rapid development of medical first-generation infrared cameras [37]. In this period, sensitivity of high-speed IRT systems reached the limiting values. Such a technological breakthrough has been made possible by the use of properly developed FPA-based photodetectors in the thermographs [38-41]. The new technique materially extended the capabilities of IRT in the biomedical sphere [5-8, 20, 21, 40, 42]. Modern high performance IRT exposed a lot of "reefs" generating a need for examination of many new details that became apparent in the thermograms measured with high accuracy and sharpness in a wide dynamic range. The recognizable new details, consequently, brought forth many tough problems to medical specialists, sending them in search of new IRT-based diagnostic methodologies and criteria. The performance of this task still remains the "problem of today" in medical IRT. In this chapter, some painful and topical problems associated with the use of modern IRT in medicine are touched on.

## 2. INTRODUCTION OF STANDARDIZED ELEMENTS INTO MEDICAL IRT METHODOLOGY

Following the centuries-old traditions well-established in medicine, a medical practitioner is usually governed by strictly specified regulations, and therefore, he needs in concrete instructions on how to act, but not in the examination and analysis of the wealth of information published in scientific journals. Taking into account this specificity, some specialists in medical IRT make efforts to work out such instructions to standardize the process of IRT examination.

Standardization is heading mainly in three directions. The most aged of them is the comparison of thermograms obtained *in situ* with those taken before and generally accepted as "correct" (exhibiting thermal patterns of some apparently healthy individual) or with reference images which display typical examples for provisional diagnosis [43-47]. "Standard" thermograms are used by the physician either in the form of ready-made pictures represented in atlas, monograph, computer database, *etc.* or in the numeric form describing temperature gradients typical for the current case. In the framework of this approach, many quantitative criteria have been proposed for recognition of different diseases. Application of these criteria usually reduces to comparison of temperature values in pathological area and in surrounding tissues or/and in the contra lateral side of body.

Second direction of standardization includes a strict definition of geometrical boundaries of human body parts to be analyzed [48]. Third direction is the definition of standards in respect of the arrangement of infrared examination including preliminary preparation of the patient. A list of standard requirements to patient preparation includes special dietary pattern, the ban against the use of cosmetic means and other creams, bowels cleaning (to give an enema), exclusion of physiotherapeutic procedures on the eve of infrared examination, *etc.* As regards examination procedure, the mentioned are patient's adaptation time before the onset of examination, temperature and humidity of air in the room, presence of powerful heat sources inside a room, imaging hardware, cross calibration of camera systems, *etc.* [43-45, 49, 50]. An apparent purpose of such requirements is elimination of the artifacts which may prevent inexperienced physicians from correct capturing of thermal images.

It may be declared a twofold attitude to infrared examination subjected to indispensable standardization. On the one hand, standardization might be welcomed because, when absolutized, it would not only facilitate the problem of data exchange between different medical users [50], but also make it possible to exercise the IRT diagnostics by the hands of junior medical personnel and, consequently, liberate the experienced physicians from routine tests. On the other hand, each sensible person ever handled medical IRT understands the complete absurdity of the latter idea. Variability of human surface thermal fields do not permit diagnose based on any mechanically executed manipulations, if the mind get not involved in the work with interpretation of thermal patterns *in situ*. Moreover, rigorous observance of the above-mentioned requirements can sometimes unreasonably extend the time of examination and raise the price of diagnostic procedure, with a consequent reduction of the number of examinees. Accumulated experience discloses such a fact that variability in individual organization of the body and the human nervous system status usually contribute into thermal pattern much more nuances and artifacts than, say, inequality of environmental parameters in the course of examinations.

Turning back to the problem of replacement of the qualified specialist by an inexperienced one, it appears that IRT leaves room for this fantasticality nevertheless. The way allowing doing so is implication of IRT to telemedicine, the technology of remote diagnosis and treatment of patients by means of telecommunications. This idea was suggested in reference [8]. The modern computerized IRT complies with telemedicine general requirements.

It is necessary to point to one more direction in medical IRT standardization. It is elaboration of different standard tests meant for diagnostics. This direction is usually realized with the help of so-called dynamic or active thermography. Dynamic tests usually include the exposure of the organism to external impacts [20, 21, 51-55]. The most simple and popular among them is the cold test equipped with the cold water for extremities to be submerged. It should be noted that cooling and moistening are used sometimes to enhance the contrast of surface thermal pattern [56, 57]. The cold stimulus as well as overheating of the organism induce the autonomic vasoconstriction and vasodilatation reflexes in normal structures, thus enhancing the thermal contrast due to differences in the vasculature of the different skin regions [20, 21, 58]. From our point of view, a closer look at the dynamic IRT data is called for. There is a need to pursue this direction in IRT in every possible way because the future of medical infrared diagnostics may well lie with the active methods of examination.

## 3. FOUR LEVELS AND TWO ADJACENT FACES OF A CUBE IN MEDICAL IRT

To date, thousands scientific works devoted to biomedical IRT have been already published. Despite this fact, muddle reigns in the mind of physicians on what they should think about this method, how indispensable is it, *etc*. Mixed opinion about IRT stems from the fact that this method, as opposed to many other medical methods, is many-sided, and physicians are lost in these "sides" as in a labyrinth. For this reason, it makes sense to sort out the papers published on medical IRT.

In concordance with the used methodological approaches and problems solved by the different authors all the publications may be classified under four main groups (levels) as follows:

1)   Examination.

2)   Correlation.

3)   Analysis.

4)   Diagnostics.

The most simple and popular level is examination. At present, it is fitted by most part of the published works. The papers [59-67] are some of the examples. This level includes inspection, observations, statement of fact complementing the clinical observations, monitoring, quantitative measurements, estimation, obtaining information about any irregularities or singularities in thermograms, finding or discovery of suspicious areas. Infrared camera serves here just as the instrument broadening physical capacities of human vision. In spite of simplicity, it is a very useful application of IRT, because an unpretentious monitoring of temperature changes, as well as finding of boundaries of thermal fluctuations offer an opportunity to get an expression of the adequacy of treatment, the time history of pathological or restorative process, *etc*.

On the second level should be placed the works of authors who aspire to demonstrate the diagnostic capabilities of IRT by means of examination of the patients with diseases known in advance. Examples of such works are given in references [68-71]. On the basis of the second-level papers, databases are formed containing "standard" (reference) thermograms typical to certain diseases. Quantitative and qualitative criteria for IRT diagnostics are also suggested in these works to be used in subsequent practice. Many publications describing the results obtained with the use of dynamic IRT tests also belong to the second level. The authors of such works often reveal from the thermograms the signs of diseases under investigation and thereupon promulgate the relevance of IRT for medical diagnostics. Vulnerability of such an approach consists in the fact that it does not allow to obtain a reliable result when the inverse problem is pending. Namely, it does not place any convincing diagnostic character at practicing physician's disposal because similar thermograms and thermal manifestations can be obtained also for many other diseases and even for natural physiological deviations of a normal organism. Most likely, the good points of these works are that, when applied to screening, they give a chance to suspect a disease and based thereon to refer the patient to supplementary examination.

The third level is concerned mainly with physiological investigations [20, 21, 72, 73]. Here the scientists have no restricted themselves to trivial description of different peculiarities found in thermograms. They endeavour to fit the obtained experimental data to physiological mechanisms responsible for these results. The authors of the third-level papers try to discover the logic pattern in the observed thermal manifestations by reconciling their data with information taken from anatomy and physiology. With the works of this level we are able both to extract substantial amount of fundamental information about human beings from the IRT data and to lay the groundwork for medical IRT diagnostics.

The works of the fourth, highest, level represent IRT in its best and most perfect form - as an independent medical diagnostic tool. In order not to put any selected work on a pedestal and not to hinder the future authors in their efforts to perfection, let's not cite any papers which could belong to this level. As applicable to this level of IRT development, the next pictorial model may be offered.

Numerous results collected to date on medical IRT may be represented as two adjacent faces of a cube (Fig. 1). One of the faces (left) carries information, broadening the capacities of human vision; generally, it reflects what the infrared camera was possible to recognize. This face gathers also the knowledge acquired from scientific analysis of the measured thermograms. Infrared data represented on this face can tell about peculiarities associated with peripheral and cutaneous blood flow, sweating, status of nervous system, *etc.* The second face (right) reflects the true diagnostic capabilities of IRT. Namely, on the basis of information selected from the materials of the first face it makes possible to carry out infrared examination using protocol allowing formulation of exact diagnosis. Such kind of selected information is called pathognomonic (or diagnostic) signs.

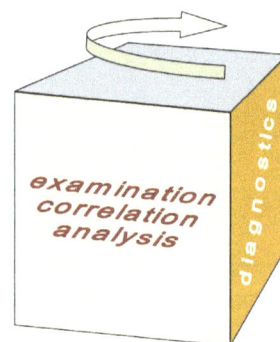

**Figure 1:** State of the art of IRT in medicine.

Fig. **1** shows for convenience that at the current state of the art in the development of medical IRT this cube presents its right face to us with the substantially smaller angle than the angle of the left face. One can count his blessings if he meets indeed a scientific article containing persuasive arguments in favour of the statement that the obtained set of diagnostic signs is sufficient to indicate the exact diagnosis or pathological deviation. Unfortunately, the last situation is a great piece of luck today. The majority of the articles devoted to medical IRT is predominantly descriptive. These works just demonstrate how powerful is such an instrument as infrared camera and pay attention of other physicians to this method calling them to turn the cube along the arrow shown in Fig. **1**. The main objective of the modern medical IRT is the speeding-up of such a rotation.

## 4. INTRICATE STANDING OF IRT IN MEDICINE

A great number of different applications of IRT in medicine utilize this method directly, namely, to measure approximate values of surface temperature and to monitor the temperature change in the course of various influences and interventions the organism is subjected to. This situation gives rise to complaints against the IRT method. Practicing physicians expect IRT to identify the diseases directly and exactly. When the physicians do not catch a thing out of IRT, they call IRT auxiliary instrument of a sort, dropping a hint that they would manage without its assistance.

A rational question at this point is: Is it possible today to manage in medicine without blood test, urine analysis or, say, roentgenography? It is clear that anyone's answer is "No". Though, the cited methods also do not put final diagnosis at physician's disposal, they just fill them in on what happened with some quantitative characteristics of the organism or its part. No X-ray picture or erythrocyte sedimentation rate value can serve as the sufficient factor verifying the disease. It follows herefrom that the doubtful reputation attached to IRT consists not in groundlessness or inferiority of the method itself, but in the fact that overwhelming majority of physicians are simply not experienced in medical IRT or have not learned to understand the infrared data so far. In this context, they consider more rational to refuse new information coming from IRT than to provide help themselves to diagnose more precisely and make the patient's recovery somewhat quicker.

Medical practice demonstrates that interpretation of infrared data is more difficult than the interpretation of results obtained by other diagnostic methods. This is, in particular, due to the fact that one can find many similar elements in the approaches to interpretation of, say, roentgenograms, MRI-tomograms, CT-tomograms, *etc.* The routine operational procedure typical to diagnostics executed with the use of the above-mentioned morphological methods is habitual to physicians because their skills here rest predominantly on both human anatomy delivered them thoroughly in medical educational institutions, and long-term practical experience obtained by medicine in interpretation of X-ray pictures. In contrast to this, IRT data look nothing like data obtained with the methods mentioned just now. Take the fact, for example, that natural asymmetry typical to viscera (unilateral location of liver, spleen, heart, *etc.*, frontal or dorsal location of kidneys, bowels, *etc.*) has little or no effect on a healthy human outer thermal pattern which remains highly symmetric. Some quantitative characteristics of this symmetry are given in [22, 74].

It is improperly to lay categorical claim to IRT to supply medicine with diagnostic information which would compare favourably in definiteness and specificity with that obtained, say, in roentgenology, at least taking into account that the development of IRT diagnostic methods, which were arisen coincidentally with the classical works of Lawson R. [2, 3], started approximately 60 years after the morphological diagnostics, arisen coincidentally with the works of Roentgen W. C. in 1895, had been started. At the same time, how arbitrarily fast would be the development of IRT diagnostics, the certain advantages of IRT over morphological methods will remain unshakeable at least on the grounds that morphological methods by virtue of general principles built in them are unable to recognize many physiological and pathological manifestations which lend themselves admirably to infrared control.

Admittedly, the medical IRT diagnostics goes at present through, probably, one of the most difficult periods in its development. On the one hand, each physician and physiologist recognizes that IRT makes possible to disclose the whole stratum of information hidden from the naked eye and related to the functioning of living organism. At that, it is important that this substantial amount of information cannot be some time extracted by other methods. On the other hand, it is not secret that many attempts in using an infrared camera have not been successful in revealing the precise diagnosis. It turned out, seemingly evident hypotheses and suggestions often were not corroborated in actual practice. We can mention the work [75] here which describes how the quite reasonable hope for positive correlation between breast cancer and extent of vascularisation in the affected breast was not justified. A long train of these practices has at length formed distrust and even enmity towards IRT in many physicians.

At the same time, here and there prevailing identification IRT with something like "futile" is a quite undeserved reproach. It should be recollected that the physicians themselves were responsible for the impending prejudice against IRT, when in due time they schemed using IRT to deal successfully, at one fling, with the problem of early recognition of breast cancer. At the earliest stage of high-performance FPA-based IRT, the prevailing view based on published data (*e.g.*, [76-79]) declared that thermography "is of minor importance for the diagnostic clarification of breast disease and for breast cancer screening; it demonstrates high diagnostic accuracy for advanced breast cancer, but is rather ineffective in indicating the presence of non-palpable lesions" [80].

Undeservedly negative attitude of some people to medical IRT is resulted also from rather poor formulation of conclusions in some separate studies. Here, we are dealing with the conclusions containing ambiguities and, as a result, leading to incorrect understanding of the essence of the authors' assertion. As an example, the work [81] might be given. Its authors carried out the titanic work, in the course of which they got involved more than 10000 women into the screening on IRT examination of breast. In five years, second examination was conducted, and the obtained results were brought into correlation with real clinical presentation. Based on results expressed in percents and making not in favour of IRT, the authors formulate the next conclusion: "Thermography is not sufficiently sensitive to be used as a screening test for breast cancer, nor is it useful as an indicator of risk of developing the disease within five years". These words can discourage any people erroneously equating "thermography" and "temperature recordings using infrared camera" from the use of IRT method for examination of mammary glands at a later time. At the same time, it is immediately obvious that the unconvincing result obtained in [81] is not a fault of infrared cameras, as such. The fault should be primarily blamed on methods and criteria which were used to separate mammary glands into two groups - healthy and affected. In the example cited above, imperfect diagnostic criteria resulted, through no fault of the authors of [81], in another deceleration of IRT at its penetration into medicine. Rather than stimulate scientists to develop new diagnostic methods and criteria in medical IRT, such formulations like those given in [81] can turn the medical scientists away from IRT at all.

IRT not infrequently had been left holding the baby if exorbitant fantasies of its users were not realized. It is useful to remind the fact that a number of physicians, which are not very much familiar with IRT, up till now do not understand that an infrared camera do not allow to see internals and to diagnose directly the gastropathy, liver or brain diseases as allowable to ultrasonic scanning, roentgenography or, say, MRI. And vice versa, from time to time, unfounded diagnostic inferences and IRT signs of diseases can be met in the internet and advertisements. This kind of infrared data representation often bears some resemblance to flourish of trumpets that is detrimental to IRT and seems impermissible.

It happens that thermographers apply the brake to themselves when they are carried away by pseudoscientific trends in medical diagnostics. To take an illustration, they persevere with the search for visualization and identification of signs employed in the traditional oriental medicine. Their efforts can unlikely be fruitful at least because medical IRT reflects solely objective processes developing in living objects, in particular in humans, whereas general principles and categories of the traditional oriental medicine are developed on entirely different basis. This opinion is supported by experimental data obtained in [82, 83]. The authors of these works have concluded that the visualization of energetic paths in the sense of meridians seems to be not possible using thermography.

The use of prehistoric criteria and approaches, when the results obtained by high-performance FPA-based infrared camera are analyzed, has also a severe impact on medical IRT. And *vice versa*, IRT method loses appreciably its popularity in fault of users who believe they can afford to formulate the correct diagnosis on the basis of thermograms obtained with low-sensitive infrared cameras of the previous generations. For instance, the diagnostic sign as "thermo-amputation" of digits, hands, toes or feet is still in practical use by some physicians working with outdated infrared cameras. "Thermo-amputation" effect appears when the infrared camera is not capable to distinguish cold distal parts of extremities on the background of other warm parts of the body. In this case, cold regions merge into the rather cold environmental background and it seems that they disappear from view (become "amputated"). As demonstrated in [84], wide dynamic range and high temperature sensitivity of modern FPA-based infrared systems make possible to discern easily such cold fingers on the room background. The use of new-generation infrared cameras enables physicians to eliminate the term "thermo-amputation" from their vocabulary and replace this "loss" by quantitative analysis of temperature changes measured in cold parts of the body.

Scepticism against the IRT method can be sometimes understood, if we address oneself, for example, to the results of the work [85]. The author of the cited article realized the extremely interesting idea by the use of IRT at occupational selection of volunteers for police and military forces. In other words, screening was made among predominantly healthy subjects. 4162 volunteers were examined, and all of them were previously, by standard medical examination, declared as healthy. As a consequence, IRT found in 30% (!) cases the presence of some diseases, which were not been detected by ordinary medical examination.

There are two hypotheses which may explain such an appreciable disagreement of the IRT results with those obtained by standard examination. One of them consists in the assumption that the historically established system of medical assessment of military personnel could be defective. However, this version is rather disputable because a mistake amounting to 30% would has long been noticed in defence establishment, and rectified. According to the second, more probable, version, over diagnosis took place when IRT examination was performed. It could had been possible if the author of [85] was using the IRT diagnostic criteria which were developed for the previous-generation of infrared cameras. The latter hypothesis is indirectly confirmed by considering the type of infrared camera used, which is the TB-03K, having frame time 3.2s and advertised temperature sensitivity 0.15°C [86] (note, the author of [85] reported the latter parameter as being 0.05°C). Despite the fact that subsequent in-depth medical examination of the potentially unhealthy subjects showed presence of disease, which did not allow or restricted of professional ability of the volunteers for service in military forces, the results of the discussed work hardly can be considered as convincing, because it is not very obvious that among the remaining 70% persons the diseases would not be found at equally detailed additional examination. As one witty man observed, there are no healthy people, they are all under examined.

## 5. IRT AS UNIQUE MONITORING INSTRUMENT AND MEDICAL DIAGNOSTIC TOOL

As far back as the seventies of the 20th century, three broad areas of diagnosis and monitoring in which thermography plays a useful part had been recognized [87]: 1) change in emissivity of the skin surface due to damage (burns, chronic ulceration or other afflictions which damaged the epidermis), 2) trouble in the vascular system itself, and 3) other circulation problems, ranging from frostbite to deep vein thrombosis. Abundance of illnesses and abnormalities affecting a human engendered abundance of attempts to look in infrared region at outer manifestations of these illnesses and states. The diversity of corresponding scientific works is wide. IRT has been used to monitor vascular reactions of peripheral vessels in diabetics [52, 88], vascular reactivity in arterial hypertension [51, 89], in vertebrology [90], to reveal outer thermal manifestations in non-verbal individuals with severe motor impairments [91], to diagnose some other developmental disorders [92], in pregnancy [54], in obesity [73], in examination of smokers [89, 93]. IRT method is actively engaged in oncology [57, 58, 94, 95], in particular, in diagnostics of breast cancer [96, 97], *etc.* Both humans and animal models participated in various physiological and clinical investigations [62, 98-101]. Ground for successful use of IRT in physiology is the fact that in many cases surface temperature represents adequately the blood perfusion, which in turn represents the physiological processes developing in the organism [20, 21, 93, 102].

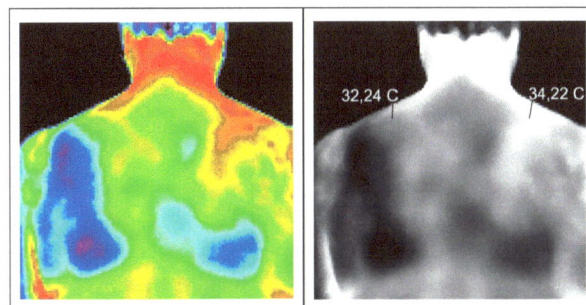

**Figure 2:** Right humeroscapular periarthritis, color and monochrome representations. Seen is a thermal-pattern asymmetry with a hyperthermic zone in the vicinity of right shoulder; the temperature excess over the symmetric region is about 2°C.

Unfortunately, overwhelming majority of physicians are grateful for small favours and content themselves just with the most trivial and entry-level information taken from the total store of opportunities presented by IRT. This simplistic approach does without mathematical analysis of thermal patterns as well as without anything but the qualitative examination of thermograms supplemented with local temperatures measured in several points. Nevertheless, it helps sometimes to learn more about the organism than it might be done without infrared camera. Some illustrations of this sort taken from our atlas [103] are presented in Figs. **2-5**.

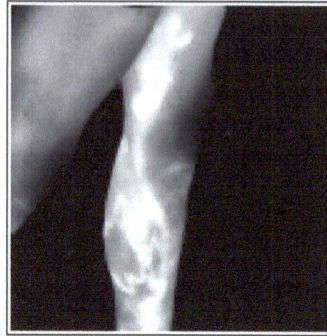

**Figure 3:** Varix dilatation in left lower extremity.

A number of specific examples illustrating in more details the use of IRT in medical applications are described below.

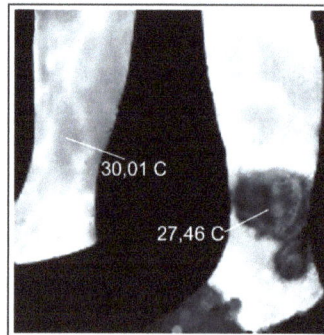

**Figure 4:** Left crus trophic ulcer. Seen is the extended hypothermic area in left shin. Temperature drop between the affected and normal parts of legs is about 2.5°C.

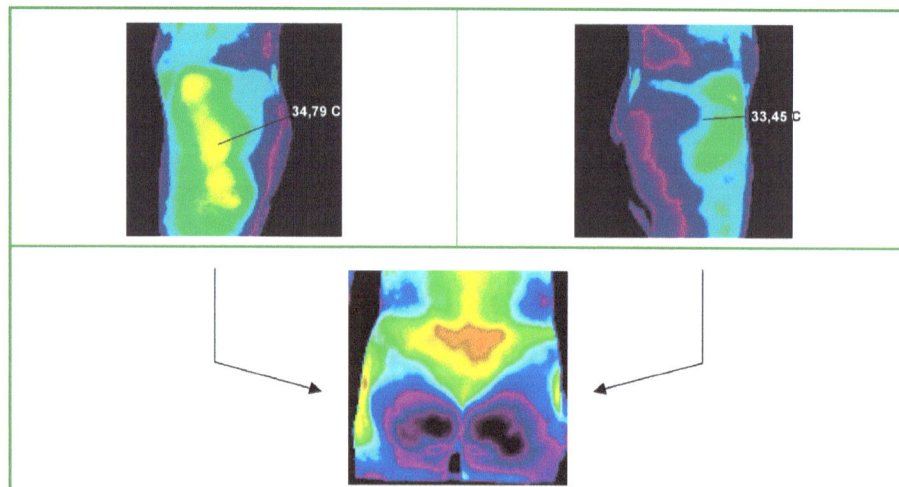

**Figure 5:** Acute arthritis of the left hip joint. Seen is a hyperthermic area in the vicinity of left hip joint with 1.3 degrees centigrade temperature excess over the symmetric region.

## 5.1. IRT in Surgery

Usage of IRT control in the course of surgical operation is a notable example of medical application of this method. Hardware implementation of the IRT in the conditions of operating room in the course of surgical operation demands of some special attributes. For example, in [18], the camera, except for the lens, was covered with conventional surgical draping to maintain the sterility of the operative site. Besides, to reduce local heating and intensity of background radiation, surgical lights near the operative site was redirected from the surgical field during the infrared measurements.

During operation, surgeon has the extremely limited time at his disposal to analyze the operative site and choose optimal tactics how to act. In these circumstances, every means is pertinent if it helps to represent the status of the region undergone surgical intervention. Any additional real-time anatomical and pathophysiological information may significantly contribute to an improved extent of tumour resection [61]. Modern high-performance IRT serves in some cases as an indispensable assistant of the surgeon.

### 5.1.1. IRT in Cardiovascular Surgery

Basing on several tens of medical cases, it is demonstrated in [104] that IRT can be advantageously used for *in situ* control in the course of surgeon operations in open heart and arteries. From the obtained results the authors of [105] inferred that cardiothermography can properly evaluate arterial bypass graft patency, the measured thermal patterns can easily be analyzed for qualitative flow-and quantitative temperature changes; myocardial protection can also be safely assessed with thermography. Plenty of IRT-assisted experimental works in cardiovascular surgery are carried out in animal models. In [105], cardiothermography was used concurrently with electrophysiological measurements in the course of surgical operations in 16 dogs with dissection of their chest. In these investigations, cardiothermography proved to be a useful tool in monitoring epimyocardial temperature changes during coronary artery occlusion and Endothelin-1 induced vasoconstriction.

### 5.1.2. Visualization of Ischemic Tissues

As reported in literature, the visual and tactile clues such as colour, mottling, and tissue turgor are currently used in the operating room for subjective assessments of organ ischemia [106]. As shown in a series of works, ischemia is amenable to IRT control which is capable to present the objective information about this phenomenon. A lot of investigations of this kind have been conducted in animal organs. In [107], the corresponding study was performed using a porcine model. Using thermal imaging and several other methods, the authors of [107] examined foci of bowel ischemia and the intestinal necrosis attributable to the ischemia. It was shown that thermal imaging was 100 percent sensitive for necrotic bowel, though it ranked below fluorescein dye and Doppler ultrasound in the positive predictive value.

In the course of investigation of renal ischemia in a porcine model, Gorbach A. M. detected variations of infrared signal arisen due to temperature fluctuations [106] and concluded that it is possible to assess the degree to which renal surface temperature reflects underlying renal ischemia. It was observed in the cited work that thermal oscillations were attenuated substantially in ischemic segments, but not in the perfused segments of the kidney. In a variety of IRT medical applications, reperfusion after prolonged cold ischemia in kidney allograft is shown [108] to be also the subject of considerable interest.

### 5.1.3. Brain Infrared Imaging

It is known from surgical practice that in the operating room some vessels represented on the surface of the cerebral cortex are immediately evident to the naked eye. At the same time, visual inspection alone cannot sensitively detect patterns of blood flow through the cortical vessels in and around tumours [18]. Real-time assessment of cerebral vessel patency and cerebral perfusion are the direct applications of the IRT. As noted in [7], IRT method is potential for localization of epileptic foci, identification of functional cortex during awake craniotomy and determination of tumour border.

At the same time, the potential of IRT as applied to surgical operations in brain still remains open to question. It is noted in [18] that at least at this stage in its development and use, intra-operative infrared imaging cannot be used to define the margins of a tumour for accurate resection while preserving contiguous functional brain tissue. By contrast, the authors of [61] revealed using an infrared camera a clear demarcation of tumour (intra-cortical metastasis of melanoma) with significant temperature differences, up to 3.3 °C, between the tumour core and the surrounding normal tissue.

### 5.1.4. Thermography in Ophthalmologic Surgery

Remote control of temperature profiles is of particular value for cases when it is difficult or impossible to apply contact temperature methods. Ophthalmology is one of a number of medical spheres which may take advantage from use of remote thermography. Considerable recent attention has been focused by ophthalmologists on IRT which proved to be particularly suitable for control of thermal processes developing in the course of ophthalmologic operations.

In [109] thermography was used as an instrument for measuring temperature of the ocular surface in order to analyze and compare three surgical procedures for cataract removal performed with the phacoemulsifier. Immediately after the introduction of phacoemulsification, developed by Kelman C. D. [110], severe surgical complications can be observed, such as burns on the cornea and the sclera. The complications are due mainly to a rise in heat in the area around the tip of the phacoemulsification probe [111-113]. It is shown in [109], that thermography may be utilized as an effective temperature monitoring instrument, which can be applied *in vivo* without requiring any change in the surgical procedure.

### 5.2. Visualization of the Pain Reflex Components

It is assumed that IRT might be helpful to visualize the pain reflex components by showing an increased or decreased skin temperature compared to the surrounding or opposite body area. The simplest way to reveal this reflex is mirroring and subtracting the left from the right body area to show temperature asymmetries [114]. To our mind, this approach has some weak points and should be handled with care. Unilateral pain is a sign of asymmetry itself. One of the regular dominant signs of asymmetry is asymmetric vascularisation [20] that cannot be associated with the pain at all. Another artifact can be arisen from quite natural behavioural direct actions of patient to rub his area of pain. The latter can intensify skin blood flow but have no relation to the pain syndrome.

### 5.3. Burns and IRT

Determination of burn wound depth at present is left predominantly to the surgeons' visual examination. At the same time, as it is highlighted in [87], it is extremely difficult to tell by visual examination whether burning damage is so deep that grafting is required or whether self-healing will take place. Delay in decision-making can be dangerous to patient's organism. IRT reveals clearly the drop in temperature, which accompanies the lowering of emissivity when the epidermis is completely destroyed, and presents an isothermal pattern which labels the areas affected. Some research directed to understand whether IRT can recognize the peculiarities inherent to skin affected by burn are based on experimental works conducted with the use of animal models (*e.g.,* domestic pigs [115]) *in vivo*. Human burned skin is also undergone detailed investigations. It was established in [116] that focal plane staring array midrange infrared systems appeared promising in determination of burn depth one to two days post-burn. This result can be considered as quite important and promising because wound depth determination often requires waiting 5 to 7 days. Researchers are pinning their hopes on the approach combining data obtained from IRT with calculation results obtained from the solution of the Pennes equation [117]. The authors of the work [118] suppose that such an approach is reliable in differentiating among *2a* degree (superficial cutaneous) and *2b* degree (deep cutaneous) burns that is of primary concern to hospital doctors.

### 5.4. IRT and Laparoscopy

The authors of the works [119, 120], using porcine model, undertook an interesting attempt to apply IRT to laparoscopic surgical procedures. For this purpose the two-channel laparoscope was used. One of the channels served for the visible image transmission, whereas the second channel was accommodated to transfer thermal image with its subsequent capture by 3-5μm infrared camera. Using this system, the problems in laparoscopic cholecystectomy, dissection of the ureter, and assessment of bowel perfusion were attacked. Thermal imaging proved to be useful in differentiating between blood vessels and other anatomic structures. Transperitoneal localization of the ureter as well as differentiation of the cystic duct and arteries were successful in all cases using the infrared system, whereas use of the laparoscopic visible system had failed. Both works [119, 120] demonstrate that IRT may potentially be a powerful adjunct to laparoscopic surgery.

While IRT, as applied to laparoscopic surgery, is discussed, the work [121] may be recollected, which is a good example demonstrating how infrared control (though, in the regime of a high temperature measurement) can render tangible assistance when the medical equipment functioning is analyzed. In [121], iatrogenic thermal injuries sometimes appearing during laparoscopic surgery using new generation vessel-sealing devices are described. It is shown that infrared camera makes possible to see the real temperature profiles provided by different active laparoscopic devices. Infrared control supplied the investigators with the quantitative data exhibiting thermal situation within and around the area of surgical intervention. The obtained results were helpful at least because they made it possible to prevent, at a later date, the accidental hand burn injuries taking place during hand-assisted surgeries.

### 5.5. The Use of IRT to Monitor and Document the Drug-Induced Actions

Among the new approaches giving an insight into pharmacological activity of drugs, IRT method is of considerable current use. Not only does this method allow investigating the action of new pharmacological means, but it also allows medical science to be provided with documentary proof of drug-induced action. With the aid of IRT, the authors of various scientific works assessed human reaction to sumatriptan [72, 122, 123] and betablockers [124] as the medicines for migraine, to L-Dopa as a medicine for Parkinson's disease [68], to dilevalol and atenolol in essential hypertension [125], to dalargin in the course of treatment of malignant tumours of the throat [94], *etc.* A set of thermograms obtained before and after taking the medicine served as the documentary evidence of the drug action.

### 5.6. IRT as a Screening Instrument

As absolutely harmless and relatively inexpensive technology, medical IRT is accepted as one of the most appropriate instrumentations for screenings all over the world. The most wide-ranging screenings, which involved tens thousands of people, were conducted in the context of projects devoted to breast cancer. The end of 20th and the beginning of 21st centuries were marked by spread of hitherto unknown infections characterized by body temperature rise: SARS (severe acute respiratory syndrome), HPAI (highly pathogenic avian influenza), the so called swine flu, *etc.* To exclude the ingress of infection into countries and their districts, sanitary and epidemiological services of the countries mounted infrared cameras at the checkpoints of airports, railway stations and some other crowded places in order to recognize and select the individuals with fever from streams of people. The area undergone infrared control was a face of the individual. Some statistical information about the latter kind of screenings was recently published. For example, in [126] it was reported that during a month 72327 visitors entered one of the Taiwan clinics. A total of 305 febrile patients (0.42%) were detected by IRT. Among them, three probable SARS patients were identified after thorough studies including contact history, radiology examinations and laboratory tests.

Inasmuch as the fever alone is not a sufficient sign of the disease, the more universal screening techniques have been appeared with intention to be used at a quarantine depot. One of them is presented in [127]. Besides infrared camera, the described system contains laser Doppler blood-flow meter to determine pulse, and 10-GHz microwave radar for the remote determination of respiration rate. This system is capable of conducting a human screening within five seconds.

A particular case demonstrating how the body elevated temperature may be revealed through the thermal pattern of face is exhibited in Fig. **6**. However, it should be taken into consideration that human surface thermal pattern can be altered significantly at various external loads applied to organism. Fast stroll as well as a carry of luggage at the airport may result in either skin cooling due to profound sweating or skin warming due to dilation of surface vessels. In [20, 21], one can find quantitative characteristics of both above-mentioned phenomena studied with the use of FPA-based IRT. It is instructively that the authors of [127], when tested their system, simulated a pseudo-infection condition exactly by a loading of healthy subjects with ergometer exercises.

It may well be that further detail IRT-assisted study of patients affected by fever as well as the use of nonstandard mathematical methods to process the obtained thermograms will make possible to find new particular signs or criteria helping to select such individuals from mass more correctly. Investigations towards this direction are in progress [128].

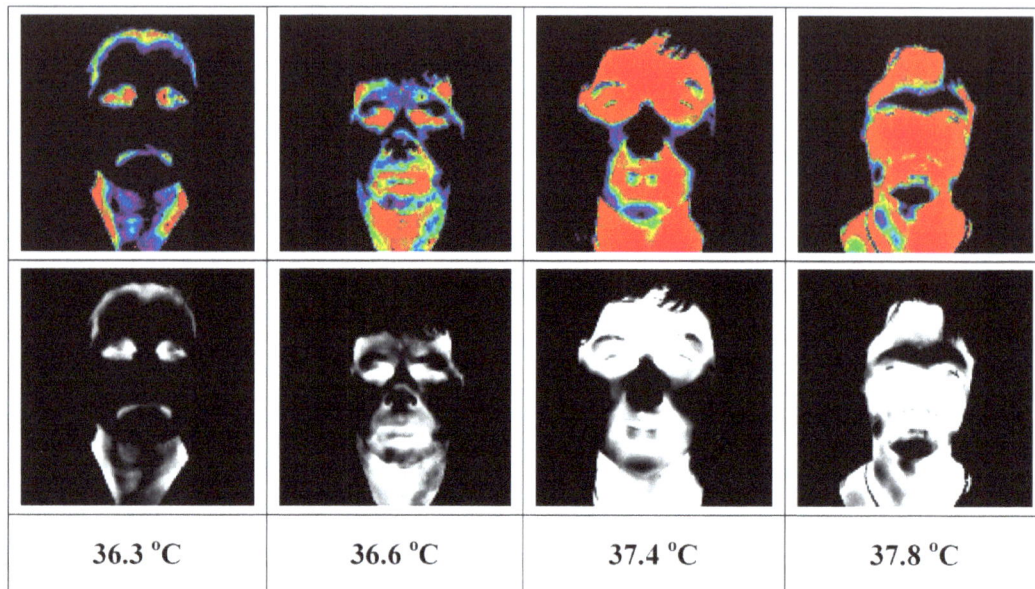

| 36.3 °C | 36.6 °C | 37.4 °C | 37.8 °C |

**Figure 6:** A set of pairs of thermograms (colour and monochrome) illustrating how the body temperature, indicated under each pair, may reflect on the thermal representation of face.

### 5.7. Breast Cancer - A Sinister Impetus to Medical IRT Diagnostic Progress

Mammology, the science devoted to study mammary glands, is a direction in medicine which, according to published materials, exploits IRT method most extensively. What is especially important, it exploits IRT for *prophylactic* diagnostics of benign and malignant tumours of the breast, predominantly in women [4]. Simultaneous use of the long-wave quantum well and two mid-wave photovoltaic infrared cameras in breast examination has been demonstrated recently a quite promising result as applied to breast cancer diagnostics [129].

To date a lot of reviews devoted to the problem in question have been published [96, 130, 131]. Possible physiological mechanisms responsible for increased intensity of infrared radiation emitted from area affected by malignant process are increased metabolism and increased perfusion within this area. Both mechanisms result in increased heat exchange between blood and superficial tissues which becomes warmer. Positive correlation between venous convective processes and presence of malignant tumour in the associated region was verified as early as in 1963 in [132].

There are two verisimilar models which have been invoked to account for such fact as the increased intensity of electromagnetic radiation at an early (as yet non-invasive) stage of oncologic disease. One of them is the so called angiogenesis [133]. It is agreed by some authors that malignant tumour, even in the very early stage and non-detectable state, prepares for itself the developed system of "forcible feeding" in the form of highly brachiferous vascularity. Malignization of tissue is followed by a release of special compositions called prostaglandins stimulating proliferation of minute vessels which are necessary for tumour growth. Another model was presented in [134, 135]. Here, the increased intensity of infrared radiation takes place not due to origination of new vessels but due to dilatation of existent ones. It is assumed that vasodilatation is induced by NO molecules produced within the region of a young tumour.

IRT diagnostics of breast cancer is not restricted sometimes by a trivial recording of surface temperature measured in the area of mammary glands. Various attempts at invoking methods of mathematical treatment of thermograms are undertaken to make diagnostics more informative, though it does not always work [75]. Such a method as the dynamic area telethermometry (DAT) [134] described briefly in the next section was also put forward as a quantitative test in breast cancer diagnostics [136, 137].

Those who launched IRT to medicine faced insuperable obstacles in early recognition of breast cancer. We still are up against a difficult problem. Even so, solving the problem of early detection of breast cancer seemingly will remain the key propelling force of medical IRT at least as long as this problem is really cracked.

## 6. IRT DIAGNOSTIC EXPEDIENTS AND THE MATHEMATICAL TREATMENTS of THERMOGRAM

### 6.1. Conventional Diagnostic Procedures

There is no need to describe in details the conventional old-time diagnostic expedients and principles accepted in medical IRT. They are well presented in many articles, reviews and monographs [43-45]. The cases demonstrated in Figs. **2-5** are typical examples of conventional use of IRT in medicine.

As an example of conventional diagnostic expedient, we refer to the cold provocation test consisting in immersion of hands or foots in a cold water (a few degrees centigrade) for several minutes and subsequent measurement of temperature recovery time using an infrared camera [52, 69, 138, 139]. As a consequence of the impaired function of peripheral vessels, sick persons usually demonstrate the more prolonged time of cold stimulated vascular reaction than healthy ones. With this test, Raynaud's syndrome, pneumatic hammer disease, peripheral atherosclerosis and some other abnormalities can be revealed.

### 6.2. Dynamic Area Telethermometry and Fast Fourier Transforms

The approach suggested by Anbar M. and consisting in Fourier analysis of telethermometry data [134] is worthy of mark to be applied to the studies of the physiology or pathophysiology of the vascular system and its neuronal control. As explained in [22], DAT needs over a thousand consecutive infrared images taken at (15-150)-millisecond intervals. The obtained images are analyzed quantitatively using fast Fourier transforms (FFT) to determine the temporal modulation of cutaneous temperature and its spatial homogeneity. The FFT calculates the amplitude of frequency components in the range 8mHz to 8Hz. It was assumed that the frequencies in the range (8-815)mHz were attributed to neuronal controlling effects, while the frequencies of 815 mHz and higher were attributed to hemodynamic effects [134].

It is believed that the DAT method invoking FFT analysis characterizes the dynamics of skin perfusion and allows the assessment of hemodynamics by a fast noncontact procedure. At the same time, it should be noted that the DAT technique needs a high quality and very stable focal plane array camera [140] and highly stable environmental conditions in the period of measurements.

### 6.3. Computer-Assisted Representations and Mathematical Treatments of Medical Thermograms

The necessity of computer processing of thermal images was recognized from early times of computerization [141]. In particular, thermal image processing is necessary to enhance the original infrared images because of the blurring, low contrast, motion artifacts, *etc.* [142-144].

One of the simplest but useful operations that one can perform to transform infrared data is three-dimensional (X, Y, T) representation of two-dimensional (X, Y, hue) thermograms. This enhances the understanding of physiological processes reflected by surface thermal pattern. As demonstrated in [20, 21], the thermogram 2D-3D transformation is informative in the studies of individual sweat glands. In [20, 21], temperature axis was directed not upwards, as usual, but downwards. The height of the displayed peaks in this case characterizes the extent of skin local cooling, due to the sweat evaporation, and intelligibly describes functional capacity of separate sweat glands. It should be noted that the 2D-3D transformation mentioned above shows to good advantage not only when it is applied to biomedical thermograms but also when applied to some other kinds of physical thermal images [145, 146].

A number of authors elaborate hardware-based and mathematical methods focused on superposition of thermal and visible images [147, 148]. They consider that the combined visual-thermal representation of infrared images helps physicians to apprehend thermal pattern of patient more easily and more adequately. Our experience in medical IRT shows that sometimes it is really necessary to have at hand both thermal and visible images of a region of interest. At the same time, superposition of these different representations on a common background, in spite of presentable view of resulted images, is suited rather not to be used in infrared examination room itself, but to be passed to outer people, among them physicians inexperienced in medical IRT.

The importance for biomedical diagnostics of such a new characteristic of thermal images as heterogeneity were first enunciated and substantiated by Vainer B. G. in [20, 21]. In these works, the correlation of the degree of heterogeneity and human physiological status as well as the necessity of elaboration of mathematical and computational methods dedicated to quantitative description of the extent of heterogeneity were highlighted. In a set of works published thereafter, some pioneering solutions of this problem were presented [24, 149-151]. Human surface thermal heterogeneity is the mirror of a deviation from homeostasis and the characteristic of a neurovascular function of the organism. The extent of thermal heterogeneity is influenced by allergic reaction, acute situational reaction, alarm reaction, anxiety reaction, *etc.*, therefore it may serve as an informative quantitative diagnostic sign in medicine.

In the same manner as in the period of sickness, overheating of the organism leads to dilatation of surface vessels and results in increase of surface thermal heterogeneity [20, 21]. Some of the vessels, the so-called vein-collectors [20, 21], are projected onto a skin as hyperthermic spots of different squares. Shown in Fig. 7 spectral distribution of such spots in terms of their effective areas is a particular case of quantitative representation of two successive states of the organism.

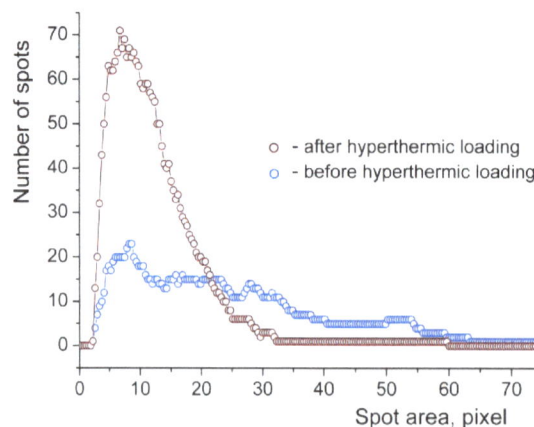

**Figure 7:** Quantitative comparison of two different states of the human organism in terms of heterogeneity of its surface thermal pattern.

Since the analysis of heterogeneity is not aimed at comparison of separate details in the thermal patterns, it results in assessment of the status of the organism as a whole. It is expected from this approach that the evaluation of quantitative characteristics of heterogeneity might contribute to implementation of idea declared in [20, 21] and based on the general principle of neurohumoral regulation. The above-mentioned idea consists in an assumption that in early diagnostics of diseases it is more reasonable to hope for quantitatively more pronounced response from integral deviations of physical characteristics in the spacious region under examination than to rely on local external manifestation of the pathology.

As methodological approaches in medical IRT become more developed, and the obtained infrared results become more correct and reproducible, artificial neural networks technology is expected to be of far-reaching importance. To date, a number of attempts have been made to elaborate this approach as applied to IRT diagnostics of breast cancer [152, 153], carpal tunnel syndrome [154], *etc.*

Further development of medical IRT should rest on intensive use of mathematical methods, in particular developed for solving the problems of different allied spheres such as tomography [155]. It is believable that through new approaches based on mathematical treatment of thermograms, the highly accurate diagnosis using thermography techniques can be achieved. We can expect a considerable step forward in medical diagnostics from solving the inverse problem of IRT. This avenue of inquiry is thorny but very fruitful, and many scientists are addressing their efforts to this problem [156-161]. Breakthrough in this area will make possible, in specific cases, to bring reliable and distinct information about biological processes which are hidden deep in the body. It prevents the patient from harmful exposure to penetrating radiation and from some other diagnostic chemical and physical agents, which have pernicious effects on the organism.

## 7. CONCLUSIONS

As far back as the beginning of the eightieths of the last century Parker R. P. argued in his review [162] that IRT should hardly be accepted as a key means playing a leading part in the diagnostics of breast cancer, but on the other hand it can:

- Serve as an informative means for early detection of bedsores,

- Tell whether a skin graft has taken,

- Be useful to monitor burns.

It is almost unreal, but no sudden changes have happened in applied medical IRT in the elapsed thirty years. Up to the present, this method has been used successfully by physicians just in the cases where the interpretation of IRT results is evident (at a level of "perfusion - no perfusion"). Physicians' indulgence towards such a situation can be considered, no doubt, as inadmissible passivity and disinclination to invoke their intellect with the intent of further development of medical IRT diagnostics.

FPA-based infrared cameras changed drastically the general principles accepted at present in conventional medical IRT diagnostics. The most key change is the demand to take into account surface temperature variations at a level of hundredths of a degree when diagnostic procedures and expedients are now under elaboration. Close adherence to this principle presents an advantage over outdated conventional approach considering just half a degree-or degree-level gradients as diagnostically significant. Conventional diagnostic principles based on the use of older-generation infrared techniques disappointed physicians since they had demonstrated many diagnostic mistakes. As it often happens in medical IRT, half a degree-or degree-level temperature gradients provide physician with redundant and needless information because in these cases the illness subjected to diagnostics becomes already self-explanatory. Unfortunately, favourable opportunities granted at present by FPA-based infrared cameras are exploited by medicine not in full measure yet.

The future of medical thermal diagnostics may well lie with all-round use of active-IRT methods based on real-time measurements of surface temperature changes and mathematical treatments of the obtained primary data. The use of modern high-performance FPA-based IRT systems just to "take a photograph" of the body and then examine this picture is the same as to use a sledge-hammer to crack nuts. As far back as 1936, in other words twenty years before the first use of thermal diagnostic imaging in medicine, the authors Hardy J. D. and Muschenheim C. exhibited in their work [163] a highly informative photograph of a forearm obtained in infrared part of spectrum and demonstrated the superficial distribution of surface veins. Was it wise to develop IRT systems with the aim to nothing more than at present expend quite a lot on each device and, at that, maintain state-of-the-art attained three fourth a century ago?

It is pertinent here to make mention of an incorrectness which is not uncommon in the literature. It consists in equalization of two terms - "infrared technique" and "infrared thermography". The authors of the book [49] quoting the work [164] give the next preferred definition of IRT: "The method used to record temperature of a body or its distribution by means of measurements of infrared radiation emitted by the surface of that body, usually for the spectral range from $0.8\mu m$ to $1.0mm$". With all due respect for the creators of this terminology, the formulation cited above can hardly be accepted for modern IRT. Method reduced to recording of temperature or its distribution, that is to the function of infrared camera (infrared system), is capable just to convert infrared radiation into visible or digitized temperature pattern. However, the features and functions inherent to modern IRT embrace also mathematical or/and computer treatment of thermal images followed by intelligent (sometimes, automated) and, normally, application-dependent analysis of the obtained and processed experimental data. In as much as the ultimate goal of IRT is the objective information about the real features of the object under investigation, but not the information represented in the form of measured temperature profile, IRT is inseparable from the methods devoted to reveal this objective information. Here, recording of temperature distribution acts as just a starting point and initial stage at these methods. The foregoing proves the next conceptually new definition of IRT: *Infrared thermography is a scientific and technical research area embracing elaboration, development and application of investigative techniques and revelation of features of material objects basing on measurement of spatial, time and energy characteristics of thermal radiation emitted or transformed by the objects under examination.* The key concepts embedded in the presented new form of IRT definition was first formulated in [20, 165].

Unfortunately, a lot of attractive nonstandard approaches which the modern computerized FPA-based IRT makes possible to realize in medicine with high sensitivity and operating speed still remain without due attention. Among them are integrated, mathematics-and computer-assisted quantitative analysis of human surface thermal patterns, search for a correlation between the results of this analysis and homeostasis state with regard to different systems of the organism, precise study of objective laws responsible for quantitative changes of heat production at different diseases, as well as those at normal and abnormal states of the organism when arisen in response to external loads, *etc.* Consequently, medical diagnostics misses the exceptionally important information that might be provided by a unique channel delivering regular diagnostically significant data which concern the local and integral thermoregulatory characteristics of a human organism. The main objective for the immediate future is to correct cardinally this situation.

## REFERENCES

[1]　Otsukay K, Togawa T. Hippocratic thermography. Physiol Measurement 1997; 18: 227-32.
[2]　Lawson R. Implications of surface temperatures in the diagnosis of breast cancer. Can Med Assoc J 1956; 75: 309-11.
[3]　Lawson R. Thermography; a new tool in the investigation of breast lesions. Can Serv Med J 1957; 8: 517-24.
[4]　Park JY, Kim SD, Kim SH, Lim DJ, Cho TH. The role of thermography in clinical practice: review of the literature. In: 9th European Congress of Thermology. 2003. Online: http: //www.comp.glam.ac.uk/pages/staff/pplassma/MedImaging/News/RoleOfThermography(Jung%20Yul%20Park).ppt
[5]　Head JF, Lipari CA, Wang F, Elliot RL. Cancer risk assessment with a second generation infrared imaging system. Proc SPIE 1997; 3061: 300-07.
[6]　Vainer B. High-resolution medical infrared thermography: new capabilities. Vrach (Physician) 1999; 2: 25-27. (in Rus.)
[7]　Ecker RD, Goerss SJ, Meyer FB, Cohen-Gadol AA, Britton JW, Levine JA. Vision of the future: initial experience with intraoperative real-time high-resolution dynamic infrared imaging. Technical note. J Neurosurg 2002; 97: 1460-71.

[8]     Vainer BG. Short-wave FPA-based IR thermographs - the advanced means for medical diagnostics, and monitoring. Bol'nichnyi List 2002; 9: 14-21. (in Rus.)

[9]     Hill DLG, Batchelor PG, Holden M, Hawkes DJ. Medical image registration. Phys Med Biol 2001; 46: R1-R45.

[10]    Wlodarczyk W, Hentschel M, Wust P, Noeske R, Hosten N, Rinneberg H, Felix R. Comparison of four magnetic resonance methods for mapping small temperature changes. Phys Med Biol 1999; 44: 607-24.

[11]    Miller NR, Bamber JC. Ultrasonic measurement of the temperature distribution due to absorption of diagnostic ultrasound: potential, and limitations. J Phys: Conf Ser/Adv Metrol Ultrasound Med 2004; 1: 128-33.

[12]    Arthur RM, Straube WL, Trobaugh JW, Moros EG. Non-invasive estimation of hyperthermia temperatures with ultrasound. Int J Hyperthermia 2005; 21: 589-600.

[13]    Jirak Z, Jokl M, Stverak J, Pechlat R, Coufalov H. Correction factors in skin temperature measurement. J Appl Physiol 1975; 38: 752-56.

[14]    Rustemeyer J, Radtke J, Bremerich A. Thermography, and thermoregulation of the face. Head & Face Medicine 2007; 3: 17(8 pp).

[15]    Petruzzellis V, Troccoli T, Candiani C, Guarisco R, Lospalluti M, Belcaro G, Dugall M. Oxerutins (Venoruton): efficacy in chronic venous insufficiency - a double-blind, randomized, controlled study. Angiology 2002; 53: 257-63.

[16]    Yahara T, Koga T, Yoshida S, Nakagawa S, Deguchi H, Shirouzu K. Relationship between microvessel density, and thermographic hot areas in breast cancer. Surg Today 2003; 33: 243-8.

[17]    Roback K, Johansson M, Starkhammar A. Feasibility of a thermographic method for early detection of foot disorders in diabetes. Diabetes Technol Ther 2009; 11: 663-7.

[18]    Gorbach AM, Heiss JD, Kopylev L, Oldfield EH. Intraoperative infrared imaging of brain tumors. J Neurosurg 2004; 101: 960-9.

[19]    Oerlemans HM, Graff MJ, Dijkstra-Hekkink JB, De Boo T, Goris RJ, Oostendorp RA. Reliability and normal values for measuring the skin temperature of the hand with an infrared tympanic thermometer: a pilot study. J Hand Ther 1999; 12: 284-90.

[20]    Vainer BG. FPA-based Infrared Thermography in Physiology: Investigation of Vascular Response, Perspiration, and Thermoregulation in Humans. Novosibirsk: Publishing House SB RAS; 2004. (in Rus.)

[21]    Vainer BG. FPA-based infrared thermography as applied to the study of cutaneous perspiration, and stimulated vascular response in humans. Phys Med Biol 2005; 50: R63-R94.

[22]    Jones BF. A reappraisal of the use of infrared thermal image analysis in medicine. IEEE Trans Med Im 1998; 17: 1019-27.

[23]    Ivanitsky GR. State of the art of thermovision in biomedicine. Uspekhi Fizicheskikh Nauk 2006; 176: 1293-320. (in Rus.)

[24]    Vainer BG. The use of infrared thermography for the investigation of thermoregulation in humans. In: Cisneros AB, and Goins BL, editors. Body Temperature Regulation. New York: Nova Science Publishers, Inc.; 2009. p 123-53.

[25]    Swain ID, Grant LJ. Methods of measuring skin blood flow. Phys Med Biol 1989; 34: 151-75.

[26]    Hardy JD. Physiology of temperature regulation. Physiol Rev 1961; 41: 521-606.

[27]    Cisneros AB, and Goins BL, editors. Body Temperature Regulation. New York: Nova Science Publishers, Inc.; 2009.

[28]    Leroy Y, Bocquet B, Mamouni A. Non-invasive microwave radiometry thermometry. Physiol Meas 1998; 19: 127-48.

[29]    Fear EC. Microwave imaging of the breast. Technol Cancer Res Treat 2005; 4: 69-82.

[30]    Vetshev PS, Chilingaridi KE, Zolkin AV, Vesnin SG, Gabaidze DI, Bannyi DA. Radiothermometry in diagnosis of thyroid diseases. Khirurgiia (Mosk.) 2006; 6: 54-58. (in Rus.)

[31]    Hardy JD. The radiation of heat from the human body: III. The human skin as a black-body radiator. J Clin Invest 1934; 13: 615-20.

[32]    Hardy JD. The radiating power of human skin in the infra-red. Am J Physiol 1939; 123: 454-62.

[33]    Steketee J. Spectral emissivity of skin, and pericardium. Phys Med Biol 1973; 18: 686-94.

[34]    Steketee J. The influence of cosmetics, and ointments on the spectral emissivity of skin. Phys Med Biol 1976; 21: 920-30.

[35]    Togawa T. Non-contact skin emissivity: measurement from reflectance using step change in ambient radiation temperature. Clin Phys Physiol Meas 1989; 10: 39-48.

[36]    Zaproudina N, Varmavuo V, Airaksinen O, Narhi M. Reproducibility of infrared thermography measurements in healthy individuals. Physiol Meas 2008; 29: 515-24.

[37]    Watson J L, Gore WG, Spears AB, Wolfe PA. A new scanning thermometer. J Phys E: Sci Instrum 1971; 4: 1029-35.

[38] Parrish WJ, Blackwell JD, Kincaid GT, Paulson RC. Low-cost high performance InSb 256x256 infrared camera. SPIE 1991; 1540: 274-84.

[39] Destefanis G, Audebert P, Mottin E, Rambaud P. High performance LWIR 256x256 HgCdTe focal plane array operating at 88 K. Proc SPIE 1997; 3061: 111-6.

[40] Kurishev GL, Kovchavtzev AP, Vainer BG, *et al.* Medical infrared imaging system based on a 128x128 focal plane array for 2.8-3.05 μm spectral range. Optoelectronics Instrumentation and Data Proc. (Autometria) 1998; 4: 5-10.

[41] Manissadjian A, Tribolet P, Chorier P, Costa P. Sofradir infrared detector products: the past, and the future. Proc SPIE 2000; 4130: 480-95.

[42] Ring EFJ, Minchhton MAB, Elvins DM. A focal plane array system for clinical infra red imaging. In: First Joint BMES/EMBS Conference "Serving, Humanity, Advancing Technology"; 1999. p 1120.

[43] Rozenfield LG, editor. Basics of Clinical Remote Thermodiagnostics. Kiev: Zdorov'ya; 1988. (in Rus.)

[44] Dragun VL, and Filatov SA. Computational Thermography. Application in Medicine. Minsk: Navuka i Technika; 1992. (in Rus.)

[45] Melnikova VP, Miroshnikov MM, Brunelly EB, *et al.* Clinical Infrared Thermography. St. Petersburg: GOI im. S.I.Vavilova; 1999. (in Rus.)

[46] Ring E, Ammer K, Jung A, Murawski P, Wiecek B, Zuber J, Zwolenik S, Plassmann P, Jones C, Jones BF. Standardization of infrared imaging. Conf Proc IEEE Eng Med Biol Soc 2004; 2: 1183-5.

[47] Schaefer G, Zhu SY, Ruszala S. Visualisation of medical infrared image databases. In: Proc. of the 2005 IEEE Engineering in Medicine, and Biology 27th Annual Conference; 2005. p 634-7.

[48] Ammer K. The Glamorgan Protocol for recording, and evaluation of thermal images of the human body. Thermology Intern 2008; 18: 125-44.

[49] Jung A, and Zuber J. Thermographic Methods in Medical Diagnostics. Warsaw: MedPress; 1998.

[50] Ring EFJ, Ammer K, Wiecek B, Plassmann P, Jones CD, Jung A, Murawski P. Quality assurance for thermal imaging systems in medicine. Thermology Intern 2007; 17: 103-6.

[51] Strangi T, Lombardi G, Braccili MP, Lo Sterzo E, Lalli A, Pennesi A. Peripheral vascular hyperreactivity in arterial hypertension. Int J Cardiol 1989; 25: S57-S61.

[52] Fushimi H, Inoue T, Yamada Y, Matsuyama Y, Kubo M, Kameyama M. Abnormal vasoreaction of peripheral arteries to cold stimulus of both hands in diabetics. Diabetes Res Clin Pract 1996; 32: 55-59.

[53] Kaczmarek M, Nowakowski A. Analysis of transient thermal processes for improved visualization of breast cancer using IR imaging. In: 25-th Annual International Conference of the IEEE EMBS; 2003. p 1113-16.

[54] Watanabe S, Asakura H, Power GG, Araki T. Alterations of thermoregulation in women with hyperemesis gravidarum. Arch Gynecol Obstet 2003; 267: 221-26.

[55] Niehof SP, Huygen FJPM, van der Weerd RWP, Westra M, Zijlstra FJ. Thermography imaging during static, and controlled thermoregulation in complex regional pain syndrome type 1: diagnostic value, and involvement of the central sympathetic system. BioMedical Engineering OnLine 2006; 5: 30(13 pp). [Online]: www.biomedical-engineering-online.com/content/5/1/30.

[56] Vainer BG. Treated skin temperature regularities revealed by IR thermography. Proc SPIE 2001; 4360: 470-81.

[57] Santa Cruz GA, Gonzalez SJ, Bertotti J, Marin J. First application of dynamic infrared imaging in boron neutron capture therapy for cutaneous malignant melanoma. Med Phys 2009; 36: 4519-29.

[58] Santa Cruz GA, Bertotti J, Marin J, *et al.* Dynamic infrared imaging of cutaneous melanoma, and normal skin in patients treated with BNCT. Appl Radiat Isot 2009; 67: S54-S58.

[59] Agarwal-Kozlowski K, Lange AC, Beck H. Contact-free infrared thermography for assessing effects during acupuncture: a randomized, single-blinded, placebo-controlled crossover clinical trial. Anesthesiology 2009; 111: 632-9.

[60] Nhan BR, Chau T. Classifying affective states using thermal infrared imaging of the human face. IEEE Trans Biomed Eng 2009. [Online]: ieeexplore.ieee.org/xpl/freeabs_all.jsp?arnumber=5338025.

[61] Kateb B, Yamamoto V, Yu C, Grundfest W, Gruen JP. Infrared thermal imaging: a review of the literature, and case report. Neuroimage 2009; 47, Suppl 2: T154-T162.

[62] Kuwahara M, Yurugi S, Mashiba K, Iioka H, Niitsuma K, Noda T. Thermography of hands after a radial forearm flap has been harvested. Eur J Plast Surg 2006; 29: 5-8.

[63] Lekas R, Jakuska P, Krisciukaitis A, *et al.* Monitoring changes in heart tissue temperature, and evaluation of graft function after coronary artery bypass grafting surgery. Medicina (Kaunas) 2009; 45: 221-5.

[64] Lindemann J, Wiesmiller K, Keck T, Kastl K. Dynamic nasal infrared thermography in patients with nasal septal perforations. Am J Rhinol Allergy 2009; 23: 471-4.

[65]  Murthy JN, van Jaarsveld J, Fei J, *et al.* Thermal infrared imaging: a novel method to monitor airflow during polysomnography. Sleep 2009; 32: 1521-7.

[66]  Sodi A, Giambene B, Miranda P, Falaschi G, Corvi A, Menchini U. Ocular surface temperature in diabetic retinopathy: a pilot study by infrared thermography. Eur J Ophthalmol 2009; 19: 1004-1008.

[67]  Wu CL, Yu KL, Chuang HY, Huang MH, Chen TW, Chen CH. The application of infrared thermography in the assessment of patients with coccygodynia before, and after manual therapy combined with diathermy. J Manipulative Physiol Ther 2009; 32: 287-93.

[68]  Dalla Volta G, Magoni M, Pezzini A. Telethermography in patients with early unilateral Parkinson's disease. Clin Autonomic Res 1999; 9: 304-5.

[69]  Coughlin PA, Chetter IC, Kent PJ, Kester RC. The analysis of sensitivity, specificity, positive predictive value, and negative predictive value of cold provocation thermography in the objective diagnosis of the hand-arm vibration syndrome. Occup Med 2001; 51: 75-80.

[70]  Ma X, He S, Wang Y. Value of infrared thermograms on diagnosis of intervertebral disc protrusion. Med J Wuhan Univ 2005; 26: 783-4.

[71]  Murray AK, Moore TL, Manning JB, Taylor C, Griffiths CE, Herrick AL. Noninvasive imaging techniques in the assessment of scleroderma spectrum disorders. Arthritis Rheum 2009; 61: 1103-11.

[72]  Dalla Volta G. Telethermography: dynamic study of the autonomic nervous system in migraine. Clin Autonomic Res 1999; 9: 294.

[73]  Savastano DM, Gorbach AM, Eden HS, Brady SM, Reynolds JC, Yanovski JA. Adiposity, and human regional body temperature. Am J Clin Nutr 2009; 90: 1124-31.

[74]  Ring EFJ. Quantitative thermal imaging. Clin Phys Physiol Meas 1990; 11,Suppl A: 87-95.

[75]  Stevens RG, Beaman SA. The use of difference of Gaussian image filtering to assess objectively the correlation between breast vascularity, and breast cancer. Phys Med Biol 1988; 33: 1417-31.

[76]  Kopans DB. 'Early' breast cancer detection using techniques other than mammography. Am J Roentgenol 1984; 143: 465-8.

[77]  Heywang SH. Status of research in the field of diagnostic imaging of the breast with particular reference to magnetic resonance tomography. Rontgenpraxis 1988; 41: 384-94. (in Germ.)

[78]  Zhou X, Gordon R. Detection of early breast cancer: an overview, and future prospects. Crit Rev Biomed Eng 1989; 17: 203-55.

[79]  Jackson VP, Hendrick RE, Feig SA, Kopans DB. Imaging of the radiographically dense breast. Radiology 1993; 188: 297-301.

[80]  Sabel M, Aichinger H. Recent developments in breast imaging. Phys Med Biol 1996; 41: 315-68.

[81]  Williams KL, Phillips BH, Jones PA, Beaman SA, Fleming PJ. Thermography in screening for breast cancer. J Epidemiol Community Health 1990; 44: 112-3.

[82]  Litscher G, Infrared thermography fails to visualize stimulation-induced meridian-like structures. BioMedical Engineering OnLine 2005; 4: 38.

[83]  Litscher G, Ammer K. Visualization of equipment dependent measurement errors, but not of meridian-like channels in complementary medicine - a thermographic human cadaver study. Thermology Intern 2007; 17: 32-5.

[84]  Vainer B. FPA-based infrared thermography systems in medicine. Vrach (Physician) 1999; 10: 30-1. (in Rus.)

[85]  Makarova TV. Application of infrared thermography for professional screening in police, and military forces. Med Thermogr 2002; 1: 41-58.

[86]  Online: http: //ultramed.com/tvk_03k.htm (2002)

[87]  Davy JR. Medical applications of thermography. Phys Tech 1977; 8: 54-61.

[88]  Mabuchi K. Clinical significance of thermography - a non-invasive, and non-contact method to evaluate peripheral circulatory function in the diagnosis of diabetic complications. Nippon Rinsho 1990; 48, Suppl: 580-7.

[89]  Lecerof H, Bornmyr S, Lilja B, De Pedis G, Hulthen UL. Acute effects of doxazosin, and atenolol on smoking-induced peripheral vasoconstriction in hypertensive habitual smokers. J Hypertens Suppl 1990; 8: S29-S33.

[90]  Zhang HY, Kim YS, Cho YE. Thermatomal changes in cervical disc herniations. Yonsei Med J 1999; 40: 401-12.

[91]  Nhan BR, Chau T. Infrared thermal imaging as a physiological access pathway: a study of the baseline characteristics of facial skin temperatures. Physiol Meas 2009; 30: N23-N35.

[92]  Coben R, Myers TE. Sensitivity and specificity of long wave infrared imaging for attention-deficit/hyperactivity disorder. J Atten Disord 2009; 13: 56-65.

[93]  Miland EO, Mercer JB. Effect of a short period of abstinence from smoking on rewarming patterns of the hands following local cooling. Eur J Appl Physiol 2006; 98: 161-8.

[94]    Maleev OV, Rozenfield LG, Kolotilov NN. Technique to increase treatment efficacy in patients with malignant tumors of the upper respiratory tract. Oncology 2002; 4: 107-8. (in Rus.)

[95]    Mikulska D. Dermatoscopy and thermal imaging: a comparative investigation of melanocytic nevi of the skin. Thermol Intern 2008; 18: 101-6.

[96]    Keith LG, Oleszczuk JJ, Laguens M. Circadian rhythm chaos: a new breast cancer marker. Int J Fertil Women M 2001; 46: 238-47.

[97]    Markel AL, Vainer BG. Infrared thermography in diagnosis of breast cancer (review of foreign literature). Terapevticheski Arkhiv (Therapeutic Archive) 2005; 77: 57-61. (in Rus.)

[98]    Stuttgen G, Eilers J. Reflex heating of the skin, and telethermography. Arch Dermatol Res 1982; 272: 301-10.

[99]    Shuran M, Nelson RA. Quantitation of energy expenditure by infrared thermography. Am J Clin Nut 1991; 53: 1361-7.

[100]   Zontak A, Sideman S, Verbitsky O, Beyar R. Dynamic thermography: analysis of hand temperature during exercise. Ann Biomed Eng 1998; 26: 988-93.

[101]   Nagashima K, Nakai S, Matsue K, Konishi M, Tanaka M, Kanosue K. Effects of fasting on thermoregulatory processes, and the daily oscillations in rats. Am J Physiol Regul Integr Comp Physiol 2003; 284: R1486-R1493.

[102]   Stoner HB, Barker P, Riding GS, Hazlehurst DE, Taylor L, Marcuson RW. Relationships between skin temperature, and perfusion in the arm, and leg. Clin Physiol 1991; 11: 27-40.

[103]   Egorov VA, Vainer BG. FPA-based Infrared Imaging Medical System TKVr-IFP/SVIT: Operating Characteristics of the Device and Atlas of Medical Thermograms. Novosibirsk: ISP SB RAS; 2004. (in Rus.)

[104]   Szabo T, Fazekas L, Horkay F, Geller L, Gyongy T, Juhasz-Nagy A. Intraoperative IR imaging in the cardiac operating room. SPIE Proc 1999; 3698: 83-7.

[105]   Geller L, Szabo T, Selmeci L, Merkely B, Juhasz-Nagy A, Solti F. Applying cardiothermography, and electrophysiology to differentiate between the ischemic, and arrhythmogenic actions of endothelin-1. SPIE Proc 1999; 3698: 88-92.

[106]   Gorbach AM, Assessment of critical renal ischemia with real-time infrared imaging. J Surg Res 2008; 149: 310-18.

[107]   Brooks JP, Perry WB, Putnam AT, Karulf RE. Thermal imaging in the detection of bowel ischemia. Dis Colon Rectum 2000; 43: 1319-21.

[108]   Gorbach A, Simonton D, Hale DA, Swanson SJ, Kirk AD. Objective, real-time, intraoperative assessment of renal perfusion using infrared imaging. Am J Transpl 2003; 3: 988-93.

[109]   Corvi A, Innocenti B, Mencucci R. Thermography used for analysis, and comparison of different cataract surgery procedures based on phacoemulsification. Physiol Meas 2006; 27: 371-84.

[110]   Kelman CD. Phaco-emulsification, and aspiration: a new technique of cataract removal. A preliminary report. Am J Ophthalmol 1967; 64: 23-35.

[111]   Polack FM, Sugar A. The phacoemulsification procedure: III. Corneal complications. Invest Ophthalmol Vis Sci 1977; 16: 39-46.

[112]   Majid MA, Sharma MK, Harding SP. Corneoscleral burn during phacoemulsification surgery. J Cataract Refract Surg 1998; 24: 1413-5.

[113]   Khodabakhsh AJ, Zaidman G, Tabin G. Corneal surgery for severe phacoemulsification burns. Ophthalmology 2004; 111: 332-4.

[114]   Devulder J. Thermograms and the vasomotor reflex. The Pain Clinic 2000; 12: 233-6.

[115]   Kaczmarek M, Nowakowski A, Renkielska A, Grudzinski J, Stojek W. Investigation of skin burns basing on active thermography. In: 23rd Annual EMBS Internatianal Conference; 2001. p 2882-5.

[116]   Hargroder AG, Davidson JE, Luther DG, Head JF. Infrared imaging of burn wounds to determine burn depth. SPIE Proc 1999; 3698: 103-8.

[117]   Pennes HH. Analysis of tissue, and arterial blood temperature in the resting human forearm. J Appl Physiol 1948; 1: 93-102.

[118]   Romero-Mendez R, Jimenez-Lozano JN, Sen M, Gonzalez FJ. Analytical solution of the Pennes equation for burn-depth determination from infrared thermographs. Math Med Biol 2009; July 17. [Epub ahead of print]

[119]   Roberts WW, Dinkel TA, Schulam PG, Bonnell L, Kavoussi LR. Laparoscopic infrared imaging. Surg Endosc 1997; 11: 1221-23.

[120]   Cadeddu JA, Jackman SV, Schulam PG. Laparoscopic infrared imaging. J Endourol 2001; 15: 111-6.

[121]   Kim FJ, Chammas MF Jr, Gewehr E, *et al.* Temperature safety profile of laparoscopic devices: Harmonic ACE (ACE), Ligasure V (LV), and plasma trisector (PT). Surg Endosc 2008; 22: 1464-9.

[122]  Parrinello G, Paterna S, Di Pasquale P, *et al.* Effect of subcutaneous sumatriptan on head temperature in migraines. Drugs Exp Clin Res 1998; 24: 197-205.

[123]  Govindan S. Imaging the effect of sumatriptan in migraine treatment. Thermology Intern 2006; 16: 60-3.

[124]  Govindan S. Imaging the effect of betablocker in migraine. Thermology Intern 2006; 16: 49-52.

[125]  Tham TC, Silke B, Taylor SH. Comparison of central, and peripheral haemodynamic effects of dilevalol, and atenolol in essential hypertension. J Hum Hypertens 1990; 4, Suppl 2: 77-83.

[126]  Chiu WT, Lin PW, Chiou HY, *et al.* Infrared thermography to mass-screen suspected SARS patients with fever. Asia Pac J Public Health 2005; 17: 26-8.

[127]  Matsui T, Suzuki S, Ujikawa K, *et al.* The development of a non-contact screening system for rapid medical inspection at a quarantine depot using a laser Doppler blood-flow meter, microwave radar, and infrared thermography. J Med Eng Technol 2009; 33: 481-7.

[128]  Ng EY-K. Is thermal scanner losing its bite in mass screening of fever due to SARS? Med Phys 2005; 32: 93-7.

[129]  Joro R, Laaperi AL, Dastidar P, *et al.* A dynamic infrared imaging-based diagnostic process for breast cancer. Acta Radiol 2009; 50: 860-9.

[130]  Keyserlink JR, Ahlgren PD, Yu E, Belliveau N, Yassa M. Functional infrared imaging of the breast. J IEEE Eng Med Biol 2000; May/June: 30-41.

[131]  Kennedy DA, Lee T, Seely D. A comparative review of thermography as a breast cancer screening technique. Integr Cancer Ther 2009; 8: 9-16.

[132]  Lawson RN, Chughtai MS. Breast cancer, and body temperatures. Can Med Assoc J 1963; 88: 68-70.

[133]  Guidi AJ, Schnitt SJ. Angiogenesis in pre-invasive lesions of the breast. The Breast J 1996; 2: 364-9.

[134]  Anbar M. Quantitative Dynamic Teletethermometry in Medical Diagnosis, and Management. CRC Press; 1994.

[135]  Thomsen LL, Miles DW, Happerfield L, Bobrow LG, Knowles RG, Mancada S. Nitric oxide synthase activity in human breast cancer. Br J Cancer 1995; 72: 41-4.

[136]  Anbar M, Eckhert KH, Milescu L. Preliminary study of women's breasts with dynamic area teletethermometry (DAT). Eur J Thermol 1998; 8: 109.

[137]  Anbar M, Brown SA, Milescu L, Babalola JA. Clinical applications of DAT using a QWIP FPA camera. SPIE Proc 1999; 3698: 93-102.

[138]  Coughlin P, Chetter IC, Kent PJ, Kester RC. Vascular surgical society of great britain, and ireland: analysis of cold provocation thermography in the objective diagnosis of the hand-arm vibration syndrome. Br J Surg 1999; 86: 694-95.

[139]  Voelter-Mahlknecht S, Letzel S, Dupuis H. Diagnostic significance of cold provocation test at 12°C. Environ Health Prev Med 2005; 10: 376-9.

[140]  Anbar M, Milescu L. Hardware and software requirements of clinical DAT. SPIE Proc 1999; 3698: 68-74.

[141]  Shimmins J, Kay MDM, James WB, Evans AL. The digitization and analysis of thermographic images. Phys Med Biol 1977; 22: 95-7.

[142]  Li Y, Hou C, Tian F, *et al.* Enhancement of infrared image based on the Retinex theory. Conf Proc IEEE Eng Med Biol Soc 2007; 2007: 3315-8.

[143]  Li Y, He R, Xu G, *et al.* Retinex enhancement of infrared images. Conf Proc IEEE Eng Med Biol Soc 2008; 2008: 2189-92.

[144]  Agostini V, Knaflitz M, Molinari F. Motion artifact reduction in breast dynamic infrared imaging. IEEE Trans Biomed Eng 2009; 56: 903-906.

[145]  Vainer BG. Narrow spectral range infrared thermography in the vicinity of 3 μm operating wavelength. In: Quantitative InfraRed Thermography 5, Eurotherm Seminar 64, QIRT'2000; 2000. p 84-91.

[146]  Vainer BG. Focal plane array based infrared thermography in fine physical experiment. J Phys D: Appl Phys 2008; 41: 065102(12 pp).

[147]  Schaefer G, Tait R, Zhu SY. Overlay of thermal, and visual medical images using skin detection, and image registration. Conf Proc IEEE Eng Med Biol Soc 2006; 1: 965-67.

[148]  Di Romualdo S, Merla A, Romani GL. Superimposition of thermal imaging to visual imaging using homography. Conf Proc IEEE Eng Med Biol Soc 2007; 2007: 3365-8.

[149]  Tarkov MS, Vainer BG. Evaluation of a thermogram heterogeneity based on the wavelet Haar transform. In: IEEE International Siberian Conference on Control, and Communications (SIBCON-2007); 2007. p 145-52.

[150]  Vainer B, Moskalev A. Heterogeneous thermograms: the methods of attack. In: QIRT2008, 9-th International Conference on Quantitative InfraRed Thermography; 2008. p 157-64.

[151] Vainer BG, Moskalev AS, and Tarkov MS. Application of infrared thermography to qualitative, and quantitative analysis of the cardiovascular system status. In: Ivanova LN, Blokhin AM., and Markel AL., editors. Cardiovascular System, and Arterial Hypertension: Biophysical, Genetic, and Physiological Mechanisms, Mathematical and Computer Modeling. Novosibirsk: Publishing House SB RAS; 2008. p 205-249. (in Rus.)

[152] Koay J, Herry C, Frize M. Analysis of breast thermography with an artificial neural network. In: 26th Annual International Conference of the IEEE EMBS; 2004. p 1159-62.

[153] Ng EY, Kee EC. Advanced integrated technique in breast cancer thermography. J Med Eng Technol 2008; 32: 103-14.

[154] Jesensek Papez B, Palfy M, Mertik M, Turk Z. Infrared thermography based on artificial intelligence as a screening method for carpal tunnel syndrome diagnosis. J Int Med Res 2009; 37: 779-90.

[155] Baddour N. Fourier diffraction theorem for diffusion-based thermal tomography. J Phys A: Math Gen 2006; 39: 14379-95.

[156] Feasey CM, Davison M, James WB. Effects of natural, and forced cooling on the thermographic patterns of tumours. Phys Med Biol 1971; 16: 213-20.

[157] Deng Z-S, Liu J. Blood perfusion-based model for characterizing the temperature fluctuation in living tissues. Physica A 2001; 300: 521-30.

[158] Kakuta N, Yokoyama S, Suzuki T, Saito T, Mabuchi K. Evaluation of infrared images by using a human thermal model. In: 23rd Annual EMBS international Conference; 2001. p 2816-2819.

[159] Deng Z-S, Liu J. Mathematical modeling of temperature mapping over skin surface, and its implementation in thermal disease diagnostics. Comput Biol Med 2004; 34: 495-521.

[160] Vainer B, Belozerov P, Baranov V. Use of IR thermography for bioheat transfer studies. In: 7th International Conference on Quantitative IR Thermography; 2004. p J.8.1-J.8.4.

[161] Wu Z, Liu HH, Lebanowski L, Liu Z, Hor PH. A basic step toward understanding skin surface temperature distributions caused by internal heat sources. Phys Med Biol 2007; 52: 5379-92.

[162] Parker RP. Medical imaging. Phys Technol 1982; 13: 70-7.

[163] Hardy JD, Muschenheim C. Radiation of heat from the human body. V. The transmission of infra-red radiation through skin. J Clin Invest 1936; 15: 1-9.

[164] Clark R, De Calcina-Goff ML. International standardization in medical thermography. Thermol Osterr 1997; 7: 85.

[165] Vainer BG. Infrared thermography monitoring in cosmetology. Innovatsii (Innovations) 2005; 7: 119-22. (in Rus.)

# CHAPTER 4

# The Use of Infrared Thermography in Livestock Production and Veterinary Field

## Petr Kunc[*] and Ivana Knizkova

*Department of Technology and Breeding Technique of Farm Animals, Institute of Animal Science, Pratelstvi 815, 104 00 Prague Uhrineves, the Czech Republic*

**Abstract:** This chapter presents the application and use of infrared thermography (IRT) in farm animals and veterinary medicine. Changes in vascular circulation in living organisms result in an increase or decrease in tissue temperature, which is then used to evaluate the situation in the given area. IRT is mainly used in veterinary medicine, primarily for diagnostic purposes, particularly in the diagnosis of orthopaedic diseases in horses. IRT can, however, be used extremely successfully for study and research in farm animals such as pigs, cattle, sheep and poultry. Areas of research include reproduction, thermoregulation, animal welfare and the milking process. Although IRT is applied less frequently in the study of the milking process, recent results show the potential of this measuring method. Generally, IRT is a suitable tool for early detection and screening for mastitis, and can also be useful for studying and evaluating the effects of various milking techniques on the teats and udders. IRT is recommended as a method that can produce important information where the possibilities of conventional diagnostic techniques have been exhausted. There are, however, certain limitations and factors that must be considered when using IRT in animals.

**Keywords:** Infrared thermography, animal practice, research, diagnosis, horse, cattle, pig, sheep, poultry, rabbit, milking process, mastitis.

## 1. INTRODUCTION

This chapter is an overview incorporating already published works of authors [1, 2]. These publications were used as the basis for this chapter and the chapter is supplemented by new information and knowledge in this area.

The thermographic method has found numerous applications mainly to architecture, civil and industrial engineering, but also in human and veterinary medicine. Infrared thermography (IRT) was used in veterinary medicine for the first time in 1965 by Delahanty and Georgi; it was at that time that proper thermograms in particular areas of horses' bodies were defined [3]. Changes in vascular circulation in living organisms result in an increase or decrease in tissue temperature, which is then used to evaluate the situation in the given area [4]. For example, heat generated by inflammation is transmitted to the overlying skin *via* increased capillary blood flow and dissipated as thermal energy. IRT proved extremely useful in enabling the diagnosis of many diseases before they could be demonstrated clinically [3]. The IRT method, however, is not particularly suitable for the diagnosis of internal diseases.

Infrared energy can be measured using an infrared camera and a specially developed analysing software program [5]. Spruyt *et al.* [6] recommended IRT as a good method to help study thermoregulation. A major advantage of the method is that it does not require direct physical contact with the surface monitored, thus allowing remote reading of temperature distribution [7].

Infrared thermography can be used in animal practice in three ways:

1. As a diagnostic tool. In these cases, IRT becomes a physiological imaging method. The difference of 1°C between two anatomically symmetric regions indicates there inflammation;

*****Address correspondence to Petr Kunc:** Department of Technology and Breeding Technique of Farm Animals, Institute of Animal Science, Pratelstvi 815, 104 00 Prague Uhrineves, the Czech Republic; Tel:+420-267009638; Fax:+420-267710779; Email: kunc.petr@vuzv.cz

**Carosena Meola (Ed)**

2.  To enhance the possibilities of physical examination. In these cases thermography can determine suspicious areas where the heat is increased or decreased;

3.  In wellness programmes. In this case, animals are monitored on a routine basis once a week. Thermography can be used to detect subclinical problems, as clinical changes occur two weeks after thermographic changes [8].

There are, however, certain limitations and other factors that need to be considered when using IRT. As a main rule, thermograms must be collected out of direct sunlight and wind currents. The surface should be free of dirt, moisture and extraneous substances. The effect of weather conditions and circadian and ultradian rhythms are also factors that need to be considered and require further investigation as part of the process of validating IRT.

## 2. APPLICATION IN HORSE BREEDING

Thermography has been found useful for the diagnosis and evaluation of several disease conditions in the horse, such as back pain, soft tissue injuries, arthritis and affections of flexor tendons, hooves and related structures. The example is shown in Figs. **1** and **2**. Greatest attention has been paid to the diagnosis of orthopaedic diseases in horses [5, 9-24]. All authors recommended IRT as a diagnostic tool [25], however, they investigated the value of IRT in the clinical examination of chronic, scintigraphically positive diseases of bony structures in the horse. The results of their study indicate that the clinical use of IRT cannot be recommended for imaging equine bones and joints during lameness examination in chronic cases.

Yanmaz *et al.* [26] present IRT as a useful imaging tool for evaluating diagnoses of certain lameness conditions in the horse, that also enables appropriate treatment to be applied before any clinical condition occurs. Eddy *et al.* [18] write that IRT has been used to evaluate a number of different clinical syndromes not only in the diagnosis of inflammation, but also in monitoring the progression of healing.

Braverman [27] used IRT for early detection of summer seasonal recurrent dermatitis (sweet itch) in horses. In horses sensitive to this disease, warmer areas in affected zones were already detected in winter. IRT was also used in the following studies: the character and duration of the pharmacological effects of intravenous isoxsurpine and the pharmacological effects of isoxpurine [28, 29], the scintigraphic, radiographic and thermographic appearance of the metacarpal and metatarsal regions of healthy adult horses treated with non-focused extracorporeal shock wave therapy [30], assessment of scintigraphic and thermographic changes after focused extracorporeal shock wave therapy on the origin of the suspensory ligament and the fourth metatarsal bone in horses without lameness [31], evaluation of skin sensitivity after shock wave treatment in horses [32], and thermographic study of *in vivo* modulation of vascular responses to phenylephrine and endothelin-1 by dexamethasone in the horse [33].

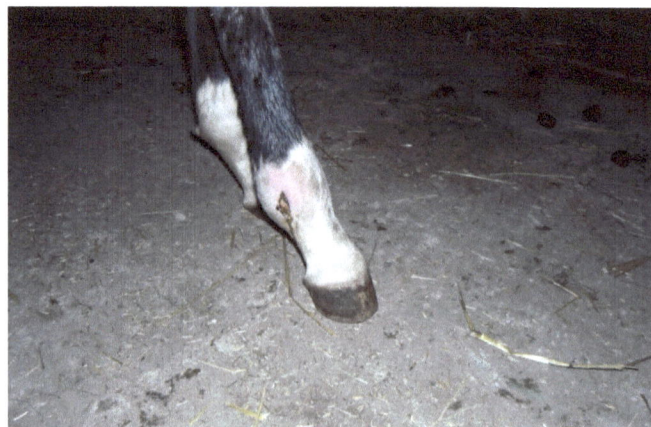

**Figure 1:** Real photo of the injured leg of a young horse.

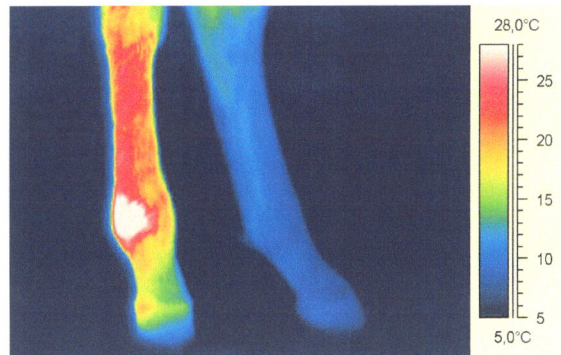

**Figure 2:** Thermogram of the injured leg of a young horse.

IRT can also be used as a method to assess the skin temperature in thermoregulatory studies in horses [4]. Ghafir *et al.* [34] studied the training effects on thermoregulation in exercising horses by means of measurements of rectal and skin temperature. The results have shown a smaller increase of rectal temperature and a greater increase of skin temperature following a training period of 6 week.

Jodkowska and Dudek [35] demonstrated the usefulness of body surface temperature measurements for the evaluation of training progress and preparation of horses for extreme exertion. Simon *et al.* [36] determined the amount of time required for the surface temperatures of thoracic and pelvic limbs in horses to return to pre-exercise temperatures after high-speed treadmill exercise, as detected using IRT imaging. There were no significant differences in surface temperatures between thermograms obtained before exercise and after 45 minutes after exercise. The study of [37] focused on thermographic evaluation of the lower critical temperature in weanling horses. They determined the heat loss of weanling horses in a cold environment by IRT to assess their thermoregulatory capacity. However, the forts on the surface of their coat at – 23°C disturbed the thermographic examination.

Marlin *et al.* [38] investigated post exercise changes in compartmental body temperature accompanying intermittent cold water cooling in the hyperthermic horse. Their study introduces a new dimension to the understanding of cold water cooling in the hyperthermic horse. Detailed dynamic changes in compartmental temperatures, and in pulmonary artery temperature in particular, have been demonstrated by measuring body temperature continuously. The pulmonary artery temperature variation clearly demonstrates that the application of cold water at 6°C to the body surface when the horse is hot does not cause vasoconstriction of the blood vessels in the skin in the initial stages, as indicated by the cyclical variations intimately reflecting intermittent pattern cooling.

## 3. APPLICATION IN CATTLE BREEDING

Hurnik *et al.* [39] studied the suitability of IRT for the detection of health disorders in Holstein-Friesian dairy cattle and concluded that it was suitable for the purpose.

Schwartzkopfgenswein and Stookey [40], using IRT, compared differences in the extent and duration of inflammation observed on hot-iron and freeze brand sites as an indicator of tissue damage and the associated discomfort to the animals. The thermographic evaluation of hot-iron and freeze brand sites indicated that both methods caused tissue damage. However, hot-iron brand sites remained significantly warmer than freeze sites at 168h after branding. In addition, hot-iron sites were significantly warmer than control sites, while freeze sites were not warmer than control sites at 168h. The prolonged inflammatory response observed in hot-iron animals indicates that more tissue damage and perhaps more discomfort are associated with hot-iron branding.

Spire *et al.* [41] used IRT to detect inflammation caused by contaminated growth-promoting ear implants in cattle. The authors found that significant differences existed between ears with contaminated implants and control ears. Cockroft *et al.* [42] described the use of IRT as an aid in the diagnosis of septic arthritis of the right metatarsophalangeal joint of Friesian heifers. IRT was able to identify the focus of inflammation accurately and provide supporting evidence for more invasive diagnostic techniques and treatment therapies.

Schaefer *et al.* [43] used IRT for identifying calves with bovine viral diarrhoea virus. They found that increases in eye temperature were more consistent than other anatomical areas. There were also significant changes in eye temperature several days to one week before other clinical signs of infection. Then the authors [44] used IRT for early detection of bovine respiratory disease in calves. Their data demonstrated that IRT was able to identify animals in early stages in illness, often several days to over one week before clinical signs were manifest.

Nikkah *et al.* [45] observed the hooves of dairy cows. Images of hooves were taken using IRT to determine the temperature of the coronary band and that of a control area above the coronary band. The authors recommend IRT as a tool for monitoring hoof health. The authors [46] also hypothesize in their study that IRT may be able to detect inflammation in the coronary band region that is associated with lameness earlier than locomotion score. Their results indicate that IRT may be a useful tool for assessing lameness in dairy cattle. The authors recommend IRT as an on-farm tool for lameness detection.

Stewart *et al.* [47] studied the possibility that pain can be detected from changes in eye temperature and heart rate variability disbudding was examined in calves. The measurements of temperatures were provided by IRT. Heart rate variability and eye temperature together may be used as a useful non-invasive and more immediate index of pain than hypothalamus-pituitary-adrenal axis activity alone.

Another area using IRT is reproduction. Hurnik *et al.* [48] studied the relationship between differences in body surface temperatures and the oestrous in Holstein-Friesian dairy cows and the possibility of using this technique to determine the onset of the oestrous cycle. Because inaccuracies were encountered in determining the oestrous cycle during the experiment, the authors did not recommend IRT for routine detection of the oestrous, but it is nevertheless completely adequate in skin temperature studies or, more precisely, in studies of body surface temperature changes. Hellebrand *et al.* [49] concluded that gravidity of heifers in their usual environment (pasture or barn) could not be determined by simple monitoring with a thermal imager, but they did find that the external pudendum temperature follows the core body temperature, and thus IRT can be utilised for oestrus climax determination.

Kozumplik *et al.* [50] used IRT in the diagnosis of inflammatory processes on the sex organs of breeding bulls. The objective was to obtain an overall thermogram of the gonads of bulls with normal and disturbed fertility, and to assess the possibility of using IRT for the diagnosis of sex organ diseases accompanied by local temperature changes. Thermograms of the scrotum showed that the warmest and the coldest zones lie at the head of the epididymis and the part of the testicles adjoining the tail of the epididymis, respectively. The authors recommend thermography as a powerful tool for the identification of initial stages of diseases of the sex organs in breeding bulls.

Kastelic *et al.* [51] found that temperature gradients were most pronounced on the scrotal surface, less in the scrotal subcutaneous tissue, and slightly negative (relative to the surface) within the testicular parenchyma. Scrotal surface temperature decreased from the neck of the scrotum to the ventral aspect to scrotum. Conversely, the ventral pole of the testis was slightly warmer than the dorsal pole. The caput of the epididymis was warmer than the adjacent testicular parenchyma, while the cauda was cooler. IRT was also used by [52, 53] to study environmental factors that influence the bovine scrotal surface temperature and to study the influence of ejaculation on the scrotal surface temperature in bulls. The authors concluded that representative temperature measurements of the scrotal surface could be taken at any time of the day except at feeding and rising. Moreover, the scrotum should be dry. Measurements are independent of the ambient temperature provided it is stable. Abrupt changes in the ambient temperature may, however, result in artefacts due to overcompensations. Thermographic measurements showed that spontaneous ejaculations, as well as electroejaculations, increased the surface temperature of the scrotum. Further, the authors [54, 55] found that insulation of the scrotal neck affected scrotal surface temperature, scrotal subcutaneous and intratesticular temperatures, and increasing testicular temperature results in defective spermatozoa, with recovery dependent on the nature and duration of the insult.

Kastelic *et al.* [56] used IRT to study the contribution made by the scrotum, testes and testicular artery to scrotal/testicular thermoregulation in bulls at two ambient temperatures. Their results supported the hypothesis that blood within the testicular artery has a similar temperature at the top of the testis compared with the bottom, but subsequently cools before entering the testicular parenchyma.

Gabor *et al.* [57] determined the effect of GnRH treatment on plasma testosterone concentrations and scrotal surface temperature, the repeatability of various morphologic, thermal and endocrine measures before and after GnRH treatment. They also examined the correlation between the total number of spermatozoa and the proportion of live and motile spermatozoa, using various morphologic, thermal and endocrine measures before and after GnRH challenge. The authors concluded that GnRH treatment significantly increased plasma testosterone concentrations and usually caused significant increases in scrotal surface temperature. Scrotal circumference and the total number of spermatozoa per ejaculate were highly correlated. Other measurements were less correlated, though with an apparent effect of ambient temperature on vide-measurements of the scrotum and assessment of scrotal surface temperature. Significant regression equations were derived for the total number of spermatozoa and the percentage of motile spermatozoa; plasma testosterone concentrations and scrotal surface temperature gradients, respectively, were the significant independent variables.

In a series of experiments with beef cattle, the authors [58] investigated IRT as a non-invasive method of evaluating thermoregulatory responses in undisturbed animals (beef cattle) on the pasture. IRT was found to be a highly reliable tool under field conditions. Coppola *et al.* [59] conclude that it is possible to evaluate the thermal status of cattle without restraint using IRT. Schaefer *et al.* [60] studied the effect of the period of fasting and transportation of slaughter animals on, inter alia, infrared heat loss of beef cattle. In three experiments, the transportation distances were 3,320 and 320km, and the period of fasting 24, 48 and 72 hours, respectively. The body surface temperature was measured by IRT, and thermograms showed that the heat loss decreased with longer fasting and transportation distances, which corresponded with the darker meat colour of slaughtered animals. A similar study was realized by [61] in beef cattle, and IRT was used successfully to detect dark-firm-dry beef.

The environment plays an extremely important part in cattle breeding. Kimmel *et al.* [62] studied the effects of evaporative cooling on heat stress of Israeli-Holstein dairy cows in summer, and used IRT to measure differences between temperature zones on their bodies. For two hours, the cows were alternatively sprinkled with water for 0.5 minutes and cooled with a flow of air ($3ms^{-1}$) for 4.5 minutes. Their rectal temperature dropped from 38.2°C to 36°C and remained unchanged for another hour. Thermograms of the cows also revealed a drop of 1.5°C in body surface temperature caused by the cooling.

As part of research into the protection of cattle against high ambient temperatures, Knizkova *et al.* [63] conducted an experiment involving thermographic monitoring of body surface temperatures immediately before and after cooling, 15, 30 and 45 minutes after cooling, and after complete drying. They also wanted to identify the warmest zones on cattle that were exposed to high ambient temperatures. Air temperatures at the time of the experiment were between 27 and 31°C. The cattle were sprinkled for 60 seconds with water by means of special nozzles mounted above. Evaluation of the thermograms showed that the warmest zones included the neck, shoulder and rib regions. After a 60-second period of cooling at the given ambient temperatures, the mean body surface temperature dropped by 1.2°C for 45 to 60 minutes. An evaluation of naturally ventilated dairy barn management using the thermographic method was performed by Knizkova *et al.* [64]. The thermal comfort of the cows in the barn was assessed by monitoring changes in their body surface temperature (Figs. **3-5**).

Microclimatic factors in the barn were modified by opening and closing sidewall plastic curtains in the barn and doors in alleys. While no changes in body surface temperature were recorded when the air temperature dropped by 3.1°C, a significant response was recorded when the air temperature dropped by 6.5°C. Significant changes in the air velocity at temperatures within the thermoneutral range influenced thermal conditions in the barn, and significant changes in body surface temperature caused by vascular responses were recorded. It is impossible to assess the thermal comfort of dairy cattle housed in barns objectively merely on the basis of the visually detectable thermoregulatory behaviour of the cows or microclimatic parameters measured in barns, because different combinations of air temperature and air velocities will result in different intensities of body surface cooling. This is reflected in variations of body surface temperatures that can be reliably monitored by IRT.

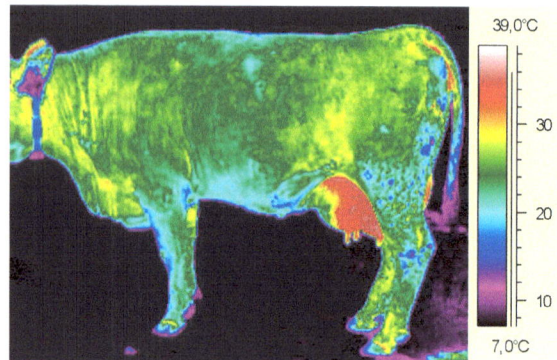

**Figure 3:** Thermal profile of dairy cows at an air temperature of 3°C.

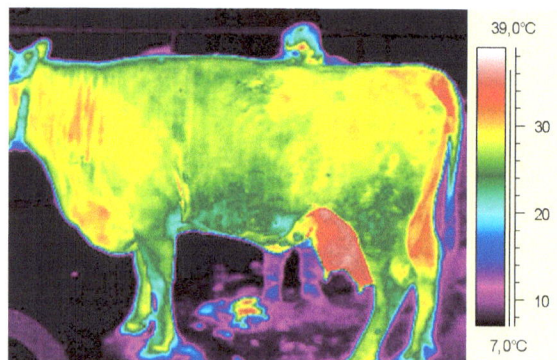

**Figure 4:** Thermal profile of dairy cows at an air temperature of 12°C.

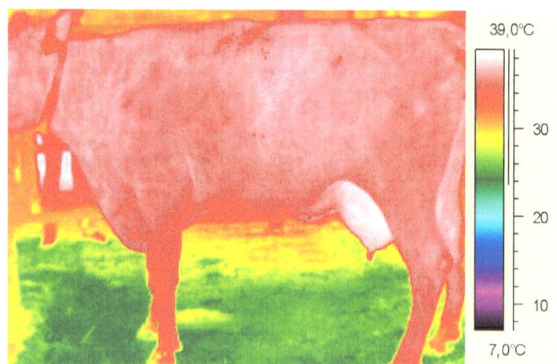

**Figure 5:** Thermal profile of dairy cows at an air temperature of 29°C.

Verkerk *et al.* [65] studied cold stress in dairy cattle with the use of IRT. This technology may prove to be a useful indicator of overall stress. Generally, researchers [66, 67] recommend IRT to detect stress in dairy cattle. IRT can detect changes in peripheral blood flow from the resulting heat loss and may therefore be used as a useful tool for measuring stress in animals. Stewart *et al.* [68] also proved the effects of a nonsteroidal anti-inflammatory drug on pain responses in dairy calves to hot-iron dehorning using IRT. Unshelm *et al.* [69] assessed damage to the joints of cattle due to housing conditions using IRT.

The IRT method can be recommended for detection of joint damage resulting from poor housing systems. The results of this study indicated that the measurement distance should be as short as possible. A distance greater than 80 cm can result in errors as a result of the colder background radiation. Stewart *et al.* [70] used IRT for evaluation of the texture suitability of floors in cow resting areas from the viewpoint of their thermal properties and with a view to various sorts of top-layer structure quality. Applied data collection of the temperature difference of the warmed resting areas with straw were, for all significance levels of 30-minute-long observations, found to correspond to the results of typical warm floors covered with mattresses filled with rubber foam and insulating mats with rubber covers. Their temperature differences were significantly higher than the results for concrete and brick floors.

Harrison *et al.* [71] used IRT in an extremely untraditional manner to predict growth efficiency. Their invention provides a method for predicting the growth efficiency of an animal by using infrared thermography by generating a predictive model, comprised of selecting a sample population from a group of animals; scanning each animal to obtain a thermographic image represented as an array of pixels providing temperature data; calculating a value of a statistical measure of the temperature data (input variable); calculating a value of a measure of growth efficiency (output variable); and determining a relationship between the input and output variables to generate a predictive model. The predictive model is then used to predict growth efficiency in an animal from the same group but not in the sample population by scanning the animal to obtain a thermographic image; calculating a value of a statistical measure of the temperature data (input variable); and solving the predictive model to provide the value of the growth efficiency of the animal. Stanko *et al.* [72] mention that the thermal camera shows itself to be a useful tool in estimating live body weight, which is of significant importance in reducing stress during progeny testing and beef production.

IRT was also tested for predicting heat production and methane production and for the detection of physiological events (*e.g.,* heat increment of feeding) in dairy cattle [73]. The results show that IRT can be successfully applied in the assessment of heat and methane production by means of analysis of feet temperature and the temperature difference between the left and right flanks, respectively. The authors write that IRT is also useful for assessing physiological responses to milking and feeding.

## 4. APPLICATION IN PIG BREEDING

In their study of osteoarthrosis tarsi deformans (OATD), the authors [74] found significant disease-related temperature differences when they used IRT to monitor the superficial temperature of the tarsus in Swedish Landrace boars. Boars with a positive finding also had a higher temperature of the tarsus. OATD was not ascertained in only ten boars of the entire group. Other authors [75] have examined the feasibility of using IRT as a method for diagnosing inflammation of the leg joints in fattening pigs. Joint inflammation in fattening pigs cannot be detected by thermography alone due to false – positive cases.

Loughmiller *et al.* [76] determined the relationship between ambient temperature and mean body surface temperature measured using IRT, and evaluated the ability of IRT to detect febrile responses in pigs following inoculation with *Actinobacillus pleuropneumonia*. IRT can be adjusted to account for ambient temperature and used to detect changes in mean body surface temperature and radiant heat production attributable to a febrile response in pigs. This conclusion was, however, disclaimed by Montanholi *et al.* [77]. They assert that IRT is not a good alternative for the detection of fever in pigs.

IRT was also used [78] to observe an increasing correlation between the incidence of meat quality defects and increasing skin surface temperature in pigs prior to stunning. Gariepy *et al.* [78] conclude that IRT can be a practical and rapid method of detecting which pigs will yield a significant proportion of meat quality defects. Schaefer *et al.* [79] studied the relationship between stress sensitivity and meat quality in pigs of known genotypes. Using a thermographic camera, the authors mapped the distribution of temperature fields on pigs immediately before slaughter and on carcasses after slaughter, and then related the temperature fields to meat quality. Although thermographic analysis failed to demonstrate any significant differences between mean superficial temperatures before slaughter and superficial temperature of carcasses in pigs of different genotypes, it was ascertained that a higher drip loss and percentage of pale, soft and exudate meat (PSE) may be expected in pigs with a lower superficial temperature.

Adamec *et al.* [80] studied the possibility of reducing heat stress on fattening pigs during the summer period by means of water evaporative cooling. Changes in body surface temperature were measured by IRT. The authors concluded that evaporative cooling decreased heat stress on pigs, and improved growth and feed conversion.

Using a thermographic system, the authors of reference [81] studied the temperature of the mammary gland skin in twenty Large White sows. The evaluation of their body surface thermograms from the last days of pregnancy showed that the area of the mammary glands was the warmest zone of the body. On the first day of lactation, the temperature in some limited areas was 39°C, and average skin temperatures in the following days of lactation ranged from 37 to 38°C. Similar temperatures were ascertained on the skin of sucking piglets or those resting close to the sows. Using IRT, the authors were able to show that the mammary gland and the piglets make up a single isothermic complex supporting the integrity of the mother-offspring biological unit. Xin [82] tested the thermal comfort of group-housed 4 – 5 week-old pigs exposed to 20, 24, 28, 32 and 36°C with air velocity at 0.1, 0.5, 1.0 and 1.5m.s$^{-1}$. The thermal profiles of pigs were obtained by IRT. His research may further elucidate the thermoregulatory responsiveness of pigs to environmental modifications.

Loughmiller *et al.* [83] write that IRT can be used to detect differences among individual pigs in mean body surface temperature associated with feed intake, growth rate and dietary energy content in more variable environmental conditions than those used with calorimetry. Development of IRT applications in swine research and production may allow direct estimates of changes in swine thermoenergetics in response to treatment or environment. These direct estimates may be useful information in traditional growth assays, growth modelling and commercial production situations. The paper by Warris *et al.* [84] describes an investigation of the value of thermal images of the surface of pigs' bodies to predict their core body temperature, using pigs being studied primarily to monitor the effects of mechanical ventilation of vehicles on the welfare of slaughter pigs; these pigs had a relatively wide range of body temperature. The mean ear temperature of the pigs was significantly correlated with the mean temperature of the pig's blood, and the mean activity of serum creatine kinase was positively correlated with the mean ear temperature. It was concluded that hotter pigs were suffering from a higher level of stress. Thermal imaging cameras are still relative expensive, but appear to be reliable under field conditions.

## 5. OTHERS

In their study Chepete and Xin [85] investigated the efficacy of intermittent partial surface sprinkling on cool caged hens at 20, 38 and 56 week of age. The body surface temperature of the hens was measured by IRT. The authors found that sprinkling once every 5 to 6min provided adequate cooling to prevent the surface temperature from rising.

Mala *et al.* [86], and Knizkova *et al.* [87] observed and compared the thermal insulation of the birthcoat in 3-day-old lambs in two, then four genetic types exposed to a cold environment and to rain simulation. Body surface temperature was measured by IRT. The lowest heat losses were recorded in the Sumavská breed. The results showed that cold resistance of postnatal lambs is influenced by breed and by birthcoat character, and that the Sumavská breed is better adapted to cold and rainy conditions (Fig. **6**) than the Merinolandschaf breed and the crossbreeds Suffolk x Merinolandschaf (Fig. **7**) and Suffolk x Sumavská. The results obtained by IRT were confirmed by biochemical and haematic analysis.

Stewart *et al.* [88] recommended IRT as a non-invasive tool for studying animal welfare. Reliable, non-invasive tools that can be used to measure acute and chronic stress during commercial practices and pre-slaughter are required. IRT fits these criteria and has great potential as a way of assessing animal welfare. For example, this fact is confirmed by the paper of [89]. The authors used IRT to detect rabbit skin zones most suitable for temperature monitoring during stress challenges. The results of this study show that IRT is a suitable method for the evaluation of superficial temperature variation in rabbits, according to the applied stressor.

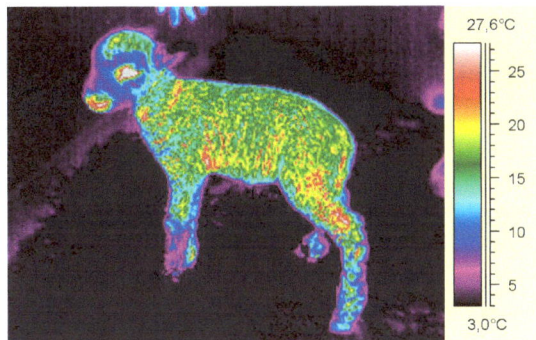

**Figure 6:** Very good thermal insulation of birthcoat in Sumavská breed lamb (air temperature 4°C).

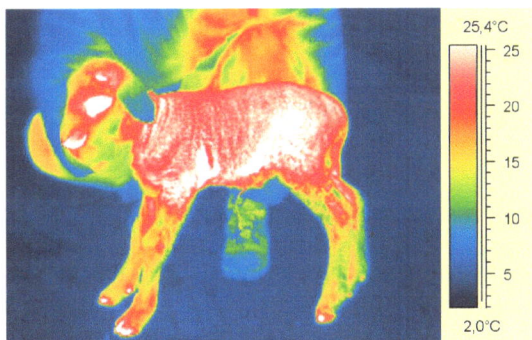

**Figure 7:** Very bad thermal insulation of birthcoat in Suffolk x Merinolandschaf crossbreed lamb (air temperature 4°C).

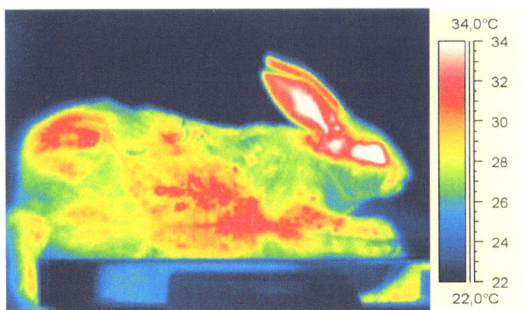

**Figure 8:** Thermal profile of a healthy rabbit.

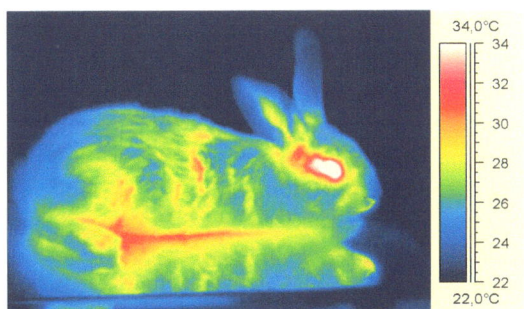

**Figure 9:** Thermal profile of a rabbit infected with *Eimeria intestinalis* oocysts (10th day of disease).

The variations related to the physiological changes during the stress reaction are evidenced in the ear pavilion and periocular area, where vasoconstriction occurs. Knizkova *et al.* [90] observed the effect of coccidiosis (*Eimeria intestinalis*) on rectal temperature, body surface temperature and performance in rabbits. They concluded that *Eimeria intestinalis* invokes changes in rectal temperature and body surface temperature (Figs. **8** and **9**). Temperatures decrease depending on the day of disease. Thermography is a suitable method for the evaluation of superficial temperature changes.

## 6. APPLICATION IN THE MILKING PROCESS

Milking is an important process in farming. Various milking routines, and the very fact of using machine milking, can affect the health and welfare of animals because an extremely sensitive organ, *i.e.,* the mammary gland, comes into direct contact with the milking machine in this process. The teats are the most stressed part of the udder, because milking changes their condition. Repeated teat compressions may cause mechanical and circulatory changes in teat tissues and hyperaemia in the teat wall [91-94]. Such changes may even lead to pathological traumatisation manifested by, for example, congestion, oedema, cracks in mucous membranes, induration. There are a number of factors in milking that influence the condition of the teats. Literary sources emphasise the importance of the milking vacuum, and also the pulsation rate, pulsation ratio and the quality of the teat cups. Assessment of the teats and udder before and after milking is usually based on visual observations. A cutimeter or a classification system [95-97] or ultrasonographic scanning [98] is used for such assessments.

Bovine mastitis infection is a widespread problem in the dairy industry. This common affliction is difficult to treat, and its effects include reduced milk quality resulting in lower milk prices, as well as reduced output and increased veterinary costs for dairy management. Early detection of mastitis can improve profits through increased milk production, decreased milk dumped due to treatment, reduced veterinary and drug costs, reduced labour costs, fewer culling and death losses, and improved quality premiums [99]. Detection of mastitis is generally provided by electrical conductivity [100] or by somatic cell counts (SCC), the California mastitis test (CMT) or bacterial isolation and identification [101]. New techniques are, however, being sought for early detection of mastitis in the dairy industry.

Thermographic measurements of the milking process have been taken by [102], who investigated the temperature responses of the udder to machine milking. This study shows that conventional milking machines may cause an increase of the teat-end temperature by 2°C. Caruolo *et al.* [103] studied the relationship between the internal and surface temperatures of the mammary gland and the temperature of milk in goats. The authors used IRT to measure the surface temperature of the udder and teats and found an increase in temperature following machine milking. This supports the findings of [104], who reported increased teat temperature after milking in 90% of dairy cows, although the evaluation of the milking used did not show any significant damage to the teats. Paulrud *et al.* [105] used IRT to evaluate milking-induced alternations in teat tissue fluid circulation and obtained similar results. The authors concluded that IRT was useful for studying and evaluating the effects of various milking techniques on teat fluid dynamics.

Kejik and Maskova [106] and Malik *et al.* [107] took udder thermograms and evaluated the relationship between the traumatised zones and the quality of the teat rubber. The thermographic study showed that milking may cause traumatisation in certain zones of the udder and teats. The authors point out that such traumatisation during the course of milking may be the cause of mastitis. Kunc *et al.* [108] and Kunc *et al.* [109] investigated the dynamics of teat temperature changes in relation to vacuum changes (40kPa *vs.* 45kPa). After evaluation of thermograms it was found that a 40kPa vacuum evoked lower teat temperatures than a 45kPa vacuum. Further, the authors [110] used IRT to monitor udder temperature responses in healthy dairy cows under standard operating conditions in an autotandem milking parlour, in which all technical specifications complied with the standard. A comparison of all thermograms showed that milking caused significant changes in teats, particularly in those that were in direct contact with the milking machine and were subject to a significant stress (Figs. **10** and **11**). Teat temperature was increased by an average of 2.62°C. Similar results were reported by [111].

Kunc *et al.* [112] compared rubber liners from two producers. The results showed that milking increased the temperature of teats. The highest values were obtained immediately after milking. This trend was recorded in liners from both producers. The differences in the temperature states of teats were, however, not significant between the producers. New liners (immediately after exchange) increased the temperature of teats more than old liners (immediately before exchange), but the differences were not significant. Schmidt *et al.* [113] found that cows with a high milk production had higher udder temperatures pre-and post-milking than low producing cows. These data suggest that IRT may have value as a diagnostic tool for assessing udder function in relation to temperature gradient changes and the level of milk production. Paulrud *et al.* [114] obtained similar results.

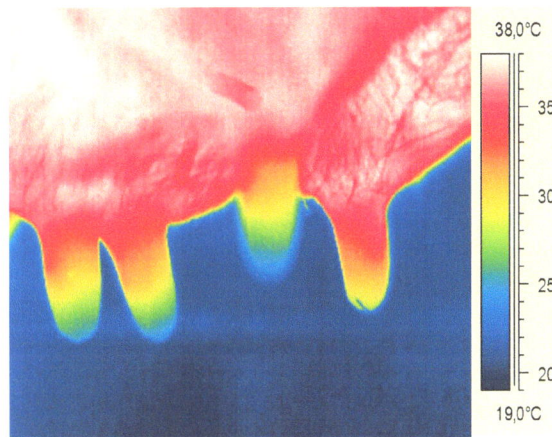

**Figure 10:** Thermal image of bovine udder and teats before milking.

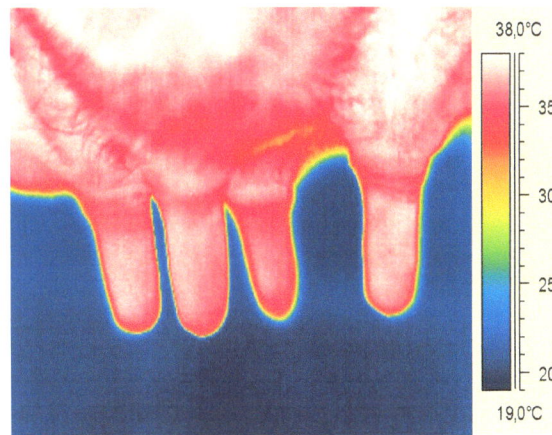

**Figure 11:** Thermal image of bovine udder and teats after milking.

Kunc *et al.* [115] studied the effect of machine milking on teats (vacuum 42.6kPa) as compared to a suckling calf by means of the thermographic method. Generally, teat temperature showed a significant increase after milking and suckling. The effect of calf suckling on the temperature of teats was dependent on age. Calves in the colostrum period (age 5 days) stressed teats significantly less than older calves (calves in the milk period, age 20 days). Further, these older calves stressed teats more than machine milking.

Berry *et al.* [116] used IRT to investigate the effects of environmental factors on daily variation in udder temperature. The authors found a distinct circadian rhythm in udder temperature and significant increase in udder temperature caused by exercise. The daily variation in udder temperature was, however, found to be smaller than the rise in temperature resulting from an induced mastitis response. They concluded that IRT had potential as an early detection tool for mastitis if it is combined with monitoring of environmental factors.

Recent studies have focused on the use of IRT to detect mastitis much earlier than was previously possible. Scott *et al.* [117] found that inflammation could be detected from temperature differences by using IRT earlier with either bovine serum albumin or somatic cell counts. The concentration of bovine serum albumin peaked at 6h post induction, whereas IRT temperature increases were evident within 1h post-induction. The authors [118] also proved that IRT showed potential as an early detection method for mastitis. This supports the findings of [119]. Kennedy [120] found that mastitis infections often caused udder surface temperature to rise before other clinical signs were observed. In experimentally induced mastitis, a rise of 2.3°C was recorded. She recommended that cows be walked past an infrared camera, which would photograph the rear of the udder. The camera and its associated computer would identify and record cows whose udder surface temperatures were higher than normal. Herd management would follow up with further assessment and possible treatment of cows flagged by the system.

Colak *et al.* [121] determined whether IRT had any merit for early detection of subclinical mastitis in dairy cows. Their results evidence that, as a non-invasive tool, IRT can be employed for screening dairy cows for mastitis. Hovinen *et al.* [122] tested a thermal camera for its capacity to detect clinical mastitis. Mastitis was experimentally induced in 6 cows with 10µm of *Escherichia coli* liposaccharide. The first signs of clinical mastitis were noted in all cows 2h postchallenge and included changes in the general appearance of the cows and local clinical signs in the affected udder quarter. Rectal temperature, milk somatic cell count, and electrical conductivity were increased 4h postchallenge. The thermal camera was successful in detecting the 1 to 1.5°C temperature change on udder skin associated with clinical mastitis in all cows. The authors believe that a thermal camera mounted in a milking or feeding parlour could detect temperature changes associated with clinical mastitis or other diseases in a dairy herd.

Only one study concerning sheep milking was found. Staletta *et al.* [123] performed a thermographic study of the ovine mammary gland during different working vacuum levels. The higher temperature for the low vacuum level could be attributable to a faster return to a normal condition of blood flow in all teat tissues.

## 7. CONCLUSIONS

The above examples prove conclusively that IRT can produce important information where the possibilities of conventional diagnostic techniques have been exhausted. There are, however, some limitations and factors that need to be considered when using IRT. Thermograms must be collected out of direct sunlight and wind drafts. Hair coats should be free of dirt, moisture and extraneous material. The effect of weather conditions, circadian and ultradian rhythms, time of feeding, milking, laying and rumination *etc.* are also factors that need to be considered and require further investigation as part of validating IRT. Infrared thermal measurements can then be used very successfully in prediction, detection and disease, in research, and in other applications in livestock science. This measurement method also has value as a diagnostic tool for assessing udder function and can be considered a useful method for indirect and non-invasive evaluation of the condition of teats and udders. Evaluation of IRT may be promising for detection of mastitis, and shows potential as an early detection method for mastitis.

## ACKNOWLEDGEMENTS

This chapter was supported by project MZE 0002701404.

## DISCLOSURE

Part of information included in this chapter has been previously published in "J. of Fac. of Agric., OMU, 2007,22(3):329-33"

## REFERENCES

[1]    Kunc P, Knizkova I, Prikryl M, Maloun J. Infrared thermography as a tool to study milking process: a review. Agr Trop et Subtrop 2007; 40: 29-32.

[2]     Knizkova I, Kunc P, Gurdil GAK, Pinar Y, Selvi KC. Applications of infrared thermography in animal production. The Journal of Agricultural Faculty of Ondokuz Mayis University 2007; 22: 329-35.

[3]     Kulezsa O, Rzeczkowski M, Kaczorowski M. Thermography and its practical use in equine diagnostics and treatment. Medycyna Weterynaryjna 2004; 60: 1137-248.

[4]     Harper DL. The value of infrared thermography in the diagnosis and prognosis of injuries in animals. In: Proceedings of the Inframation; 2000: Orlando, USA; pp. 115 – 22.

[5]     Embaby S, Shamaa AA, Gohar HM. Clinical assessment of thermography as a diagnostic and prognostic tool in horse practice. In: Proceedings of Inframation; 2002, Orlando, USA; pp. 30 – 6.

[6]     Spruyt P, Ghafir Y, Art T, Lekeux P. La thermographie infrarouge dans l etude de la thermoregulation. Revue de la litterature. Ann Med Vet 1995; 139: 413–8.

[7]     Speakmen JR, Ward S. Infrared thermography: principle and applications. Zoology 1998; 101: 224–32.

[8]     Head J, Dyson S. Talking the temperature of equine thermography. Vet J 2001; 162: 166–7.

[9]     Purohit RC, McCoy MD. Thermography in the diagnosis of inflammatory processes in the horse. Am J Vet Res 1980; 41: 1167-74.

[10]    Palmer SE. Use of the portable infrared thermometer as a means of measuring limb surface temperature in the horse. Am J Vet Res 1981; 42: 105-8.

[11]    Pick M. Erste Ergebnisse thermographischer Untersuchungen zur Lahmheitsdiagnostik beim Pferd mit Hilfe eines Infrarottthermographen. Tieraztl Prax 1984; 12: 229-38.

[12]    Marr CM. Microwave thermography – a non-invasive technique for investigation of injury of the superficial digital flexor tendon in the horses. Equine Vet J 1992; 24: 269-73.

[13]    Wieland M. Gegenuberstellung thermographischer und knochenszintigraphischer Begunde beim Pferd. Dissertation thesis, University of Zurich, Switzerland, 1992.

[14]    Auer JA, Wieland M, Plocki KA, Lauk KHD. Thermographische und ultrasonographische Untersuchungen über den Einfluss instrumentierter Touchierstangen auf die distalen Vordergliedmassen beim Pferd. Pferdeheilkunde 1993; 9: 41-57.

[15]    Denoix JM. Diagnostic techniques for identification and documentation of tendon and ligament injuries. Veterinary Clinics of North America: Equine Pract 1994; 2: 365-407.

[16]    Turner TA. Thermography as an aid in the localization of upper hindlimb lameness. In: Proceedings of 15th Meeting on Equine Welfare and Sports Medicine, Bonn, Germany, 1996; pp. 632-4.

[17]    Weil M, Litzke LF, Fritsch R. Diagnostic validity of thermography of lameness in horse. Tierarztl Prax.Ausg G Grosstiere Nutztiere, 1998; 26, 346–54.

[18]    Eddy AL, van Hoogmoed LM, Snyder JR. The role of thermography in the management of equine lameness. Vet J 2001; 162: 172-81.

[19]    Kowalik S, Derobek-Gilowska A, Studzinski T. Poor performance in horses – causes and diagnosis. Madycyna Weterynaryjna 2002; 58: 103-7.

[20]    Tunley BV, Henson FMD. Reliability and repeatability of thermographic examination and the normal thermographic image of the thoracolumbar region in the horse. Equine Vet J 2004; 36: 306-12.

[21]    Allmers E, Hollenberg C, Heldt S, Meyer W. Routine use of an infrared thermometer in measuring the body surface temperature pattern of horse limb. Tieraztl Prax 2005; 33: 181-7.

[22]    Ciutacu O, Tanase A, Miclaus I. Digital infrared thermography in assessing soft tissues injuries on sport equines. Bulletin of the University of Agricultural Sciences and Veterinary Medicine in Bucharest 2006; 63: 228-33.

[23]    Fonseca BPA, Alves ALG, Nicoletti JLM, Thomassian A, Hussni CA, Mikail S. Thermography and ultrasonography in back diagnosis of equine athletes. J Equine Vet Sci 2006; 26: 507-16.

[24]    Toth P.Examination methods of the equine thoracolumber and sacral region. Literature review. Magy Allatorvosok 2006; 128: 515-23.

[25]    Lauk HD, Kimmich M. Comaprasion of scintigraphic and thermographic findings in the horses. Pferdeheilkunde 1997; 13: 329.

[26]    Yanmaz LE, Okumus Z, Dogan E. Instrumentation of thermography and its applications in horses. J Anim Vet Adv 2007; 6: 858-62.

[27]    Braverman Y. Potential of infra-red thermogrpahy for the detection of summer seasonal recurrent dermatitis (sweet itch) in horses. Vet Rec 1989; 125: 372-4.

[28]    Harkins JD, Munday GD, Stanley S, *et al.* Character and duration of pharmacological effects of intravenous isoxsuprine. Equine Vet J 1996; 28: 320-6.

[29]    Harkins JD, Tobin T. The pharmacological effects of isoxspurine. Pferdeheilkunde 1996; 12: 428-30.

[30]   Verna M, Turner TA, Anderson KL. Scintigraphic, radiographic, and thermographic appearance of the metacarpal and metatarsal regions of adult healthy treated with nonfocused extracorporeal shock wave therapy – A pilot study. Vet Ther 2005; 6: 268-76.

[31]   Ringer SK, Lischer CHJ, Ueltschi G. Assessment of scintigraphic and thermographic changes after focused extracorporeal shock wave therapy on the origin of the suspensory ligament and the fourth metatarsal bone in horses without lameness. Am J Vet Res 2005; 66: 1836 42.

[32]   Valdern NM, Weishaupt MA, Imboden I, Weistner T, Lischer CJ. Evaluation of skin sensitivity after shock wave treatment in horses. Am J Vet Res 2005; 66: 2095-100.

[33]   Comelisse CJ, Robinson NE, Berney CA, Eberhart S, Hauptman JE, Derksen FJ. Thermographic study of *in vivo* modulation of vascular responses to phenylephrine and endothelin-1 by dexamethasone in the horse. Equine Vet J 2006; 38: 119-26.

[34]   Ghafir Y, Art T, Lekeux P. Infrared thermography in the study of thermoregulation in the horse: Training effects. Ann Med Vet 1996; 140: 131-5.

[35]   Jodkowska E, Dudek M. Significance of body surface temperature measurements in exploitation of horses. Madycyna Weterynaryjna 2003, 59; 584-7.

[36]   Simon EL, Gaughan EM, Epp T, Spire M. Influence of exercise on thermographically determined surface temperatures of thoracic and pelvic limbs in horses. Journal of the American Veterinary Medical Association 2006; 229: 1940-4.

[37]   Autio E, Heiskanen ML. Thermographic evaluation of the lower critical temperature in wealing horses. J Appl Anim Welf Sci 2007; 10: 207-16.

[38]   Marlin DJ, Scott CM, Casas RI, Holah G, Schroter RC. Post exercise changes in compartmental body temperature accompanying intermittent cold water cooling in the hyperthermic horse. Equine Vet J 1998; 30: 28-34.

[39]   Hurnik JF, De Boer S, Webster AB. Detection of health disorders in dairy cattle utilizing a thermal infrared scanning technique. Can J Anim Sci 1984; 64: 1071-3.

[40]   Schwartzkopfgenswein KS, Stookey JM. The use of infrared thermography to assess inflammation associated with hot-iron and freeze branding in cattle. Can J Anim Sci 1997; 77: 577-83.

[41]   Spire MF, Drouillard JS, Galland JC, Sargeant JM. Use of infrared thermography to detect inflammation caused by contaminated growth promotant ear implants in cattle. J Am Vet Med Assoc 1999, 215: 1320-4.

[42]   Cockroft PD, Henson FMD, Parker C. Thermography of a septic metatarsophalangeal joint in a heifer. Vet Rec 2000; 26: 258-60.

[43]   Schaefer AL, Cook N, Tessaro Sv, *et al.* Early detection and prediction of infection using infrared thermography. Can J Anim Sci 2003; 84. 73 – 80.

[44]   Schaefer AL, Cook NJ, Church JS, *et al.* The use of infrared thermography as an early indicator of bovine respiratory disease complex in calves. Res Vet Sci 2007; 83: 376-84.

[45]   Nikkah A, Plaizier JC, Einarson MS, Berry RJ, Scott SL, Kennedy AD. Infrared thermography and visual examination of hooves of dairy cows in two stages of lactation. J Dairy Sci; 88: 2749–53.

[46]   Munsell BA, Beede DK, Domecq JJ, *et al.* Use of infrared thermography to non-invasively identify lesions in dairy cattle. J Anim Sci 2006; 84,Suppl.1/J Dairy Sci 2006; 89, Suppl 1: 143.

[47]   Stewart M, Statford KJ, Dowling SK, Schaefer AL, Webster JR. Eay temeparture and heart rate variability of calves disbudded with or without local anaesthetic. Physiol Behav 2008; 93: 789-97.

[48]   Hurnik JF, Webster AB, De Boer S. An investigation of skin temperature differentials in relation to estrus in dairy cattle using a thermal infrared scanning technique. J Anim Sci 1985; 61: 1095-102.

[49]   Hellebrand HJ, Brehme U, Beuche H, Stollberg U, Jacobs H. Application of thermal imaging for cattle management. Proceedings of 1st European Conference on Precision Livestock Farming; 2003: Berlin, Germany; pp. 761-3.

[50]   Kozumplik J, Malik K, Ochotsky J. Vyuziti termograficke metody k diagnostice zanetlivych procesu lokalizovanych na pohlavnich organech plemeniku. Vet Med Czech 1989; 39: 305-7.

[51]   Kastelic JP, Coulter GH, Cook RB. Scrotal surface, subcutaneous, intratesticular and intraepididymal temperatures in bulls. Theriogen 1995; 44:147-52.

[52]   Kastelic JP, Cook RB, Coulter GH, Wallins GL, Entz T. Environmental factors affecting measurement of bovine scrotal surface temperature with infrared thermography. Anim Reprod Sci 1996a; 41: 153-9.

[53]   Kastelic JP, Cook RB, Coulter GH, Saacke RG. Ejaculation increase scrotal surface temperature in bulls with intact epididymides. Theriogen 1996b;.46: 889-992.

[54]   Kastelic JP, Cook RB, Coulter GH, Saacke RG. Insulating the scrotal neck affects semen quality and scrotal/testicular temperatures in the bull. Theriogen 1996c; 45: 935-42.

[55] Kastelic JP, Cook RB, Coulter GH. Scrotal/testicular thermoregulation and the effects of increased testicular temperature in the bull. Veterinary Clinics of North America: Food Animal Practice 1997a; 13: 271-82.

[56] Kastelic JP, Cook RB, Coulter GH. Contribution of the scrotum, testes and testicular artery to scrotal/testicular thermoregulation in bulls at two ambient temperatures. Anim Reprod Sci 1997b; 45: 255-61.

[57] Gabor G, Sasser RG, Kastelic, JP, *et al.* Morphologic, endocrine and thermographic measurements of testicles in comparison with semen characteristic in mature Holstein-Friesen breeding bulls. Anim Reprod Sci 1998; 51: 215-24.

[58] Gerken M, Barow U. Methodical investigation into thermoregulation in suckler cows under field conditions. Proceedings of the 49th Annual Meeting of EAAP; 1998: Warsaw, Poland; pp. 179.

[59] Coppola, CL, Colier Rj, Enns RM. Evaluation of thermal status cattle using infrared technology. Proccedings, Western Section, American Society of Animal Science 2002; 53: 1-3.

[60] Schaefer AL, Jones SDM, Tong AKW, Vincent BC. The effects of fasting and transportation on beef cattle. 1. Acid-base-electrolyte balance and infrared heat loss of beef cattle. Livest Prod Sci 1988; 20: 15-24.

[61] Tong AKW, Scheafer AL, Jones SDM. 1995. Detection of poor quality beef using infrared thermography. Meat Focus International, 1995; 4: 443-5.

[62] Kimmel E, Arkin H, Berman A.Evaporative cooling of cattle: transport phenomena and thermovision. Paper Am Soc Agric Engin 1992; 92-4028: 14.

[63] Knizkova I, Kunc P, Novy Z, Knizek J. Evaluation of evaporative cooling on the changes of cattle surface body temperatures with use of thermovision. Ziv Vyr 1996; 41:433-9.

[64] Knizkova I, Kunc P, Koubkova M, Flusser J, Doležal O. Evaluation of naturally ventilated dairy barn management by a thermographic method. Livest Prod Sci 2002; 77: 349-53.

[65] Verkerk G, Webster J, Bloomberg M, Barrell G, Tucker C, Matthews L. Minimising impact of adverse environments on stock. Dexcelink Winter, 2004:15.

[66] Stewart M. Webster JR, Schaefer AL, Cook NJ, Scott SL. Infrared thermography as a non-invasive tool to study animal welfare. Animal Welfare, 2005; 14: 319-25.

[67] Stewart M, Webster JR, Verkerk GA, Schaefer AL, Colyn JJ, Statford KJ. Non-invasive measurement of stress in dairy cows using infrared thermography. Physiol Behav 2007; 92: 520-5.

[68] Stewart M, Stockey JM, Stattford KJ, *et al.* Effects of local anestethic and a nonsteroidal anti-inflammatory drug of pain responses of dairy calves to hot-iron dehorning. J Dairy Sci 2009; 92: 1512-19.

[69] Unshelm J, Reinhart E, Platz S. Untersuchungen uber das Erfassen haltungsbedingter Schaden beim Rind mit Hilfe der Infrarottthermometrie. Tierarztl Umsch 1992; 47: 516-21.

[70] Lendelova J, Pogran S, Knizkova I, Kunc P. Cubicle lying structures and their thermo-technical comparison. Scientiarum Polonorum Acta – Architectura 2006; 5: 81-90.

[71] Harrison HJS, Scott SL, Christopherson RJ, Kennedy AD, Tong AKW, Schaefer AL. The use of infrared thermography in live animals to predict growth efficiency: 2004: Pub.No: WO/2004/105474.

[72] Stanko D, Brus M, Hocevar M. Esstimation of bull live weigh through thermographically measured body dimension. Comput Electr Eng 2008; 61: 233-40.

[73] Montanholi YR, Odongo NE, Swanson KC, Schenkel FS, McBride BW, Miller SP. Application of infrared thermography as an indicator of heat and methane production and its use in the study of skin temperature in responses to physiological events in dairy cattle (*Bos taurus*). J Therm Biol 2008; 33: 468-75.

[74] Sabec D, Lazar P. Erste Ergebnisse beruhrungsloser Temperaturmessungen mittels eines Infrarotthermometers am Sprunggelenk des Schweines mit Osteoarthrosis tarsi deformans. Dtsch Tieräztl Wschr 1990; 97: 43-4.

[75] Savary P, Hauser R, Ossent P, Jungbluth T, Gygax L, Wechsler B. Evaluation of infrared thermography as a method for diagnosting inflammation of leg joints in fattening pigs. Dtsch Tieraztl Wschr 2008; 115: 324–9.

[76] Loughmiller JA, Spire MF, Dritz SS, Fenwick BW, Hosni MH, Hogge SB. Relationship between mean surface temperature measured by use of infrared thermography and ambient temperature in clinically normal pigs and pigs inoculated with *Actinobacillus pleuropneumonia.* Am J Vet Res. 2001; 62: 676–81.

[77] Montanholi YR, Dewulf J, Koenen F, Laevens H, de Kruif A. Infrared thermography is not suitable for the detection of fever in pigs. Vlaams Diergen Tijds 2003; 72: 373–9.

[78] Gariepy C, Amiot J, Nadai S. Antemortem detection of PSE and DFD by infrared thermography of pigs before stunning. Meat Sci 1989; 25: 37–41.

[79] Schaefer AL, Jones SDM, Murray AP, Sather AP, Tong, AKW. Infrared thermography of pigs with known genotypes for stress susceptibility in relation to pork quality. Can J Anim Sci 1989; 69: 491-5.

[80]   Adamec T, Kunc P, Knizkova I, Dolejs J, Toufar O. By using evaporative cooling in summer session, it is possible to improve growth and feed conversion in pig fattening. In. Proceedings of the 48th Annual Meeting of EAAP; Vienna, Austria; 1987: pp.229.

[81]   Kotrbacek V, Nau HR. The changes in skin temperatures of periparturient sows. Acta Vet Brno, 1984; 54: 35-40.

[82]   Xin H. Assessing swine thermal comfort by image analysis of postural behaviors. J Anim Sci 1999; 77, Suppl.2/.Dairy Sci., 82, Suppl.2: 1–9.

[83]   Loughmiller JA, Spire MF, Tokach MD, *et al.* An evaluation of differences in mean body surface temperature with infrared thermography in growing pigs fed different dietary energy intake and concentration. J Appl Anim Res 2005; 28: 30–7.

[84]   Warris PD, Pope SJ, Brown SN, Wilkins LJ, Knowles TG. Estimating the body temperature of groups of pigs by thermal imaging. Vet Rec 2006; 158: 331-4.

[85]   Chepete HJ Xin H. Cooling laying hens by intermittent partial surface sprinkling. Transaction of the ASAE 2000; 43: 965–71.

[86]   Mala G, Knizkova I, Kunc P, Matlova V, Knizek J. Resistance of early postnatal lambs from two genetic types to cold environment and rain. Ann Anim Sci. 2004, 1: 169–71.

[87]   Knizkova I, Mala G, Kunc P, Knizek J. 2005. Resistance of early postnatal lambs from four genetic types to cold environment and rain. In: proceedings of the 12th International Congress ISAH, Warsaw, Poland, 2005; pp.271 –3.

[88]   Stewart M, Webster JR, Schaefer AL, Cook NJ, Scott SL. Infrared thermography as a non-invasive tool to study animal welfare. Anim Welfare 2005, 14: 319–25.

[89]   Luzi F, Ludwig N, Gargano M, Milazzo M, Carenzi C, Verga M. Evaluation of skin temperature change as stress indicator in rabbit through infrared thermography. Ital J Anim Sci 2007; 6: 769.

[90]   Knizkova I, Kunc P, Vadlejch J, Makovcova K, Langrova I. Is thermal profile changed in rabbit infected with *Eimeria intestinalis*? In: Proceedings of 14th Congress in Animal Hygiene, Vechta, Germany, 2009; (in press).

[91]   Hamann J. Stimulation and teat tissue reaction. Kieler Milchwirtschaftlich Forschungsberochte, 1992; 44: 339–47.

[92]   Isaksson A, Lind O. Teat reactions in cows associated with machine milking. J Vet Med 1992; 39: 282–8.

[93]   Burmeister, JE, Fox LK, Hillers J, Hancock DD. Effect of premilking and postmilking teat disinfectants on teat skin condition. J Dairy Sci 1998; 81: 1910–6.

[94]   Zecconi A, Hamann J, Bronzo V, Moroni P, Giovannini G, Piccini R. Relationship between teat tissue immune defences and intramammary infections. Adv Exp Med Biol 2000; 480: 287–93.

[95]   Neijenhuis F. Teat end callosity classification system. In: Proceedings of 4[l]h International Dairy Housing Conference, ASAE; 1998; pp 117–23.

[96]   Rasmussen MD, Larsen HD. The effect of post milking teat dip and suckling on teat condition, bacterial colonization and udder health. Acta Vet Scan 1998; 39: 443–52.

[97]   Neijenhuis F, Barkema HW, Hogeveen H, Noordhuizen JPTN. Classification and longitudinal examination of callused teat ends in dairy cows. J Dairy Sci 2000; 83: 2795–804.

[98]   Neijenhuis F. Teat condition in dairy cows. Ponsen & Looijen BV, Wageningen, the Netherlands: 2004.

[99]   Wiltis S. Infrared thermography for screening and early detection of mastitis infections in working dairy herds. In: Proceedings of Inframation 2005, Las Vegas, USA; pp 1–5.

[100]   Norberg E. Electrical conductivity of milk as a phenotypic and genetic indicator of bovine mastitis. Livest Prod Sci 2005; 96: 129–39.

[101]   Timms L. Mastitis Diagnosis. Western Dairy Digest 2004; 5: 10–1.

[102]   Hamann J. Infection rate as affected by teat tissue reactions due to conventional and non-conventional milking systems. Kieler Milchwirtschaftlich Forschungsberochte 1985; 37: 426–30.

[103]   Caruolo EV, Jarman RF, Dickey DA. Milk temperature in the claw piece of the milking machine and mammary surface temperature are predictors of internal mammary temperature in goats. J Vet Med 1989; 37: 61–7.

[104]   Eichel H. Temperature of teat skin in dairy cows milked in piped milking parlor. Monatshefte fur Veterinarmedizin 1992; 47: 193–5.

[105]   Paulrud O, Clausen S, Andersen PE, Bjerring M, Rasmussen MD. Infrared thermography to evaluate milking induced alterations in teat tissue fluid circulation. J Anim Sci 2002; 80/J Dairy Sci 2002; 85: 84.

[106]   Kejik C. Maskova A. Thermographic measurements of teat surface temperature during machine milking. Ziv Vyr 1989; 35: 225 – 30.

[107]   Malik K., Maskova A, Vevoda J. A thermovision study of the rubber teat cup function. Plastics and Rubber-Special Issue 1989; 16: 49-51.

[108] Kunc P, Knizkova I, Koubkova M. The influence of milking with different vacuum and different design of liner on the change of teat surface temperature. Czech J AnimSci 1999; 44: 131 – 4.

[109] Kunc P, Knizkova I, Koubkova M. Flusser J, Dolezal O. Thermographic observation of the mammary gland responses to machine milking in dairy cows. Physiol Res 2000a; 49: 21.

[110] Kunc P, Knizkova I, Koubkova M., Flusser J, Dolezal O. Machine milking and its influence on temperature states of udder. Czech J Anim. Sci 2000b; 45: 1–15.

[111] Barth K. Basic investigations to evaluate a highly sensitive infrared-thermograph-technique to detect udder inflammation in cows. Milchwissenschaft 2000; 55: 607-9.

[112] Kunc P, Knizkova I, Koubkova M., Flusser J, Dolezal O. Comparison of teat rubber liners by means of temperature states of teats. Res Agri Eng 2000c; 46: 104–7.

[113] Schmidt S, Bowers S, Dickerson T, Graves K, Willard S. Assessments of udder temperature gradients pre-and post-milking relative to milk production in Holstein cows as determined by digital infrared thermography. J Anim Sci 2004; 82/J Dairy Sci 2004; 83: 460–1.

[114] Paulrud O, Clausen S, Andersen PE, Bjerring M, Rasmussen, MD. Infrared thermography and ultrasonography to indirectly monitor the influence of liner type and overmilking on teat tissue recovery. Acta Vet Scan 2005; 46: 137–47.

[115] Kunc P, Knizkova I, Koubkova M. Teat stress by calf suckling and machine milking. Folia Vet 2002; 46: 48–9.

[116] Berry RJ, Kennedy AD, Scott SL, Kyle BI, Schaefer AL. Daily variation in the udder surface temperature of dairy cows measured by infrared thermography: potential for mastitis detection. Can J Anim Sci 20003a; 83, 687–93.

[117] Scott SL, Schaefer Al, Tong AKW, Lacasse P. Use of infrared thermography or early detection of mastitis in dairy cows. Can J Anim Sci 2000; 70: 764–5.

[118] Berry RJ, Kennedy AD, Scott SL, Fulawka D, Hernandez FIL, Schaefer AL. The potential of infrared thermography as an early detection method for mastitis. Seasonal effects on predictability. J Anim Sci 2003b; 81/J Dairy Sci 2003; 86: 85.

[119] Willits S. Infrared thermography for screening and early detection of mastitis infections in working dairy herds. In: Proceedings of Inframation 2005, Las Vegas, USA; pp. 1–5.

[120] Kennedy A. Mastitis detection using infrared thermography. Western Dairy Digest, 2004; 5: 15.

[121] Colak A, Polat B, Okumus Z, Kaya M, Yanmaz LE, Hayirli A. Early detection of mastitis using infrared thermography in dairy cows. J Dairy Sci 2008; 91: 4244–8.

[122] Hovinen M, Siivonen J, Taponen S, *et al.* Detection of clinical mastitis with the help of a thermal camera. J Dairy Sci 2008; 91: 4592–8.

[123] Staletta C, Murgia L, Caria M, Gianesella M, Pazzona A, Morgante M. Thermographic study of the ovine mammary gland during different working vacuum levels. Ital J Anim Sci 2007; 6: 600.

# Concluding Remarks to Part II – Section 1

This Section I of Part II was concerned with application of infrared thermography in the medical field; the first chapter was devoted to human medicine and the second one to veterinary.

Chapter 3 reports on the IRT's state of art in medicine with methodological approaches and a variety of applications such as in the diagnosis of breast cancer, in ophthalmologic surgery, in cardiovascular surgery, in the visualization of ischemic tissues and in many others. As illustrated in this chapter, infrared thermography roused soon interest, since the middle fifties, in the medical community owing to the Hippocrates' criterion to diagnose diseases in humans from temperature features. But, there was also wide skepticism and loose of interest mainly due to lack of well assessed procedures and difficulties in the data interpretation also because of the poor sensitivity of old devices. Renewed interest is today justified by the availability of computerized FPA high performance cameras. However, the infrared thermography is still not adequately exploited in medicine because there are still difficulties in the interpretation of thermograms.

What is the situation in veterinary?

As reported by the authors of Chapter 4, infrared thermography is useful in veterinary medicine, primarily for diagnostic purposes, particularly in the diagnosis of orthopaedic diseases in horses. It was also demonstrated that it could be successfully used for study and research in farm animals such as pigs, cattle, sheep and poultry. Areas of research include reproduction, thermoregulation, animal welfare and the milking process. As evidenced by the authors, IRT is a suitable tool for early detection and screening for mastitis, and can also be useful for studying and evaluating the effects of various milking techniques on the teats and udders. Not with standing this, it is scarcely applied in the study of the milking process.

Of course, the application of IRT to veterinary medicine is particularly useful to predict inflammation since, contrary to human beings, animals cannot explicitly communicate any symptom before the illness has become important.

The following Section II reports on the use of an infrared imaging system for monitoring the conservation conditions of fruits and vegetables. Indeed, the climate control is mostly important to elongate the store life, but also to assure the quality of products for the human beings healthiness.

# CHAPTER 5

# Application of Infrared Thermography in Fruit and Potato Storage

## Klaus Gottschalk[*]

*Leibniz-Institut für Agrartechnik, Potsdam-Bornim e.V. (ATB), Max-Eyth-Allee 100, 14469 Potsdam, Germany*

**Abstract:** An infrared camera is used to investigate the conservation condition of fruits and vegetables. Crops like apples or carrots are exposed to mechanical impacts that are caused by post harvest treatments like washing or grading. The resulting intensification of respiration causes a temperature decrease on the moist surface which is under evaporation; this is detectable with infrared thermography (IRT). In fact, IRT can be exploited to monitor current temperature distributions, to measure low local temperature differences and to visualize the influences of airflow within a wide area of a store. An appropriated climate control is crucial to maintain the quality and to enhance the shelf life of stored agricultural products like fruit, vegetable and potatoes. Hence, an infrared camera can help improve the climate control performance. The camera can be embedded in a wired system for data exchange and remote control. A number of well placed cameras distributed in the store enhance the visibility, and hence the controllability of the store. The investment cost can be paid back within several years if the IR camera system can help to improve the climate control for reducing mass loss (shrinkage) and prolonging shelf life of the crop.

**Keywords:** Infrared thermography, fruit, vegetable, potato storage, store climate control, heat transfer, mass transfer, infrared emissivity, fruit surface, transpiration, calibration, temperature data logging, air flow control, economical effect, ecological effect.

## 1. INTRODUCTION

Parameters like heat, moist, mechanical impacts *etc.* are influencing the quality and shelf life of agricultural products as fruit, vegetable, and potato. Temperature is one of the most important parameters which can be measured by means of simple thermometers or infrared (IR) cameras. The measurement of temperature is therefore a useful method to investigate and monitor the property and state of the produce. For example, transpiration and metabolic heat production as well as heat and water exchange with the surrounding air can be determined by measuring the surface temperature of the produce. Hence, infrared thermography (IRT) is suitable for monitoring current temperature distributions, measure of low local temperature differences and visualization of influences of environmental conditions, (Fig. **1**).

An infrared camera can also be used to check the storage temperatures. A stationary applied thermographic camera system allows monitoring fruit or potato tuber surface temperatures. If the IR camera is embedded in a process control system a certain region of the store can be regulated to a desired temperature value which is called the 'set-point'. The set-point value is defined accordingly to the optimal temperature for long-term storage specific to each stored produce like fruit, vegetable, or potato, (Table **3**).

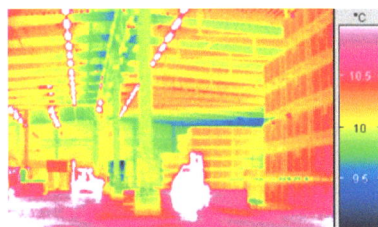

**Figure 1:** Thermographic view of a potato store house during filling with boxes.

*Address correspondence to Klaus Gottschalk: Leibniz-Institut für Agrartechnik, Potsdam-Bornim e.V. (ATB), Max-Eyth-Allee 100, 14469 Potsdam, Germany, Email:kgottschalk@atb-potsdam.de

Best control set point is the desired inner box temperature of the stored produce, but if a thermographic system is used for measurements of temperatures, only the temperature from the 'visible' part of the object surface is measurable. However, the thermogram can be used to visualise temperature fluctuations in the store caused by air flow movements and helps to detect spots or zones of insufficient or excessive ventilation.

## 2. APPLICATIONS

Infrared Thermography is being used advantageously in engineering, human and animal medical research and in many other fields. It was originally developed for military use to detect objects, but soon it became interesting also for civil purposes like detection of heat losses in buildings or detection of defective heat development in electrical facilities. The main advantage of thermography is its practical use for monitoring temperature changes. Amongst its many applications thermography is also used for early detection and prediction of infections of animals and humans.

The ability to measure temperature differences on the surface of the skin with a resolution of 0.08K or below allows infrared thermography to detect instantly (with a reaction time below 1 second) a potential elevated body temperature from fever of a person. If the temperature difference exceeds a critical value (for example, 1K higher than the average temperature of a healthy person) the checked person can be sent for further examination. In the same way thermography is applied for the detection of fire in waste bunkers of residue combustors [1]. For such problems complete technical control solutions are offered by specialized companies.

Measurements of temperatures of plant leaves and parts using infrared ray receptive sensors were already done over 40 years ago [2, 3]. A number of newer studies have shown the possibility to examine plants using thermography [4, 5].

Infrared image analysis has been applied to determine physiological disruptions in harvested plants [6] to investigate the transpiration behaviour of intact plants at the pre-harvest phase [7] and to determine the produce quality during the post-harvest phase [8, 9], (Fig. **2**). Temperature differences caused by diseases and wind influences on the leaves of standing cereal plants were observed with a thermal imaging system [10, 11]. Also, an application possibility in dairy farming is discussed [12].

Further applications were performed to investigate the water transport in plant leaves [13, 14] and to detect heat development in individual cells [15].

**Figure 2:** Thermogram of potatoes in different conditions; sprouting of tubers could not be well detected with an IR camera.

## 3. POSSIBILITIES AND LIMITS OF THERMOGRAPHY

Some advantages exist in using thermography to determine temperatures because it can be done:

-   Without disturbing the objects (non-invasive measurements);

-   From far distances in contactless way;

-   With images and visualization effects (coloured pictures of isothermals);

- With immediate visualisation of temperature changes and fluctuations (temperature gradient changes);

- With producing sequences of consecutive images at arbitrary time steps (videos);

- For a wide area within the view angle of the camera;

- With high resolution and accuracy;

- With capability to detect local heat sources or freezing spots (important in stores) even in the initial state.

The main disadvantages in using thermography for measurements of temperatures in a fruit store is the fact, that only a two-dimensional image of the observed area can be obtained. Hidden objects cannot be observed unless a heat transport takes place from the hidden object to the visible object in the foreground. Heat conduction and heat transfer processes are transporting the heat from the background to the viewed surface with a typical time delay.

It is important to take care when interpreting the thermograms because reflections from IR radiation sources in the neighbourhood may appear on the surface of the observed object causing measuring errors. However, when applying thermography in a storeroom the risk for getting measuring errors of that kind is mostly insignificant because well absorbing surfaces with emissivity close to 1.0 (like potato tubers, for example) have negligible reflections.

When applying thermography in a potato box store, for example, only the visible part of the boxes can be observed. The possibility to obtain a three dimensional view depends on the position and on the viewing angle of the camera. Anyhow, the image is a 2 dimensional projection of the 3 dimensional object on the image screen, like a photograph. Therefore, it is not possible to 'look inside' a filled potato box to determine the temperatures of the tubers inside the box. Microbial infection may cause abnormal heat production of hidden tubers inside the heap, but it remains undetected, unless the heat is transported to the foreground surfaces due to convection and heat conduction. Uncharacteristic temperature changes may result and can be detected by a well calibrated high resolution thermographic system. Such phenomena can be well identified by analysing sequences of thermograms. It is important to note that biological tissues have a poor transparency for IR radiation.

Heat from hidden objects can be detected by the thermography camera only if heat is transported by convection or is conducted through the biological material to the foreground surfaces resulting in a surface temperature change. Hidden potatoes in a box, for example, may have significant different temperatures, (Fig. **3**).

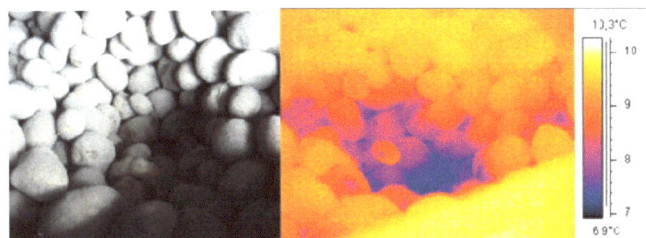

**Figure 3:** Thermogram of potatoes in a box; the upper layers are removed; the tubers inside are still moist, appearing with reduced temperatures.

## 4. INVESTIGATION OF FRUIT AND VEGETABLE WITH IR THERMOGRAPHY

Thermal imaging may be a useful tool for quality maintenance of fruit and vegetable. Evaporation of water from surfaces of transpiring fruit or vegetable reduces the temperature of the surface remarkably. Therefore, transpiration activities can be detected by thermal imaging of temperature differences between fruit surface and ambient air. Profiting this physical effect the following possibilities arise to determine [8, 9]:

- the transpiration intensity of produce at harvest time for the analysis of pre-harvest conditions;

- the transpiration intensity of produce at the postharvest stage for the evaluation of climatic stress;

- local transpiration differences of plant parts for the assessment of a freshness status;

- the transpiration intensity of produce to determine ripeness, mealiness, differences between varieties;

- local transpiration differences between plant parts for the evaluation of mechanical loads during the harvest and in the post harvest period;

- local transpiration differences between plant parts for the assessment of microbial infestation;

- the air flow influences (forced air, natural convection) to exposed produce.

Mechanical damages on fruits can also be detected because localised temperature decrease can be recognized on the thermogram. Mechanical impacts on apples affect bruising, *i.e.,* defects of cells below the skin are detected by the infrared camera which may not be visible to naked eye. This is possible since the respiration process locally enhances.

The results of thermographic investigations made by Linke *et al.* [8] could be summarized as:

- Analysis of pre-harvest conditions is possible, *e.g.,* for carrots mean temperature differences of 1.5-5.0K were determined at 20°C, 50% RH ambient air condition.

- Assessment of climatic stress on all kind of produce can be performed.

- Freshness status (fruit and stem parts of apples, cherries, strawberries, asparagus, …) can be determined using transpiration differences between plant parts.

- Conditions of produce at post-harvest stage can be distinguished.

- Mechanical impact on fruit can be assessed.

- Microbial infestation can be evaluated for some cultures.

- Air flow conditions on display sites can be analysed.

It could be shown that changes of temperatures can be below 0.1K, and therefore the change in transpiration of apples, for example, influenced by bruising is not measurable (in such a case). Since the transpiration process depends on variety and ripeness, apples can be differentiated. The maturity of apples of one variety can be estimated by thermal imaging. Different varieties are distinguishable, if the ripeness of apples is similar (Fig. **4**).

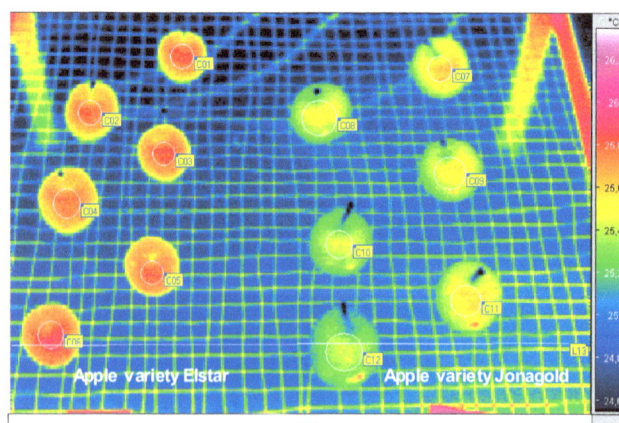

**Figure 4:** Thermogram of apples to detect different varieties; left, Elstar var., right Jonagold var., [15].

The effect of processing fruit and vegetables may be visible on a thermogram (Fig. **5**). Washing or cleaning fruits is affecting the transpiration of horticultural products which is also an indicator for freshness. Therefore, thermal imaging can be utilized to determine qualitatively freshness and shelf life of horticultural products. For different effects of washing processes, after passing certain falling steps when conveying and handling, and repetitions it is possible to classify the effects of mechanical impacts on the produce. For example, during the washing process the natural protective layer of the skin of carrots is removed. It follows that the transpiration resistance reduces significantly. On the thermogram these effects become visible because the surface temperature decreases of about 1.5-2K compared to unhandled carrots (Fig. **5**).

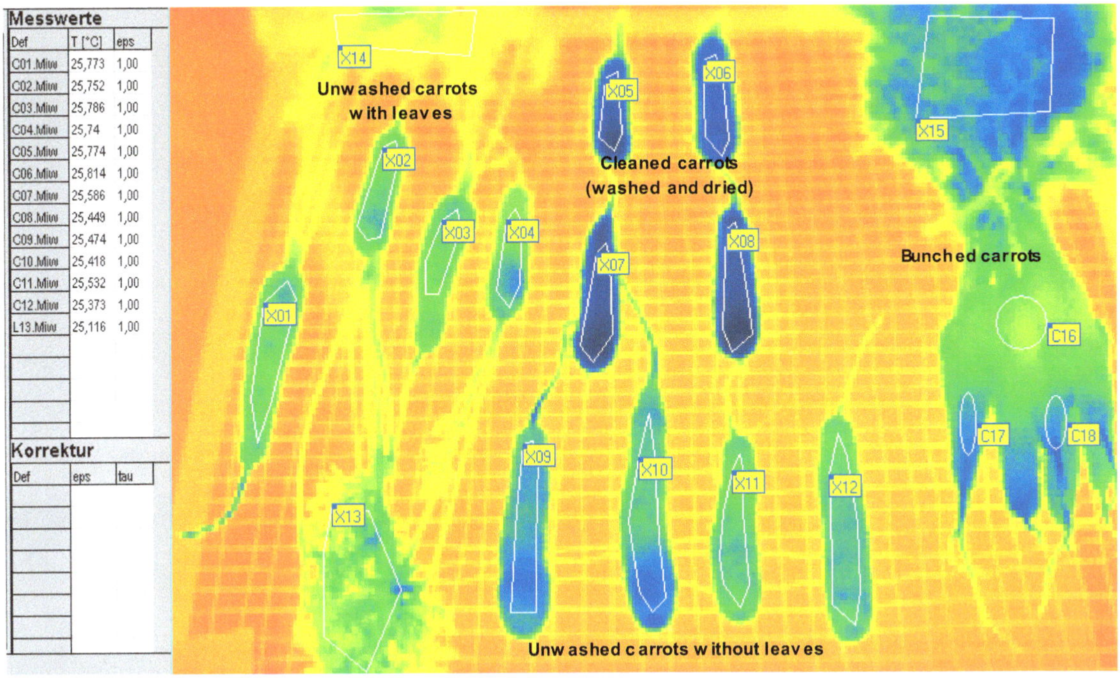

**Figure 5:** Thermogram of washed and unwashed carrots [16].

For investigations of such kind the usage of a thermal imaging with a temperature resolution ≤0.1K is necessary. It can also be achieved after calibration of the thermographic system (see below).

## 5. ANALYSIS OF THE INFRARED DATA

For a better comparison of thermal imaging and conventional measuring method, some basics of thermography are illustrated. The infrared (IR) thermographic camera is measuring indirectly the thermal radiation of a body or area. Up-to-date camera systems discussed here are using a microbolometer array mounted on a silicon substrate. For example, the resolution of such a bolometer array producing the digital image is typically 320×240 pixels or 384×288 pixels. In modern relatively inexpensive cameras uncooled bolometers are used. The performances of such types of cameras are approved as sufficient for the purposes discussed here. Cooled detectors are more expensive but have higher sensitivities because the thermal signal to noise ratio (SNR) is higher.

The most important specification for sensitivity of thermal cameras is the Noise Equivalent Temperature Difference (NETD). Typical values for uncooled detectors are 0.08K at 20°C and 1K at 600°C.

Each bolometer element, equivalent to one pixel, is absorbing the radiation which affects the bolometer element to heat up. The bolometer element is formed like a bridge of an electrical resistance. This resistance is thermal sensitive. Its characteristic is equal to a NTC (negative thermal resistance) with a typical value of approx -2%. It means that the electrical resistance changes by -2% per 1K temperature change. The camera types of interest here have sensitivities within an IR wavelength range of about 8 to 13μm (long wavelength).

The physical term for the IR radiation is the radiant flux $\Phi$ which is measured in $Wm^{-2}$. It is the power density of the radiation, *i.e.*, the power of the radiation divided by the total radiating area of the radiation source. The radiant flux depends on the temperature $T$ as well as on the emissivity $\varepsilon$ of the surface area $A$ according to Stefan-Boltzmann's law:

$$\Phi = \frac{P_{rad}}{A} = \sigma \varepsilon T^4 \qquad (1)$$

with the natural proportionality constant $\sigma = 5.67032 \cdot 10^{-8} \cdot Wm^{-2}K^{-4}$ (Stefan-Boltzmann's constant). The emissivity $\varepsilon$ is a dimensionless value between 0.0 and 1.0. A value of 0.0 signifies that the body is not emitting any thermal radiation whereas a value of 1.0 represents an ideal radiating body (a 'blackbody') which is radiating (and absorbing) completely the thermal energy, (Fig. **6**).

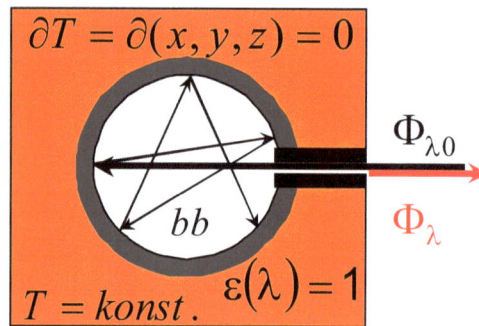

**Figure 6:** Principle of the 'black radiator' or 'black body'.

The emissivity is an object parameter and expresses the relation between the radiation of a real body ($\varepsilon < 1$) and that of a black body ($\varepsilon = 1$) at the same temperature. A 'selective' radiator is dependent on the wavelength $\lambda$, (Fig. **7**). Generally, the emissivity is dependent on the temperature $T$, the material (*e.g.,* fruit material, wood, steel, *etc.*), the surface properties and the radiation angle of the examined object:

$$\varepsilon (\lambda, T, \dots) < 1 \qquad (2)$$

The dependence on the wavelength of the emissivity may be neglected when using the thermography within the limited wavelength range for the purposes discussed here.

It is important to note that to all measuring areas or points the corresponding emissivity (Table **1**) has to be assigned in the recorded thermogram (*e.g.,* see Fig. **21**).

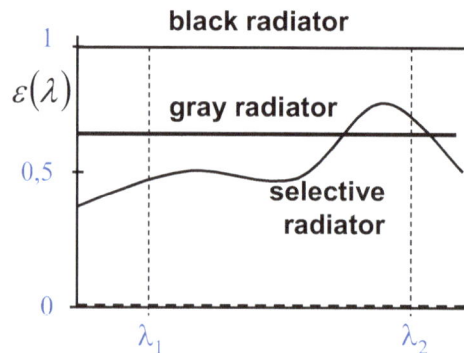

**Figure 7:** Emissivity of different types of radiators.

**Table 1:** Emissivity of several materials.

| Material | Emissivity $\varepsilon$ | Reference Source |
|---|---|---|
| Potatoes | 0.85-0.92 | [17-20] |
| Wood | 0.87-0.91 | [17-21] |
| White writing paper (paper marker) | 0.953 | [22] |
| Blackened reference sheet | 0.92 | [23] |
| Polyimide (PI) resin film | 0.94 | |
| Copper, polished | 0.04 | |

$\lambda$-range: 8…14µm.

## 6. CALIBRATION OF THE INFRARED SYSTEM

A special attention must be paid to the calibration of the infrared measuring system. As mentioned above, the temperature of the surveyed areas is calculated from the radiation flux density within a defined wavelength range, (Equation 1). The direct computation of the temperature using (Equation 1) leads to incorrect values with deviations of up to 2K. However, the measurements can be corrected using a reference temperature and applying a compensation calculation. The deviations can be reduced to below 0.1K with these calculations.

Recorded temperature values from an infrared thermographic camera can systematically differ from conventionally measured ('real') values using thermometers. Additionally, there is a temperature dependence of the IR measured data caused by internal heating of the camera. Modern cameras have a temperature compensation feature to eliminate these effects. This feature has mostly limited capabilities, therefore external calibration methods are needed to gain higher precision (see below). Referring to the manual of the camera is obvious. It could be stated that the difference between IR and conventionally measured values increases linearly along decrease of the absolute temperature (dots in Fig. **8**; the ideal relation should be like the solid thin line).

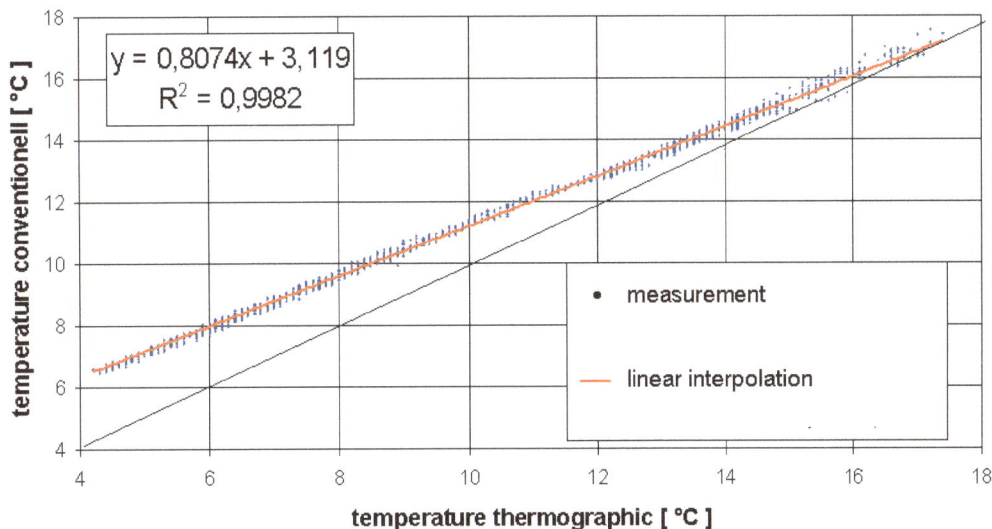

**Figure 8:** Determination of regression line to calibrate temperature measurements.

These results can be obtained during lab-experiments in a climate chamber, *e.g.,* during cooling of a potato box with forced air flow. Such results can also be verified in a real free convective ventilated store during a long-term experiment, for example during the cooling period (see Fig. **12**). IR measured data in a big box potato store may show more fluctuations compared to lab experiments.

For potato or fruit cool storage a limited temperature range is of practical interest, such as 5 to 25°C ambient or store temperature. Within this range linear temperature dependency can be assumed. Therefore, the linear regression approximation is a straight forward method to compensate these dependencies. In order to be able to apply the linear regression calculation, measured data should be recorded during an experiment time period covering the full range of the potential ambient temperatures (dots in Fig. **8**).

From these data linear correlation calculation can be carried out using Gauss' Error Mean Square method. The derived straight linear regression line can now be used to compensate the average deviation. For example, in Geyer *et al.* [22] an initial deviation of 1.37 ±0.58K between thermographically and conventionally measured temperatures was found which could be compensated to finally 0 ±0.09K. Thus, this method can be used to obtain high precision which is necessary for climate control in the store. It is recommended to repeat the calibration procedure regularly to check long term effects on the precision.

A blackbody can easily be constructed (Figs. **9** and **10**). The blackbody is a well isolated chamber heated inside (Fig. **9**) with a thermostat to assure a regulated temperature at a constant (the 'reference') value. The emissivity of the hole is 1.0 per definition (Equation 2). Its temperature is recorded simultaneously with the IR camera. In the data logging software the measuring area is marked around the hole and the maximum temperature of this area is recorded to be sure that exactly the blackbody radiation emitted through the hole is acquired. The inside temperature of the blackbody is measured by a high precision temperature sensor like a Pt100-type or Pt1000-type platinum resistance temperature detector which has online data transfer capability. The sensor is regularly calibrated with a Reference Temperature Calibrator (RTC). Such types of sensors can meet tolerance Class B ($R_0$: ± 0.12%) or Class 2B ($R_0$: ±0.24%); ($R_0$=100 or 1000 Ohms). Also, they have a nonlinear characteristic according to DIN EN 60751.

**Figure 9:** Practical design of a black radiator (blackbody) with metal corpus and insulation walls.

**Figure 10:** Practical design of a black radiator; view inside with heating element and thermostat.

Manufacturers of thermographic systems are offering 'blackbody radiation sources' as references with defined precision, calibrated emissivity, bigger target surface instead of just a hole, and adjustable features.

Instead of using a blackbody a 'reference plate' is also approved. The reference plate (Fig. **11**) is an aluminium plate covered with a matt black film of defined emissivity (Table **1**). The plate may be heated using a regulated Peltier element and its temperature is measured by a high precision temperature sensor and recorded. Using two references, the regression line (see above) can be constructed and any interpolation temperature within the defined range can be calculated.

The references must be placed within the camera view of the observed region and their values must also be recorded. The corresponding areas of the reference plate are marked in the thermogram (*e.g.*, see Fig. **21**). For these purposes the average temperatures of the marked areas calculated by the software should be used instead of the maximum temperature, as discussed above about the blackbody. To get the 'lower' temperature reference one reference plate attached with a high precision sensor is sufficient if it is placed near a 'cold spot'. The temperature difference between two references, the 'lower' and the 'higher', should cover at least the temperature range of the area (store) which needs be controlled, because extrapolation to values outside the defined range may not be precise enough.

**Figure 11:** Aluminum plate attached with matt blacked reference film and high precision temperature sensor.

The proceeds of the temperatures of an uncompensated camera and of a reference plate are shown in Fig. **12** for a time period of a few days.

The measured temperatures of the reference plate, both the IR (solid light line; Fig. **12**) and the conventional (solid dark line; Fig. **12**) should be always the same but in uncorrected mode a deviation of about -1-2K is characteristic for the IR values. It can be seen that the measuring error is relatively high within the interesting temperature range of about 4.6°C for cold storage (green solid line in Fig. **12**). Therefore, the need for a correction of the IR data is apparent.

Instead of attaching a Peltier element for thermoelectric heating or cooling the reference plate to a defined reference temperature it was practically approved to calibrate the system 'on-the-run' for some time period after starting experiments or climate control to get the needed two (upper and lower) temperature references. After a sufficient long time interval it is possible to analyze statistically the IR and conventional temperature data measured on the reference plate for deriving the regression line (Fig. **13**).

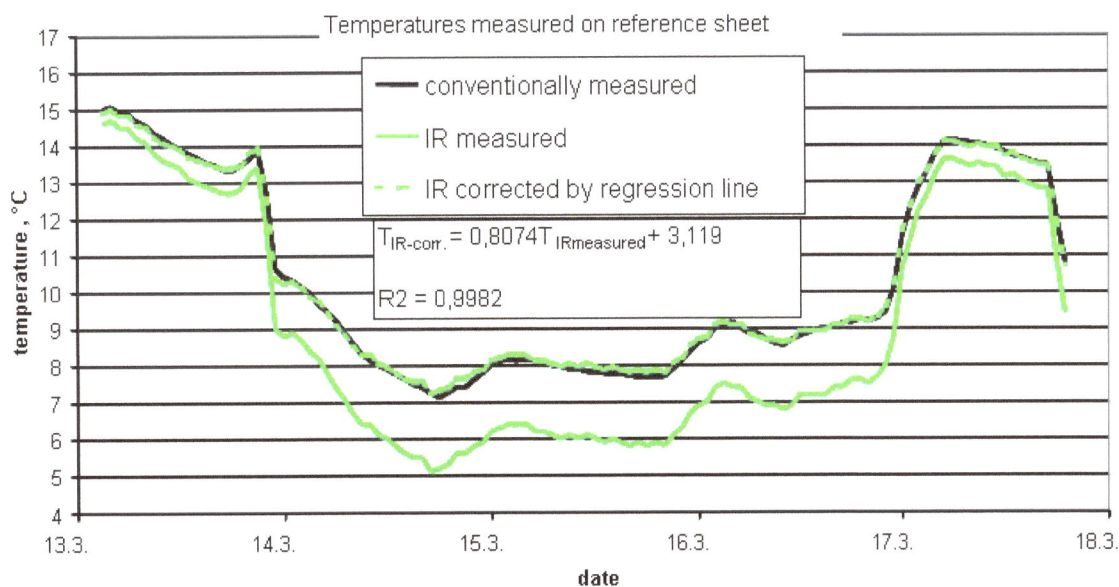

**Figure 12:** Non-calibrated and calibrated temperature measurements.

The possibility to refer to only one reference temperature may exist if the temperature measuring range is very narrow (approx. $\leq 20K$). The regression line has to be determined firstly for the specific camera. Afterward, there is only need to shift the regression line to the reference temperature point but the slope of the regression line remains practically independent on the temperature.

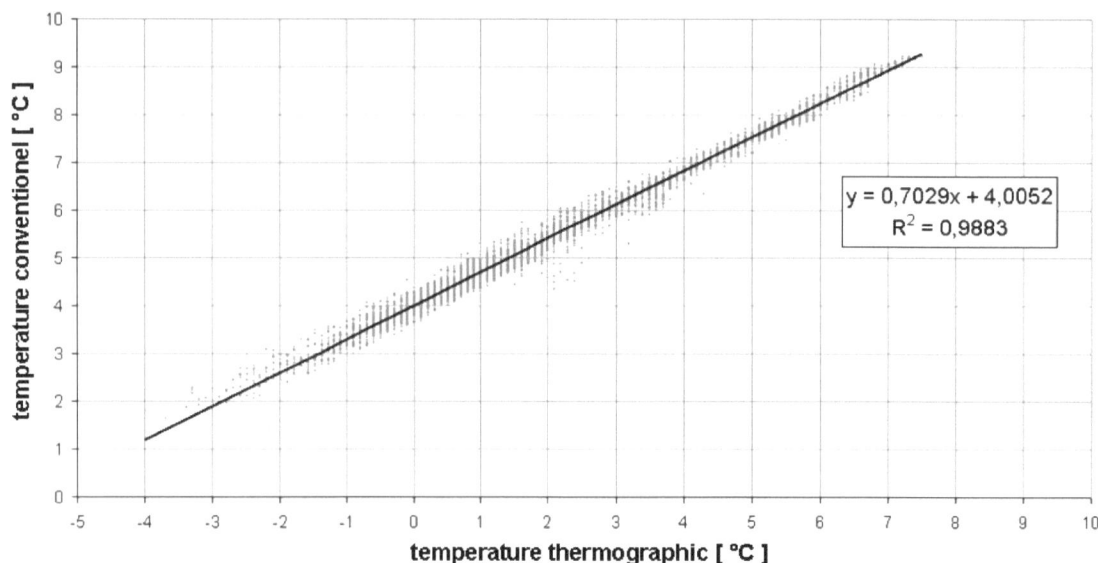

**Figure 13:** Determination of regression line to calibrate temperature measurements.

Trying to correct the data by simply compensating the error with the average of the deviations of measurements for the whole time period could not be approved and is therefore not recommended. Additionally, the uncorrected IR values are dependent on the temperature. The error increases with lowering the temperature, (Fig. **8**). This temperature dependency may exceed the precision error given by the manufacturer of the camera. The simple method by correcting with the average deviation is not able to compensate the temperature dependency. An average temperature difference of $1.37\pm0.58K$ still remained.

The IR camera has an internal compensation feature. For this feature some parameters must be set using the camera software, among others the operating ambient temperature and the air humidity. These settings should be updated regularly. There is need for an additionally temperature/RH sensor placed near the camera position. The measured values must be recorded online and fed into the software for up-to-date correction. Practically, this additional compensation increases the precision slightly. Only a slight shift of the regression line (see above) should be recognized. For ease of operation the standard setting of the camera is mostly sufficient, (Table **2**). With this method the regression line should be corrected only by an almost constant shift of approx. 1K, *i.e.,* without remarkable dependency on ambient temperature and on humidity.

**Table 2:** Example of standard parameters set with the control software for the IR camera (FLIR) [23].

| Version | ThermaCAM Researcher 2001 |
|---|---|
| Distance | 10 |
| Distance Unit | Meter |
| Emissivity | 0,92 |
| Transmission | 0,00 |
| Ext. Optics Temperature | 20,00°C |
| Ext. Optics Transmission | 1,00 |
| Reference Temperature | 0,00°C |
| Relative Humidity | 50,00% |

The camera parameters must be set according to the manual of the camera. Most important are ambient temperature and rel. air humidity (RH). Normally, this should be done regularly during the operations because climatic condition in the store room is changing. It should be checked if this is done automatically by the software. Practically, it may be found that the influence is insignificant and can be neglected; setting the parameters to standard values may be sufficient then, Table **2**.

During real operation of the camera in a potato store the compensation method can be performed in real time during operation, (Fig. **13**). The precision of IR and conventional reference temperature data can then be increased from 3.23±0.67K to an average of 0 ±0.32K [22].

## 7. POTATO STORAGE

### 7.1. State of the Art

Infrared cameras are offered as an ideal solution for assuring that temperature tolerances are not exceeded throughout manifold operations. But the application of thermography as a temperature data recording system or as a climate control element in store houses for agricultural products is a new field in practice.

Thermography has recently found application in observing temperature distribution in potato storage boxes to examine the potato temperatures dependency on different filling conditions during ventilation of the boxes [24]. In that way, the local temperature distribution of potato tubers in a box can be used as an indicator for effectiveness of ventilation of tubers for different filling or emptying trials of the box.

Fruit, vegetable and potato are living objects with biological activities like respiration and metabolic heat production [25]. To maintain storability for a long time it is necessary to reach and keep the 'optimal' storage condition which includes appropriate temperature, air humidity and air flow in the store room. The conditions should be defined according to the need of the produce which is dependent on the purpose, the kind of the produce, and on the variety of the produce. Potatoes with purpose for fresh consumption (table potato) need normally a constant storage temperature of 3.5°C at ≥95%RH. The adequate temperature should be fixed dependent on variety and purpose (Table **3**).

For most varieties this is the condition for maintaining the dormancy period of the potato tuber. The dormancy is the period of reduced metabolic activity, hence suited for prolonging storability [25]. The metabolic activity produces continuously heat which is locally distributed inside the potato bulk. The metabolic heat production causes augmentation of temperature within the bulk which can be observed. The heat is accumulating in the void areas between the tubers and must be removed frequently to reduce risk for local overheating. It can be done by ventilating regularly the bulk. Automatic control is switching the fans for this purpose. For stores working with the free convective ventilation principle only the dampers can be controlled. Temperature differences between ambient air and heated air in the bulk are driving the air ventilation naturally (Fig. **14**). The ambient air (and the air in bulk or surround the boxes) can only be refreshed by damper control for this type of ventilation.

A number of different storage principles exist for modern potato stores (Table **4**) [26].

Free convective ventilation systems are working without fans (Fig. **15**, right). The air flow is induced by buoyancy forces due to air density differences. The air density is dependent on temperature and humidity. Warm air and more humid air are lighter than cold or lower humid air. The resulting air density differences cause the buoyancy forces to drive the air. If warm potatoes are put into the store and cold air is entering the store penetrating the opened dampers from outdoor ambience, the temperature differences are initiating the air flow in 'naturally' way and the potatoes are cooled until temperature and humidity equilibrium is reached. Also, metabolic heat production is positively self-inducing air flow in the same way. The temperature differences can easily be detected using an infrared camera.

A temperature difference of 2K is inducing an air flow rate of $30 m^3 \cdot ton^{-1} \cdot h^{-1}$ at a bulk or box filling height of more than 1m. From other calculations a continuous up-flow due to respiration (metabolic) heat is assumed as 0.50 to $0.75 m^3 t^{-1} h^{-1}$ in average [27].

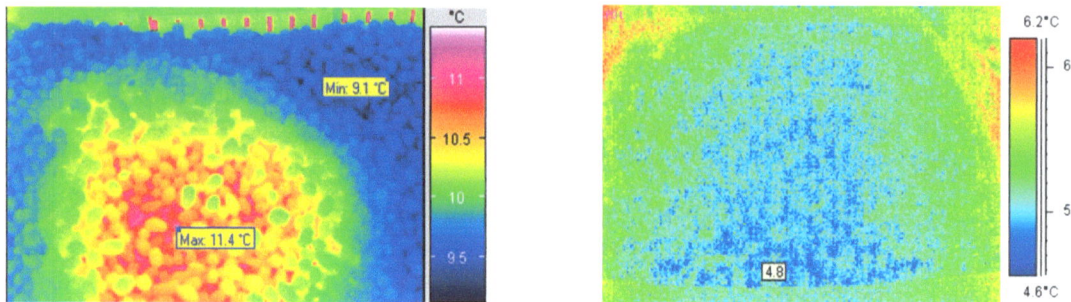

**Figure 14:** Potatoes in a box, left: harvest in August, right: before emptying the store in April; the inner parts of the tuber bulk have remarkable temperature differences to the ambient air.

**Table 3:** Storage conditions for different purposes of potatoes.

| Purpose | Long Term Storage Temperature |
|---|---|
| Seed potato | 2. 4 °C |
| Fresh potato | 5. 6 °C |
| Chips | 7. 10 °C |
| French fries | 6. 8 °C |
| Granulates | 7 °C |
| **General conditions** | |
| $CO_2$-concentration | < 2 % |
| $O_2$-concentration | 20. 21 % |
| Relative humidity | > 95 % |

**Table 4:** Potato storage principles.

| Type of Storage | Ventilation Principle | Ventilation Type | Air Flow Driving Forces in Bulk | Inlet System [1] | Type of Inlet Air | Air Conduct Elements |
|---|---|---|---|---|---|---|
| storage as bulk | free ventilation | ventilation due to buoyancy (max. 1.5m bulk height) | under-pressure due to buoyancy | open | fresh air | damper, door |
| | forced ventilation | forced ventilation | under-pressure or over-pressure or equal pressure | open or closed | fresh air mixed air recirculated air | bottom or top floor ducts |
| storage in boxes | free ventilation | free convective ventilation | over-pressure due to buoyancy | open | fresh air | damper |
| | | other types of free ventilation | | | | |
| | combined ventilation | optional combinations of elements of free and forced ventilation | | | | |
| | forced ventilation | room ventilation | underpressure or equal pressure | open or closed | fresh air mixed air recirculated air | slotted wall throw ventilation wall ventilation |
| | | forced ventilation | over-pressure | boxes closed on the sides | | slotted wall bottom floor ducts |

1)   open     = inlet/outlet *via* dampers.
     closed   = inlet/outlet *via* fans only.

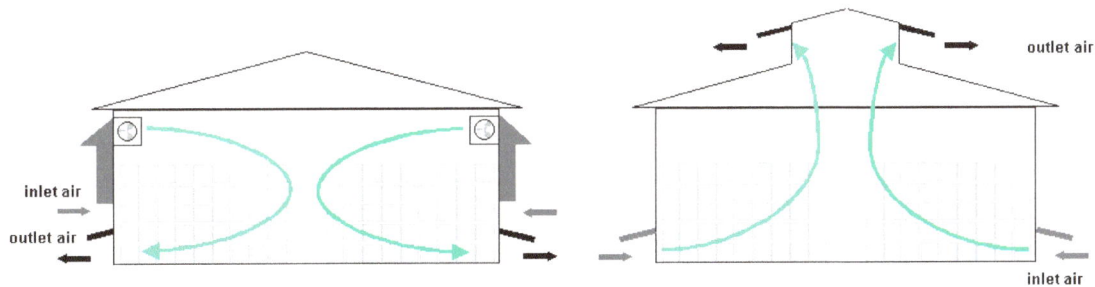

**Figure 15:** Two different ventilation systems for box storages; left: forced convection, right: free convection.

Box storages with free convective ventilation are operated since about the 70th [26, 28]. The experiences with these systems are different; some are working sufficiently, others not. These systems are severely dependent on the outdoor conditions like site positions, climatic conditions, construction and storage management. Windy store sites are more effective.

## 7.2. Problem Outline for Box Storage of Potatoes

Potato stores for boxes in Mid-Europe have an average capacity of 2 to 4 kilotons. A newer big store with a storage capacity up to 16 kilotons is operating successfully as free convective ventilation system [28].

One big hall is specially constructed to contain piles of up to 7 boxes with heights up to 8.5m placed on a ground area of approximately 5000m$^2$, (Fig. **16**).

The store is logically, but not physically, divided into 6 sections. Each of this section is independently controlled in automatic way. If necessary any section may separated with moveable curtains to prevent influences from neighbour sections during special ventilation or treatment, like cooling, drying or sprout inhibitor applications.

Since free convective ventilated stores are working without additional electrical cooling or ventilation, keeping a well adapted climatic control with free convective ventilation may arise to a great problem for stores of that size. The free convective ventilation becomes inefficient if appropriate cold outdoor air is not available for ventilation, esp. during warmer climate periods even in winter season. The desired temperature should be maintained at any box in the store during long term storage up to 5 to 8 months in the desired range of storage temperature, *e.g.,* 4 to 5°C for table potatoes.

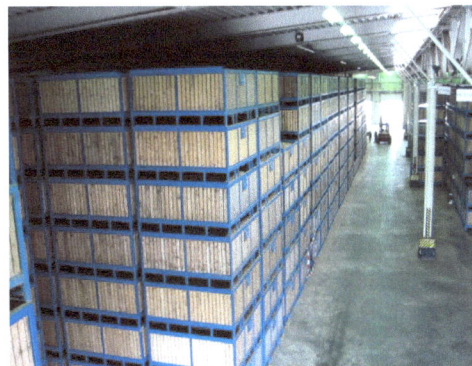

**Figure 16:** A view along a section of potato boxes in a big store.

It is found that potato temperatures diverge continuously above the normally existing temperature difference of approximately 1.5-2.0K along the height of 1 meter to 8.5 meters during the storage period. At the end of the storage period in April it may happen that the potatoes are stored too cold at the lowest level and too warm at the topmost level of the stack, (Fig. **17**).

Early sprouting due to dormancy break may occur in the upper levels if environment air is too warm for an extended period. Deficient ventilation strategies will lead to additional shrinkage and mass loss above the expected 2 to 3% mass loss caused by natural physiological respiration.

Contemporary conventional climate control is not much effective to achieve an equalize temperature distribution in the store. Traditional temperature sensors distributed in the boxes are acquiring local information about the state of the environment of the produce, like temperatures, air humidity, gas concentration, *etc.* The climate control observes only the average temperature from a number of sensors distributed in several potato boxes.

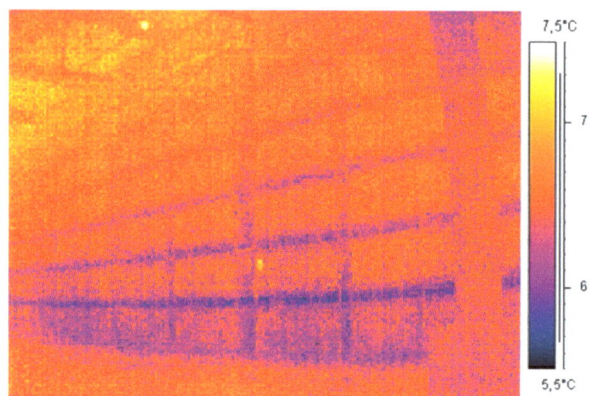

**Figure 17:** A section of the potato store after filling with boxes. The temperatures are lower at bottom due to evaporation of moist potatoes.

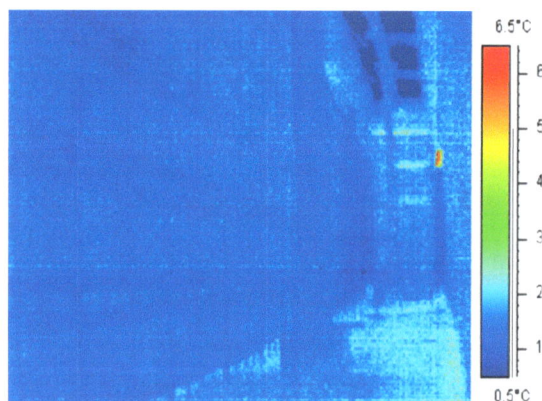

**Figure 18:** A section of potato boxes after complete cooling. The temperatures are equalized.

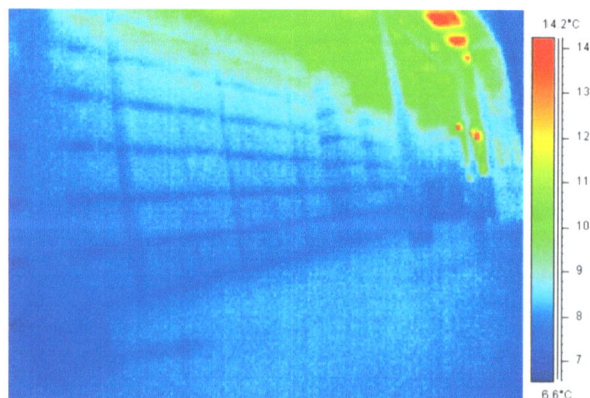

**Figure 19:** A section of potato boxes after the winter period. Warm air is penetrating through the roof dampers.

An infrared imaging system can be applied as a climate control component in a potato box store [22]. It is able to record a field of view over a comparably wide area for detection of local surface temperature differences in the storage (see Fig. **23**).

The task of such a system is to improve climate control by monitoring the temperature distribution for an extended area of the store. The use of thermography in a free convective ventilated box store for potatoes can help avoiding high temperature differences, which is a typical problem in such types of stores, (Figs. **17, 18** and **19**).

### 7.3. Thermal Behaviour of Stored Potatoes

There are a few alternative methods to ventilate a potato store house. The so called "cellar-effect" follows intuitively the traditional way to ventilate a store and is supposed to be a specific but adequate ventilation method for free convective ventilation storage [29]. Following this idea only the roof dampers are opened. It is assumed that the cold outside air is 'falling' when penetrating the roof dampers and 'sinking' down through the stack to the floor due to the buoyancy forces and is transporting away the heat of the potatoes. The sinking cool air is crossing the rising warm air which is enriched with humidity and therefore becomes lighter than dry air. More effectiveness is reported in the case of simultaneous opening of roof and floor dampers [30, 31], but generally there is low knowledge about the real airflow distributions for the different ventilation types. By means of thermography it is now possible to monitor the temperature changes on the potato and box surfaces of the stack and to detect air flow fluctuations.

According to IR measurements the hypothesis of the 'cellar-effect' could not be proven; see below. The presumption that upward flow of warm and humid air is crossing the fresh cool air is not or not totally achieved. The air flow is developing vortices instead and the warmed humid air cannot flow crossways the inlet air and pass the roof dampers without disturbances.

### 7.4. Temperature Data Logging in the Store

The infrared camera should be mounted on a mechanical stable position and adjusted to the object. The camera should not vibrate or displace and the environment should be clean, dry and in constant climatic condition, if possible. If this cannot be achieved a housing may be an option. Sufficient ventilating of the camera is necessary. In Fig. **20** it can be seen that a tube is attached against dust and moisture. The aperture of the tube corresponds to the view angle of the camera (approx. 45 ).

**Figure 20:** The IR camera mounted for test purposes in the store and protected against dust and moist.

For data acquisition the camera can be networked into a system of data logging equipment with additional conventional sensors for validation and calibration purposes. The loggers and sensors can be wired using the standardized RS 484 data exchange protocol. A wiring of the camera with optical fibre cable is compulsory to ensure high speed data transfer and to prevent electrical disturbances for long distance data exchange. An Ethernet wiring is also approved for distances below 100 meters.

Monitor software is useful for logging the conventional and infrared-recorded data from the stationary installed measuring system and for transmitting the data online to the office. The software should permit to monitor the state of the conventional climate control system which comprises acquired weather data, store temperatures, and damper operation activities. The relevant data can then be downloaded for analysis and backup. It is obvious that the camera should be featured with remote control capabilities.

Different ventilation scenarios like warming up, cooling and airflow development in the store can distinctly be visualised as motion picture sequences ('video'). Materials with low heat capacity and low dimensions are reacting very fast to temperature changes that arise when airflow direction and temperature are changing, or, for example, when fresh air is passing through the ventilation damper openings. Because of its small heat capacity, paper is reacting almost instantly upon temperature changes. Air flow fluctuations can therefore be visualised with thin paper sheets or card boards. For these purposes, a number of simple paper sheets can be fixed beside the boxes for capturing temperature changes with the IR-camera as a monitor. Front sides of wooden boxes and the paper sheets are reacting faster to temperature changes than potato tuber surfaces. It was found that there is mostly no need to place extra paper markers on wooden surfaces, since they both have almost equal response time on air temperature changes, (Fig. 21). The paper markers may be placed in empty spaces for visualization of air flow movements between or above the box stacks, *etc.*

It was approved that wooden boxes are also suitable for visualization of air temperature changes since wood is reacting fast upon temperature fluctuations.

Placing the camera exactly in a position that the view is penetrating a gap between boxes (Fig. 21) can make it possible to measure temperatures of object surfaces in the background. For example, it is possible in that way to measure the temperatures of the floor behind the box stack at high distances. Pointing a spot near the floor dampers makes it possible to measure the inlet air without using a conventional sensor including wiring or wireless data transfer (green solid line in Fig. 22). These data can be implemented into flow control.

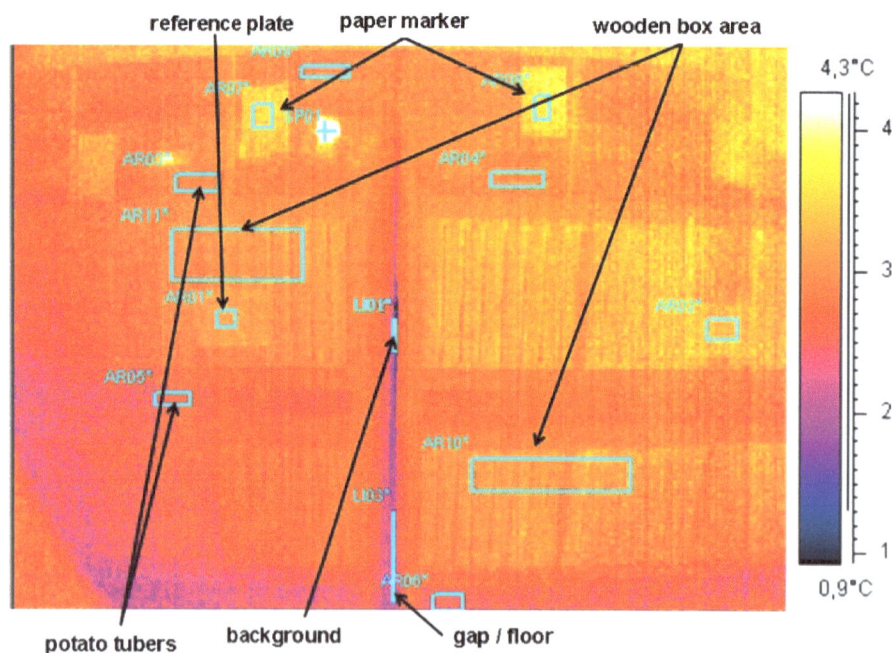

**Figure 21:** Thermogram of a section of a box store with marked areas for temperature measurements.

As an example, a test run in two modes, dampers closed and floor dampers opened, is demonstrated in Fig. **22**. Measurements-conventional and IR-are made at several points: inside a potato box, outside a potato box (tuber surfaces and wooden box plank), paper marker, and air. As long as the dampers are closed the temperatures are stable during the period of 17 hours. At the time the floor dampers are opened cold air is penetrating into the store. The temperature of the air passing the floor near the dampers is decreasing quickly. If the IR-camera is positioned in the way to be able to view through a gap between the box stack, as discussed above, the floor temperatures can be measured. The floor temperature corresponds nearly to the inlet air temperature. Anyway, best recommendation is to place a card board near the damper position which is permanently visible to the camera. The measuring area must be marked exactly by the software, (Fig. **21**). The camera software should allow defining arbitrary marking areas, like points, bars, rectangles, circles or ellipses, or polygons.

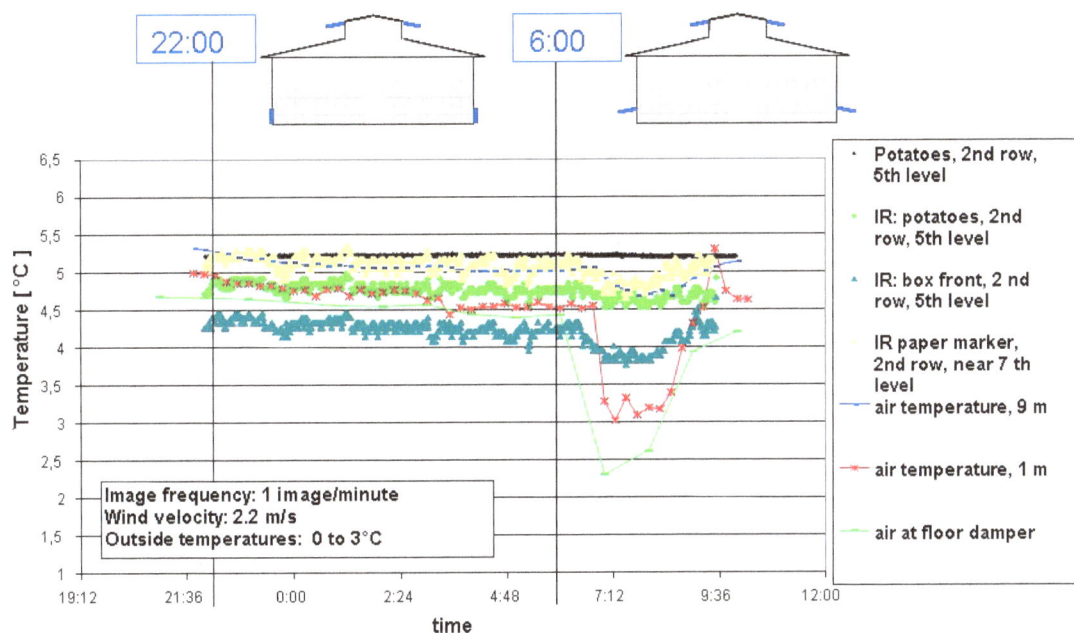

**Figure 22:** Proceed of temperature at different measuring points, conventional and IR during ventilation tests in two modes.

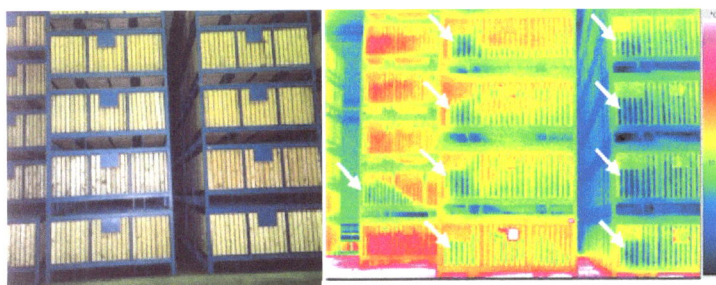

**Figure 23:** A section of potato boxes during cooling. The temperatures are not yet equalized.

## 7.5. Conventional Climate Control

The main tasks for climate controllers in potato bulk stores are to keep the storage climate within an appropriate condition for quality conservation of the produce. The internal climate of potato bulk stores is controlled by using outdoor air only if mechanical cooling systems are excluded to avoid extra energy costs [31]. In conventional storage systems electrically driven fans are used to ventilate the air (Fig. **15**, left). The need to maintain high quality produce and to meet the tendency to global climate warming during the recent decades is the motivation to use artificial cooling. Therefore, there is a trend to implant mechanical cooling systems in the recent years despite of their high power consumptions and investment costs.

The most important quality factors for vegetable and potatoes are freshness (reduced mass/water loss), health (dry storage) and absences of sprouts (cool storage). To avoid energy costs, cooling and heating systems are used only temporarily in some storage facilities, during a period when needed and applicable. Some cooling systems are mobile. When not using such systems, the store climate is regulated with outdoor climate only.

Harvesting time and storing time for potatoes takes place about in September to October (in Mid-Europe). In this time mostly there are cool nights and sometimes cool days to obtain fresh cool air to ventilate the potato bulk or boxes directly.

Under Mid-European climatic conditions, the following objectives are to be fulfilled ('rules' to regulate the climate [32]:

-   Dry moist potato tuber surfaces as fast as possible, if necessary;

-   Cool down the potatoes as fast as possible after filling the store;

-   Keep potato bulk cool at constant temperature during the whole storage time;

-   Keep the tubers dry, *i.e.,* avoid water condensation on the potatoes to prevent spread of microbial diseases;

-   Warm up the potatoes as fast as possible after storage time;

-   Minimize mass loss during the whole storage time;

-   Minimize energy consumption during the whole storage time.

To achieve these aims, a well adapted control algorithm has to be implemented in the control equipment [33]. Conventionally, the climate controller manages the fans and the damper openings to get inlet air flow to the potato bulk (Fig. **15**, left). The controller follows the current outdoor climate, *i.e.,* outdoor air temperature and rel. humidity. In practical stores the outdoor air humidity may also be handled by the controller, if this feature is available. The climate controller takes the following values into account: the average potato temperature, the inlet air temperature, and the inlet air humidity, if available in practical stores. In stores with ventilation systems the outlet air may be mixed with recirculated air to obtain the most suitable inlet air condition. Humid air above 85% RH helps to reduce mass loss or shrinking of the tubers.

The basic control 'rules' for cooling and storing potatoes are defined as:

*if*   cooling air is needed

and

*if*   cool air is available

*then* ventilate (open dampers and switch on fans)

*else*  do not ventilate (close dampers and switch off fans).

These rules are refined more to obtain the most appropriate air conditions for ventilation, *i.e.,* to reach appropriate inlet air temperature and air flow velocity into the bulk and, if control is more extended, to reach best suited inlet air humidity to reduce mass loss.

This is achieved by additional features for controlling dampers to regulate inlet air and recirculated air.

### 7.6. Air Flow Control in a Free Convective Ventilated Store Using IR Thermography

Main advantages of using an IR-camera in a store are:

-   Quick response to temperature changes;

-   High sensitivity to temperature differences and fluctuations;

-   Visibility of air-flow movements when ventilating with air of variable temperatures passing through top and/or bottom dampers;

-   Capability to measure at far distances without direct contact to the observed object;

-   Possibility of on-line monitoring and control without distribution of a high number of sensors.

The strategy for controlling the air-flow and the storage climate is as follows [22]. The wooden box surfaces are sufficient good indicators for the ambient air temperature. Any changes of these temperatures, which are due to air-flow movements, become visible on the IR images and temperature fluctuations can be determined with high accuracy. Adequate damper movements are controlled, based on these temperature changes. The invisible parts of the box stack, mainly the potatoes inside the boxes, can only be derived from a specific number of sensors for temperature and air humidity. A way to keep the number of these sensors to a minimum, the temperature distribution inside the boxes can be modelled for predicting the temperature changes dependent on the changes of the surface ('visible') temperatures; see below.

Investigations showed that controlling the top dampers alone mostly have an effect on the air temperature at the top region of the store only, while controlling the bottom damper alone or both bottom and top dampers, gives the possibility to regulate the whole store climate. The presumed effect that controlling the top dampers alone it should be possible to influence the whole climate efficiently could not be verified. The IR camera is a useful and effective tool for such kind of investigations, [22].

The surface temperatures of the wooden box walls and the potato tuber surfaces are changing quickly upon air flow movement with fresh air. The tuber surfaces are reacting a little slower. This can almost instantly be seen from IR-image movie-sequences. However, temperatures inside the boxes are changing very slowly, with some delay, depending on the temperature difference inside to outside (ambience) of the box stack. The wooden surfaces can easily be taken as an indicator for the air temperature. Metabolic heating due to respiration activities of the tubers results in warming up of the tubers with approx. 0.8K per day and the tuber surface temperature changes are remarkable at the following conditions:

-   Dampers are closed;

-   Outdoor temperatures are equal or higher than the inside temperatures;

-   Air exchange stops instantly when closing the dampers;

-   Dampers are closed and outdoor temperature is lower (or frosty).

The temperature decreasing rate during cooling is depended on:

-   Outside air velocity (wind);

-   Temperature difference to outside air;

-   Damper movement activity and opening time;

- Opening the dampers on roof top which leads to a low decrease rate of temperature mostly on the top of the stack (mainly useful in Jan/Feb);

- Opening the dampers on top and bottom which leads to a faster temperature decrease rate;

- Opening the dampers on bottom only which leads to temperature decrease almost on bottom of the stack;

- Mastering the warming problem to cool down sufficiently the upper region of the stack which may caused by solar heating of the roof during sunny days.

The control efficiency is therefore strongly dependent on the outdoor condition, mainly on the air temperature difference inside to outside. During warmer periods it is difficult to keep the inside temperature constant at the setpoint value, (Fig. **19**). Perhaps, only for a few hours during the night it may be possible to use cooler outdoor air for re-cooling the stack.

The desired (setpoint) average storage temperatures can be achieved and maintained, but during prolonged storage time the temperatures at 1m and 8.5m height of the box stack may diverge significantly. At the end of April it can be observed that the potatoes at the lowest level of the stack are somewhat too cold (below the desired storage temperature) and at the highest level too warm (above the desired storage temperature). Particularly in the upper levels the risk arises that the tubers start sprouting prematurely and mass loss increases remarkably due to increased transpiration. Anyway, during the long term storage at the desired temperature an almost constant temperature difference between bottom and top of approx. 2K is observed and regarded as acceptable in practice.

Using a model for the heat transfer from outside a heap of potatoes to inside the heap (or potatoes in box) can be calculated [34, 35]. The boundary conditions of the model are set to the ambient air temperature and humidity. The temperatures correspond to the thermographic data measured on the surface of the concerning box or stack of boxes, (Fig. **24**). In this way the capability is gained to predict 'hidden' temperature values for an appropriate climate control.

An approximate prediction of inside temperatures can be achieved more easily when the time shift of temperature changes from outside (surface) to inside is measured, using conventional sensors for temperatures placed inside the bulk, (Fig. **25**). The heat transfer into the box is delayed by a few hours and the temperature can be predicted from the outside (surface) temperature, measurable with the IR-camera, by taking the time shift into account.

After a validation period, the conventional sensors are dispensable. Air flow control is then based only on the IR images involving the model calculation results for the temperature distribution inside the boxes.

## 8. HEAT AND MASS TRANSFER IN FRUITS

Thermal imaging (thermometry), using an electronic infrared camera, is a method to determine the drying process when drying off the moisture from a fruit surface [15, 36, 39].

Cooling, drying and storing of fruit or potatoes is accompanied with mass loss, *i.e.,* loss of water which means loss of weight, loss of quality, pecuniary loss, and loss of freshness. Most of interests are the physical properties of the transport and convective transfer phenomena of vapour between fruit body and ambient air, when ventilating fruits in a storeroom. The air with certain velocity, temperature and humidity, and also the fruit state (condition, type, variety, maturity, *etc.*) are the influencing parameters on the transpiration process.

The primary effects on the moisture loss in a fruit or potato tuber are the diffusion processes in the ambient air and inside the produce material. The environmental conditions can easily be obtained under test using temperature and humidity sensors. Air flow velocity around the produce is another parameter. The air flow may be induced by natural convection or by forced convection using fans.

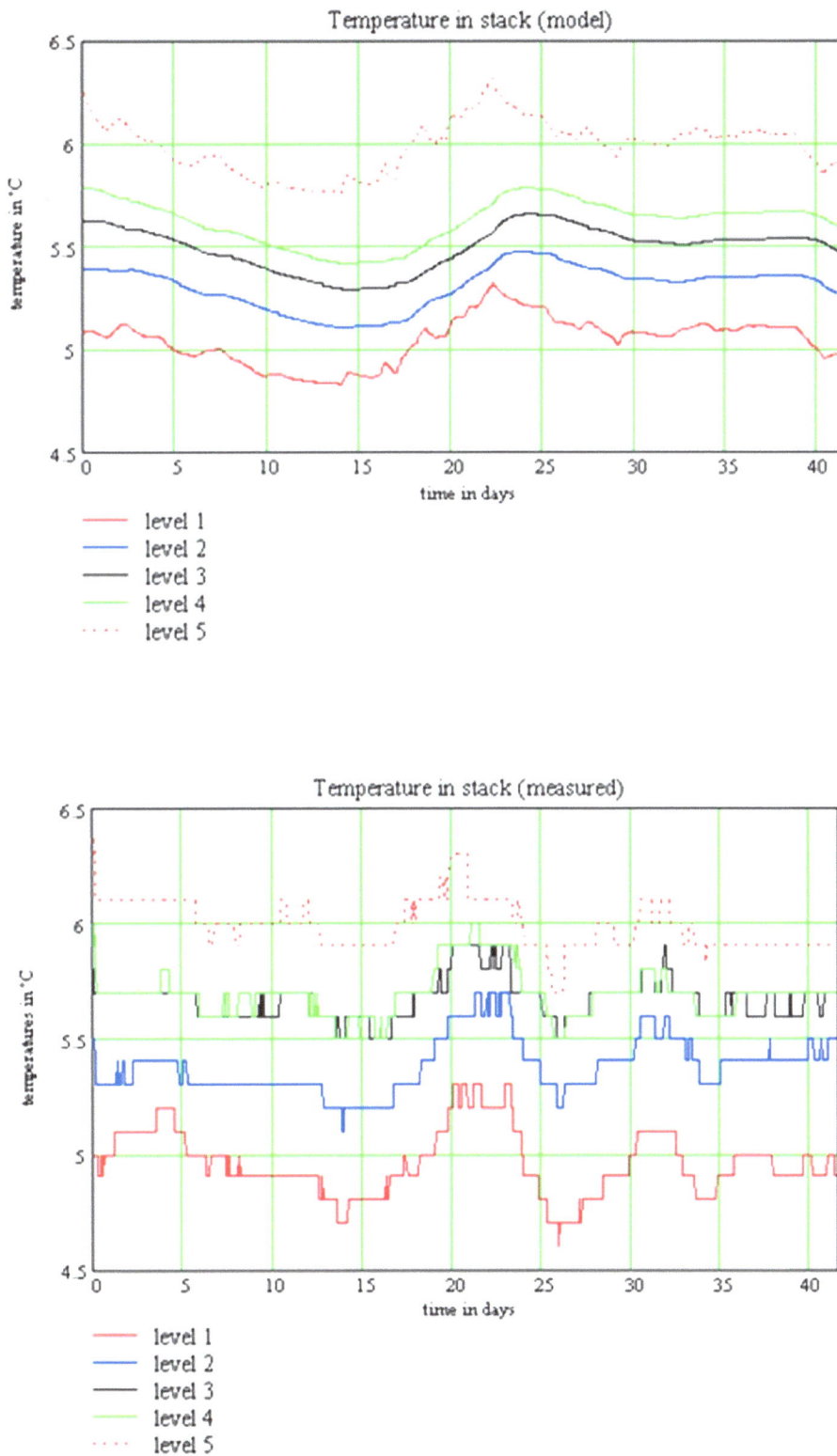

**Figure 24:** Temperature proceed of potato tubers inside a box, simulated (top) and measured (bottom).

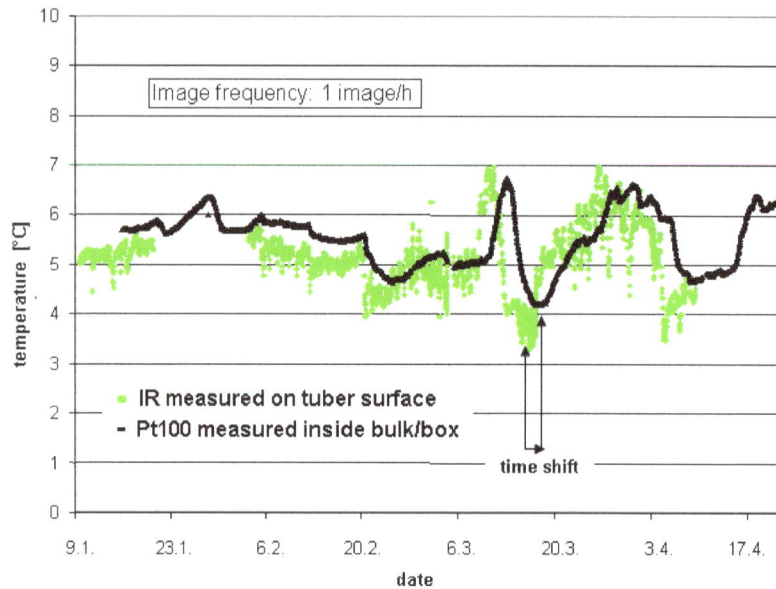

**Figure 25:** IR measured temperatures (dotted line) and conventionally measured temperatures inside a stack of potato boxes; the temperatures (peaks) are shifted in time.

## 8.1. Mass Transfer

Investigations were made by a number of authors to determine transpiration coefficients when ventilating a bulk of fruits with conditioned air of fixed properties (*e.g.,* air rate, temperature, humidity). Experimental studies of the transpiration processes were worked out before, *e.g.,* in referecne [37]. Although, it is still of interest how the air flow is developing around a fruit body when ventilating directly the body and how the transpiration and heat exchange process is influenced by the air flow pattern. Transpiration is a process on which diffused water to the surface is evaporating. As a result, the surface temperature decreases significantly and can be measured with an IR-camera. The evaporated water is diffusing from the surface of the fruit to the ambient air penetrating a boundary layer of humid air. The thickness of the boundary layer is dependent on the velocity of the air passing the fruit surface. The amount of vapour passing the boundary layer is expressed as mass flow density, *i.e.,* the flow of mass (water vapour) per time unit and traversed area (units: kg s$^{-1}$m$^{-2}$).

The course of the mass flow density gives an approximation of the boundary layer thickness $\delta$ at air flow velocity equal to zero, *i.e.,* if pure diffusion takes place. Hence, the convection term is zero but arises when forced air flow develops. At air flow velocity $w>0$ the boundary layer thickness $\delta$ decreases, which causes the mass transfer coefficient $\beta$ to increase. This process is represented by the Sherwood number:

$$Sh = \frac{\beta\delta}{D} \tag{3}$$

from which the mass transfer coefficient $\beta$ can be determined if $\delta$ is known or can be approximated using the thermal boundary layer thickness $\delta_T$ and diffusion coefficient $D$.

The air humidity ($x_A$) can easily be measured at a far distance $r_A$, while $x_{R0}$ close to the surface can be assumed to be at the saturation point for the vapour-air mixture within the boundary layer, as long as the surface is wet *i.e.,* covered with a thin water film.

The definition for $\beta$ is:

$$\beta = \frac{D_{air}\left(\frac{\partial x}{\partial r}\right)_{surface}}{x_{s0}-x_\delta} \tag{4}$$

($x_{s0}$ air humidity at saturation point on the surface, $x_\delta$ at saturation point at the boundary layer distance). For the example stated here, the ratio:

$$\frac{D_{air}}{\beta} = \frac{\delta}{Sh} \tag{5}$$

is ca. $1.20 \cdot R_0$ or $1.22 \cdot R_0$ when neglecting free convection; ($R_0$ = fruit radius).

For laminar air flow around a sphere, the Sherwood (*Sh*) number is found [38] to be:

$$Sh = 0.644 Re^{1/2} Sc^{1/3} \tag{6}$$

using the Schmidt number $Sc = v/D$ and $v$ the kinematic viscosity, $D$ the diffusion coefficient (for air). By knowing the velocity field w*(s)* around the object, the mass transfer coefficient can be determined.

The air flow velocity $w$ is enclosed in the Reynolds number:

$$Re = w R_0 v^{-1} \tag{7}$$

Using the air flow velocity distribution around the fruit (sphere), measured (Fig. **28**) or modelled, we can obtain the (local) mass transfer coefficient $\beta(\varphi)$ for the angle $\varphi$ along the path half around the fruit $0 \leq \varphi \leq 180°$.

With this method and applying Equation 5, $\beta$ was found as approx. $1.8 \cdot 10^{-3}$ m·s$^{-1}$, at the zenith of the fruit ('local', at $\varphi = 90°$), while measured in the range of $2.5-3.3 \cdot 10^{-3}$ m·s$^{-1}$, [26].

The water moist is evaporating from the fruit surface to the ambient air. The mass flux density $j$ of the water vapour is calculated as:

$$\boldsymbol{j} = \beta\, \rho_{\boldsymbol{air}}\left(\boldsymbol{x_{s0}} - \boldsymbol{x^\delta}\right) \tag{8}$$

with $x_\delta$ the abs. humidity of the air at some distance (boundary layer) from the surface and $x_{s0}$ the moist on the surface at surface temperature $T_{s0}$ (IR-measured). The humidity at the boundary is identified with the humidity of the ambient air.

In the example given in Fig. **26**, an amount of 80μg of surface water (moist) was dried off within ½ h, ventilating with air at 24°C, 40% rH, and $w = 0.15$m·s$^{-1}$. The mass loss rate is then $2.2 \cdot 10^{-5}$ kg s$^{-1}$ m$^{-2}$. Calculating Eq (8) leads to $j = 2.4 \cdot 10^{-5}$ kg s$^{-1}$ m$^{-2}$ which is close to the result of the experiment taking $\beta \approx 6.4 \cdot 10^{-3}$ m s$^{-1}$.

## 8.2. Heat Transfer

The methodology for the heat transfer is analogous to the mass transfer, if the Nusselt number *Nu* is taken [39] which is similar to the Sherwood number, *i.e.,*:

$$Nu = 0.664\, Re^{1/2} Pr^{1/3} \tag{9}$$

for an air flow around a sphere with the Prandtl number $Pr = v/\alpha$. For gases, *Pr* is approx. 0.7.

Analogously to the definition of the Sherwood number we have:

$$Nu = \alpha \frac{\delta_T}{k} \tag{10}$$

with $\alpha$ the thermal transfer coefficient, $k$ the thermal conductivity coefficient, and assuming the thermal boundary thickness $\delta_T$. Practically we can assume $\delta_T \approx \delta$, *i.e.*, approximately to the boundary layer thickness of the mass transfer.

Also, we find the definition for $\alpha$ as:

$$\alpha = -k_{air} \frac{\left(\frac{\partial T}{\partial r}\right)_{surface}}{\frac{T_{s0}}{T_{\delta T}}} \qquad (11)$$

The temperature $T_{s0}$ is the value on the moist surface of the fruit at saturation point, *i.e.*, water is condensed and the air closest to the surface is saturated with humidity (RH = 100 %), at the temperature $T_{s0}$.

The proceed of temperature at three different positions (upwind, centre, downwind) along the surface of the moist peach is shown in Fig. **27**. The evaporation of surface moisture lowers the temperature down to wet bulb temperature, if the heat transfer is adiabatic. In reality, this process is non-adiabatic, since the heat of the object (fruit body), its heat capacity and the heat conduction inside the fruit hinders the surface temperature to drop down finally to the wet bulb temperature.

As soon as the surface is partially dried off, the temperature increases again to the ambient air temperature value. The movement of a 'drying zone' from the upwind side of inlet air-flow direction to the opposite downwind side can be well recognized, (Figs. **26** and **27**). The temperature distribution development on the surface is dependent on air flow direction and velocity, as well as on temperature and air humidity. The drying time, *i.e.*, the time until the surface moist is totally removed, can be well recognized as the point when all local temperatures on the fruit surface have converged to the ambient or fruit temperature.

The air flow velocity can precisely be measured at high spatial resolution using a hot-wire anemometer, from which the Reynolds number $Re$ (Fig. **28**) and finally the mass transfer coefficient $\beta$ along the fruit surface can be calculated. In Fig. **27** we see the magnitude of the air flow velocity around the fruit (sphere) according to the fluid dynamic theory. The velocity is low at the front of the fruit at the upwind side ($\varphi = 0$ degree), is increasing to its maximum value at about $\varphi = 90$ degree and decreases again in downwind direction ($\varphi = 180$ degree). On the other hand, the mass transfer coefficient is high at the front and decreases continuously against the downwind side. This is in agreement with the movement of the drying zone (Fig. **26**) on which the drying process is faster at the front ($\varphi = 0$ degree) and decreases continuously along the flow path on the surface which we can see on the thermogram.

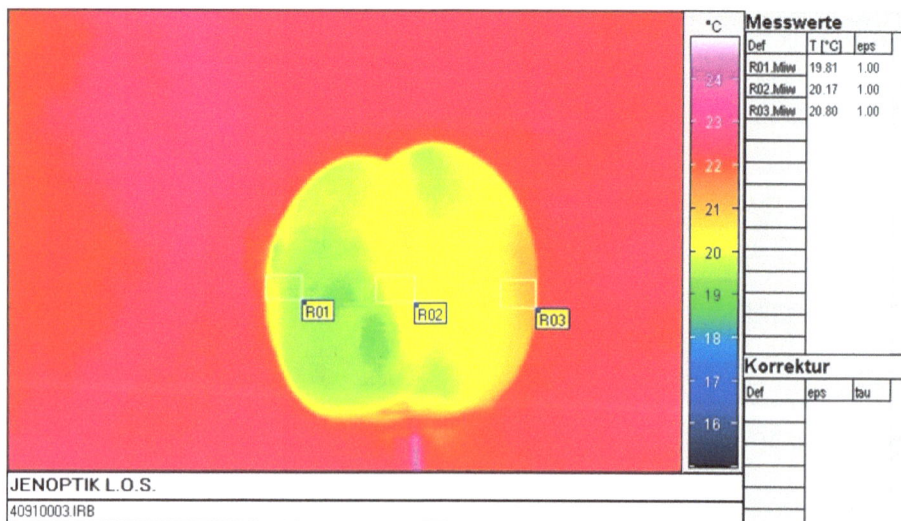

**Figure 26:** Temperature distribution on the surface of a peach during drying off surface moisture using infrared thermography.

The primary effects on the moisture loss in a fruit or potato are the diffusion processes in the ambient air and in the produce material. The environmental conditions can easily be obtained under test. The local mass transfer coefficient $\beta(\varphi)$ distribution along the surface and the total mass transfer coefficient $\beta = \pi^{-1} \cdot \int \beta(\varphi) \, d\varphi$, $(0 \leq \varphi \leq \pi)$ ($\varphi$ in radiant) can be calculated (Fig. **29**) [39].

It is important to note that the infrared thermographic camera is not measuring the wet bulb temperature of a drying surface but the temperature caused by a *non-adiabatic* evaporating process (Fig. **27**) [39]. It means that during the evaporation heat is diffusing from inside the fruit to the surface which augments the wet-bulb temperature (ca. $15^{\circ}C$ in our example). The non-adiabatic effect can also be seen (Fig. **27**) on the different and changing slopes of the temperature curves during the evaporation and re-warming.

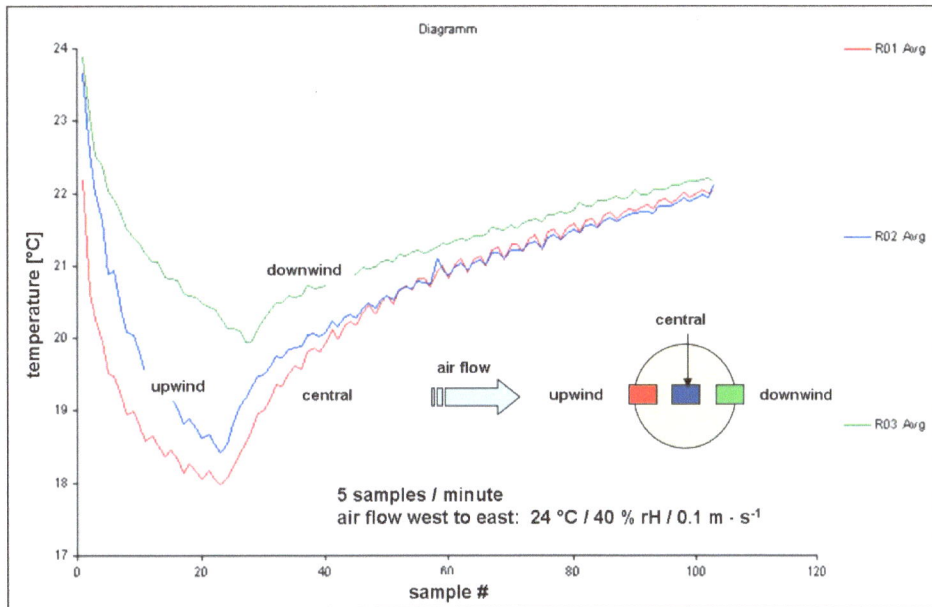

**Figure 27:** Proceed of temperature at three local points on the surface of a peach during drying off surface moisture.

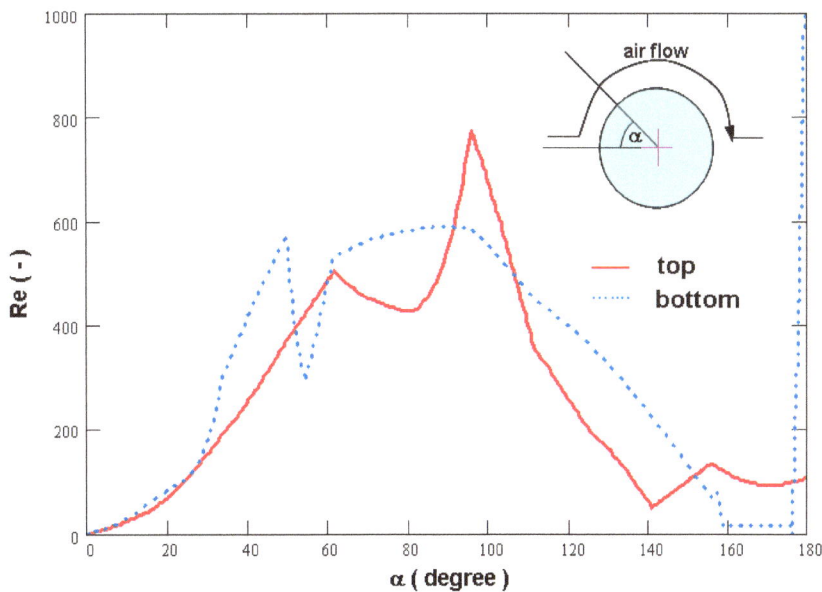

**Figure 28:** Measured proceed of Reynolds number along the surface of a peach during drying off the surface moisture.

MASS TRANSFER COEFFICIENT

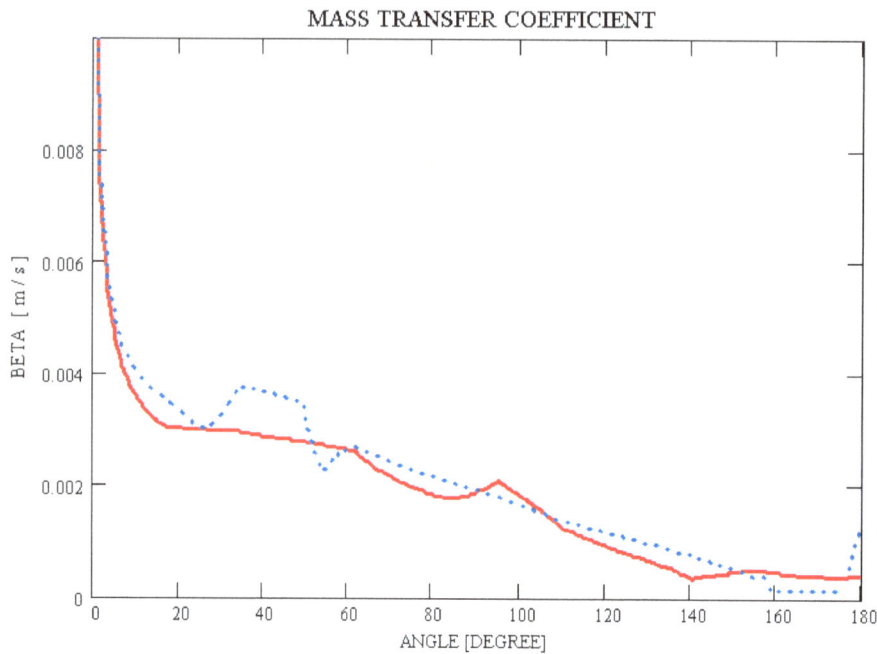

**Figure 29:** Measured course of the mass transfer coefficient $\beta(\varphi)$ along the surface of a spherical fruit.

## 9. ECONOMICAL AND ECOLOGICAL EFFECTS

By the introduction of innovative climatic control technology in connection with free convective ventilation in the store the energy expenditure is reduced. An optimal guidance for climate regulation reduces quality losses in the store. This leads likewise to saving relevant environment within the pre-aged range, since the cultivation can be reduced accordingly.

Free convective ventilation in connection with innovative control is therefore characterised by the following aspects relevant to environment. Lowering power requirement and preservation of resources as reducing losses of the stored produce makes a contribution for reduction the emission of climatic relevant gases. For example, approx. 1 to 2kWh per ton or approx. 15,000 to 30,000kW per hour of energy in total would be needed per year for a potato store with 15 kilotons capacity. Energy in this order of amount can be saved as operating cost when the store is working with the free convective ventilation principle. This saving corresponds to 150 to 300GJ primary energy for each year. The emission reduction of climatic relevant gases amounts for this regarded 15kiloton-potato store to approx. 10 to 20tons $CO_2$-Equivalent per year (Table **5**).

**Table 5:** Assessment of environmentally relevant impacts for a conventional operated potato storehaus of 1ktons capacity [22].

| Ventilation Need h/a | Primary Energy Need GJ | Greenhouse Effect kg $CO_2$-equiv. | Acidification kg $PO_2$-equiv. | Eutrophication kg $PO_4$-equiv. | Photo-oxidant kg $C_2=C_2$-equiv. |
|---|---|---|---|---|---|
| 180 | 28 | 1950 | 5.0 | 0.41 | 0.38 |
| 300 | 35 | 2380 | 5.7 | 0.46 | 0.40 |
| 600 | 52 | 3470 | 7.5 | 0.58 | 0.44 |
| 900 | 70 | 4560 | 9.2 | 0.70 | 0.48 |

Fruit, vegetable and potato are respiring during their lifetime. Therefore, mass loss is not avoidable during storage and ventilation. Ideally, a mass loss of about 2.5% during a long-term storage period of 6…8 months is inevitable. Total mass loss or shrinkage of potato tubers can increase to 5 to 10% or more for normally operated and managed storage [26]. A well adapted regulation of the store climate can alternatively reduce the loss by about 1 to 1.2%. When assuming an average market price for potatoes as ca. 80€ per ton, monetary reduction of loss amounts to ca. 10,000€ for a 15 kiloton store which is filled up to 80%, for example. Thus, investment in a suitable control system like IR camera equipment may be paid back within a few years for a big store.

Wide application fields were found for thermographic cameras the recent years. The development of high sensitive infrared sensors at low cost has extended the market for infrared systems. Handhold cameras are offered now at low-cost for simple usage like checking objects for abnormal heat development *etc.* But the suitability of such type of cameras is limited, or unsuited, for on-line control inside a storehouse.

Higher expenditure is needed for a thermographic system to apply as on-line control of storehouse climate for fruit, vegetable and potato because such a system must accomplish following features:

- Stationary mounted.

- Long time reliable and stable.

- Mechanical stable.

- Extended operating range min.-10°C to 50°C.

- Stable housing and humidity resistive (IP 54) or industrial housing IP 65 with internal cooling.

- Software controllable, *i.e.,* data exchange and remote control capability.

- Resolution min 320 × 240 pixels.

- Spectral range about 8-14μm.

- Resolution min 2K (after careful calibration min. 0.2K).

A housed camera needs a special (and high-priced) window made of an IR-transparent material like the monocristalline semiconductors Germanium (Ge) or Zinc Selenide (ZnSe). A very thin film (~20 to 40μm thickness) of Polyethylene (PE) was also proved as a window under laboratory conditions but it is mechanical unstable and causes reflections for IR radiation.

Since the required temperature measuring range is limited for usage in a store (approx. -5 to +25°C) to invest in a more expensive high range camera is not necessary. A completely filled box store mostly leave not much room to place the camera at a sufficient distance from the object, in order that the camera can get a view to cover a sufficient wide area of the observed objects. Therefore, a wide angle lens (≥ 45°, horizontal) is mostly necessary for stationary climate control which expands the investment costs. A camera with wide angle lens is taking the biggest part of the investment cost; about 70% can be assumed.

In summary, following additional equipment is needed to make the system suitable for stationary use for climate control.

- Wide angle lens.

- Solid mounting device.

- Data cable, copper wired or glass fibre.

- Rugged PC with software and sensor signal converters.

-   Extra housing when used in dusty and humid environment plus IR-transparent window.

-   Data modem and internet access for remote control.

The control computer must also be reliable and robust against dusty and humid environment if placed in the storehouse (rugged PC). A more sophisticated system may consist of a higher number of cameras distributed in the storehouse. For data exchange and remote control the cameras must be networked and the software must comprise features to process the complete system.

## 10. CONCLUSIONS

Thermal imaging technique is an up-and-coming technology for monitoring temperature/climate changes in potato stores. An infrared image supplies a lot of information about the temperature distributions and fluctuations in a store offering the chance to measure simultaneously temperatures of potatoes and the fluctuations of the ambient air.

However, the interpretation of infrared images requires special experiences. The displayed temperature values of an IR image are not complied with the real valid temperatures if the system is not correctly calibrated. Selected parts of interest of the IR image (the thermogram) must be set up in a defined session. The appended software takes into consideration the thermal radiation of the different materials (emissivity) and the object distance from the camera.

The pre-adjustment of the software is not sufficient for monitoring the real temperatures on the IR image. At potato store temperatures of 5°C average, the IR camera normally measured temperatures approx. 2.4K lower than the conventional sensors. Additionally, there is a temperature dependence of the thermographically measured values. A linear regression straight line, calculating the correlation between conventionally and thermographically measured temperatures on a reference sheet, results in a good adjustment probability. After adjusting thermographical temperatures by means of the regression line, thermographical and conventional temperature measurements correlate very well if the conventionally measured temperature is acquired from an identical surface.

It can be said that in opposition to the conventional measuring technique, the thermal imaging technique resolves the existing temperature difference, thus, lowest temperature differences can be visualized by the IR-Camera.

A potato box store may be ventilated even when outside air is below $0^{\circ}$C. For conventional control this method implies risk for freezing the potatoes placed closely to the inlet air dampers. In conventional controlled storehouses are the sensors placed inside a limited number of boxes. These sensors are measuring the warmer parts of the potatoes inside the boxes. Therefore, unsuitable too low temperatures cannot be detected in time. But a surface temperature monitoring IR-camera can do it. Lower inlet air temperature may be allowed because the air is warming up instantly when approaching the boxes and the tubers are reacting inertly. The lowest temperature limit is allowed on the tuber surface for approx. 1 hour to avoid sweetening or freezing the tubers. The thermographic analysis show that ambient air condition can be regulated to much lower temperatures than intuitively potato surface temperatures would allow. Thermographic images show appropriate temperatures on the potato surfaces. Low temperature differences can be controlled by moving the top and bottom dampers, according to the temperature fluctuations, dependent on outside wind velocity and free convective, natural flow, and can be determined by the thermography system.

Economical prospects of success for using the thermal imaging measuring system as temperature controller depend to a large extent on the relative exquisiteness of the procedure compared with competing conventional processes. Because of the relatively high investment costs for the infrared camera, additionally required equipment, and staff requirement and skill, this procedure is only in the interest of big stores at actual prevailing circumstances. But in these cases, storehouses with optimised climate control can compensate a single camera system investment, running and maintaining costs when shrinkage (mass loss) of the potato tubers is reduced by 1.2% during a period of about 7 years [22].

The range of possible application of thermography may grow in the future because costs for imaging systems are expected to decrease the next following years. Operability, accuracy, reliability *etc.* is also expected to be improved in the near future.

## REFERENCES

[1]     Monteith JL, Szeicz G. Radiative temperature in the heat balance of natural surfaces. Q J Roy Meteorol Soc 1962; 88: 496-507.

[2]     Tanner CB. Plant temperatures. Agronom J 1963; 55: 210-1.

[3]     Danno A, Miyazato M, Ishiguro E. Quality evaluation of agricultural products by infared imaging method. Memoirs Fac Agric 1978; 14 (23): 123-38.

[4]     Hellebrand HJ, Linke M, Beuche H, Herold B, Geyer, M. Horticultural Products Evaluated by Thermography. AgEng2000, University of Warwick, UK, July 2-7, 2000, Abstracts Part 2, p. 26-27, Paper No. 00-PH-003

[5]     Hellebrand HJ, Beuche H, Linke M. Determination of Thermal Emissivity and Surface Temperature Distribution of Horticultural Products, In: Proceedings of the 6th International Symposium on Fruit, Nut, and Vegetable Production Engineering (Zude, M., Herold, B., and Geyer, M., Herausg.), Potsdam, Germany, S. 2002; 363-8.

[6]     Inoue Y. Remote detection of physiological depression in crop plants with infrared thermal imagery. Jap J Crop Sci 1990; 59 (4): 762-8.

[7]     Inoue Y, Kimball BA, Jackson RD, Pinter PJ, Reginato RJ. Remote estimation of leaf transpiration rate and stomatal resistance based on infrared thermometry. Agricultural and Forest Metereology 1990; 51: 21-33.

[8]     Linke M, Beuche H, Geyer M, Hellebrand HJ. Possibilities and limits of the use of thermography for the examination of horticultural products. Agrartechnische Forschung 2000; 6: 110–4.

[9]     Linke M, Beuche H, Geyer M, Hellebrand HJ. Einsatzmöglichkeiten und Grenzen der Thermografie, Landtechnik, 2000; 55 (6): 428-9.

[10]    Nilsson HE. Remote sensing and image analysis in plant pathology. Canadian Journal of Plant Pathology 1995; 17: 154-66.

[11]    Hellebrand HJ, Herppich WB, Beuche H, Dammer KH, Linke M, Flath K. Investigations of plant infections by thermal vision and NIR imaging. International Agrophysics, 2006; 20:1-10.

[12]    Brehme U, Ahlers D, Beuche H, Hasseler W, Stollberg U. Is there a possibility of clinic application on infrared-thermography for diagnostics in oestrus detection in dairy cows? Academic Publishers: Proceedings 3rd Research and Development Conference of Central and Eastern European Institutes of Agricultural Engineering, Hungarian Institute of Agricultural Engineering, Gödöllö (H), 2003; 11.-13.09.2003, pp. 137-42.

[13]    Zohar O, Ikeda M, Shinagawa H, *et al.* Thermal imaging of receptor-activated heat production in single cells. Biophysical Journal 1998; 74 (1): 82-9.

[14]    Jones HG. Thermographic studies of stomatal conductance over leaf surfaces. Plant, Cell and Environment 1999; 22: 1043-55.

[15]    Linke M, Geyer M, Hellebrand H-J. Postharvest Behaviour Affected by Local Different Transpiration Resistances, In: M Zude, B Herold, M Geyer (Eds.) Proceedings of the 6th International Symposium on Fruit, Nut, and Vegetable Production Engineering, Potsdam, Germany, 2002, p. 291-6.

[16]    Hellebrand HJ, Linke M, Herold B. Bruises and ripeness of apples studied by thermal imaging. ASTEQ – Action Group Meeting, Assisi, Italy, 12.-14.06.2000, Proceedings Vol. 1, p.11.

[17]    King WJ. Emissivity and absorption, in: RC Weast Ed, Handbook of Chemistry and Physics. CRC Press, Boca Raton (Florida), 1987; E: 393-E-395.

[18]    LaRocca AJ. Artificial Sources, in: GJ Zissis Ed. Sources of Radiation, vol. 1 of "The Infrared & Electro-Optical Systems Handbook" (8 volumes), exec. eds. Accetta JS, Shumaker DL. ERI Ann Arbor (Michigan) and SPIE Optical Engineering Press Bellingham (Washington), 1996; 49-135.

[19]    Schuster N, Kolobrodov VG. Thermal imaging. Wiley-VCH Berlin Weinheim New York, 2000; 339: 59

[20]    Hellebrand HJ, Beuche H, Linke M. Thermal imaging. A promising high-tec method in agriculture and horticulture. In "Physical Methods in Agriculture-Approach to Precision and Quality" edited by J.Blahovec and M. Kutílek, Kluwer Academic/Plenum Publishers (ISBN 0-306-47430-1), New York 2002, p. 411-27.

[21]   Lutz P, Jenisch R, Klopfer H, Freymuth H, Richter E, Petzold K. Lehrbuch der Bauphysik, B.G. Teubner Stuttgart, 4. Auflage 1997.

[22]   Geyer S, Gottschalk K, Hellebrand HJ, Schlauderer R, Beuche H. Infrared-Thermography for Climate Control in Big Box Potato Store. Landtechnik 2004; 2: 96-7.

[23]   Flir Systems: www.flir.com/

[24]   Hoffmann T, Maly P, Fürll Ch. Ventilation of Potatoes in storage Boxes. Agricultural Engineering International: the CIGR Ejournal. Manuscript FP 06 014. Vol. IX. May, 2007.

[25]   Gottschalk K, Ezekiel R. Storage. In: Handbook of Potato Production, Improvement, and Postharvest Management. Gopal, J.; Khurana, S.M.P. (Hrsg). Food Products Press, The Haworth Press Inc, Binghampton, NY, USA, 2006, S. 489-522.

[26]   Schuhmann P. Stand und Probleme der Lagerung und Belüftung. Kartoffelbau, 50 9/10, 1999; 380-3.

[27]   Schierhorn H. Nachlese zur 25 jährigen Anwendung des Systems der Freien Konvektionslüftung nach Schierhorn von 1971 –1996. Belüftung von Kartoffeln. Vorträge zu Jahrestagung 1997, 33 –38. KLAS-Verband, Buchedition Agrimedia, ISBN 3-86037-081-2, 1998.

[28]   Kern AD, Krumbiegel E, Pötke E, Schuhmann P, Qualitätsschonende Lager-und Lüftungssysteme. Der Kartoffelbau 2002; 53: 226-33.

[29]   Pötke E, Hauschild W, Kern A. Verbesserte Auftriebslüftung in Behältern durch Zuluftluken am Boden" Kartoffelbau 50 9/10, 1999; 384-7.

[30]   Schorling E. Raumbelüftung bei Großkistenlagern. Kartoffelbau, 2001; 52(5): 204-8.

[31]   Maltry W. Elementare Wärme-und Stofftransportvorgänge in lagernden Kartoffeln. Belüftung von Kartoffeln. Vorträge zur Jahrestagung 1997, S 9 – 24, KLAS-Verband, Buchedition Agrimedia, ISBN 3-86037-081-2, 1998.

[32]   Gottschalk K. Möglichkeiten und Grenzen der Lagerklimaautomatisierung für Kartoffellager. Belüftung von Kartoffeln. Vorträge zur Jahrestagung 1997, 45–61, KLAS-Verband, Buchedition Agrimedia, 1998; ISBN 3-86037-081-2.

[33]   Gottschalk K, Nagy L, Farkas I. Improved climate control for potato stores by fuzzy controllers. Computers and Electronics in Agriculture 2003; 40: 127-40.

[34]   Gottschalk K. Mathematical modelling of the thermal behaviour of stored potatoes and developing of fuzzy control algorithms to optimise the climate in store houses. Proceeding: 2nd International IFAC/ISHS Workshop on Mathematical and Control Applications in Agriculture and Horticulture in Silsoe, UK, 12-15. Sept. 1994, Acta Horticulturae No. 406, ISHS 1996, p 331 –9.

[35]   Gottschalk K, Christenbury GD. A model for predicting heat and mass transfer in filled pallet boxes. 1998 ASAU Annual International Meeting, Paper No.983157, 1998.

[36]   Fito PJ, Ortolá MD, De Los Reyes R, Fito P, De Los Reyes E. Control of citrus surface by image analysis of infrared thermography. J Food Eng 2004; 61(3): 287-90.

[37]   Linke M. Modelling and Predicting the Postharvest Behaviour of Fresh Vegetables. Mathematical and Control Applications in Agriculture and Horticulture (Munack and Tantau, Eds.), Pergamon Press, Oxford, UK, 1997; 283-8.

[38]   Baehr HD, Stephan K. Heat and Mass Transfer, 2006. Springer-Verlag, Berlin Heidelberg. 1994.

[39]   Gottschalk, K. Surface drying of fruit and potatoes. Landtechnik 2011, 66 (1): 26-9.

# Concluding Remarks to Part II – Section 2

The use of an infrared imaging device for monitoring the environmental conditions for foodstuff conservation was illustrated in Section 2 of Part II. From the reported data, it is evident the convenience of using infrared thermography to control the climate of the store house and the respiratory activity of harvests. The author also supplied information on the location of infrared cameras in the store house and on the procedures for data treatment and interpretation.

Of course, the climate conditions are fundamental for the correct conservation of foodstuff for preventing loss of mass and of freshness with obvious economical profits. However, the conservation of foodstuff is a crucial task not only from the economic point of view, but primarily from the wellbeing perspective. In fact, the temperature and the humidity level are key parameter to be accurately set for avoiding bacterial proliferation. Not with standing the benefits, the investment costs are often considered prohibitive for stores not enough large. Fortunately, as pointed out in Chapter 2 (Part I), the technology is evolving fast while costs are becoming more affordable, and then, it is expected an increasing use by also small stores.

Now, the eBook goes on an excursion into applications of infrared thermography to the industrial field for comprehension of thermo-fluid-dynamics phenomena and for materials non destructive inspection. The use of an infrared device for climate control will be reconsidered in Section 4 but for the living comfort and to prevent degradation of artworks.

# CHAPTER 6

## An Overview on IR Thermography for Thermo-Fluid-Dynamics

## Giovanni M. Carlomagno[*]

*Department of Aerospace Engineering, University of Naples Federico II, Piazzale Tecchio 80, 80125 Napoli, Italy*

**Abstract:** This chapter deals with the exploitation of thermographic measurements in complex fluid flow configurations for the evaluation of the maps of wall convective heat fluxes. The measurement of convective heat fluxes is always performed with a heat flux sensor described by a suitable heat transfer model, where proper temperatures must be measured. By correctly choosing the measuring sensor, infrared (IR) thermography can be exploited to resolve convective heat flux maps in complex fluid flows with both steady and transient techniques. Compared to standard transducers, an IR camera appears very valuable because it is non-intrusive, has a high sensitivity (down to 20mK), has a low response time (down to 20µs), and is fully two-dimensional (from 80k up to 1M pixels) allowing for accurate detection of heat flux variations and better evaluation of errors due to radiation and tangential conduction within the sensor. After mentioning the first historical steps, this chapter reviews the most useful heat flux sensors, thermal restoration of data and several examples of convective heat transfer measurements in complex fluid flows, ranging from natural convection to hypersonic regime, which were performed by the author and his co-workers.

**Keywords:** Infrared thermography in thermo-fluid-dynamics, heat flux sensors, convective heat transfer, heated thin foil, thin film, thin skin, rotating surfaces, jet impingement, jet instability, thermal restoration, image restoration, hypersonic flow.

## 1. INTRODUCTION

Measuring heat fluxes in thermo-fluid-dynamics requires both a *thermal sensor* (with its related physical model) and *temperature transducers*. In standard techniques where temperature is measured by thermocouples, resistance temperature detectors (RTDs) *etc.*, each transducer yields either the heat flux at a single point, or the space-averaged one. Hence, in terms of spatial resolution, the sensor itself has to be considered as *zero-dimensional*. This makes measuring particularly troublesome whenever temperature, and/or heat flux, fields exhibit high spatial variations.

Instead, the infrared camera is a truly *two-dimensional* transducer since it allows measurements of temperature maps even in the presence of relatively high spatial gradients. Accordingly, also the heat flux sensor becomes two-dimensional, as long as the necessary corrections are applied.

When compared to standard techniques, the use of an infrared (IR) camera as a temperature transducer in convective heat transfer measurement appears useful from several points of view. In fact, since the IR camera is fully two-dimensional (up to 1M pixels), besides producing a whole temperature field, it allows for easy evaluation of errors, which are due to radiation and to tangential conduction. Further, it is non-intrusive (*e.g.*, allows removing the conduction through thermocouple or RTD wires), it has high sensitivity (down to 20mK) and relatively low response time (down to 20µs). As such, IR thermography can be effectively employed to measure convective heat fluxes in complex fluid flows with both steady and transient techniques.

The earliest attempts to measure heat transfer coefficients arose in the hypersonic regime and were performed by using scanners operating in the middle IR band (3-6µm) of the infrared spectrum, at the time called *short wave IR band*. In particular, the AGA Thermovision 680SWB camera was employed by

---

**\*Address correspondence to Giovanni M. Carlomagno:** Department of Aerospace Engineering, University of Naples Federico II, PiazzaleTecchio 80, 80125 Napoli, Italy. Email: carmagno@unina.it

**Carosena Meola (Ed)**

Thomann and Frisk [1] to measure the heat flux distribution on the surface of an elastomeric paraboloid in a hypersonic wind tunnel at *Mach number M* = 7. The thin-film sensor (see Section 2) was used to determine convective heat transfer coefficients, which showed good agreement with data already obtained with different techniques, and was encouraging in view of using IR systems for heat flux measurements.

Once the method was proven to work, efforts were mainly oriented towards comprehension of potential errors sources which could affect measurement accuracy and especially to developing devices which could simplify the IR camera use. Compton [2], at NASA Ames, realized that the bottleneck of IR thermography was data acquisition, storage and processing. In fact, the heat flux map had to be computed on a pixel-by-pixel basis from temperature readings, which were generated at rates of about 88,000 data points per second. The solution was devised in the automation of data processing and this concept finally brought to the systems, which are currently in use.

In 1976, the Arnold Engineering Development Center (AEDC) was embarked on a large scale research program to develop IR camera capability of extensive heat transfer testing in hypersonic regime [3]. In particular, the von Karman facility was dedicated to hosting an IR imaging system for test series that prolonged for a long time period. To assess the accuracy of the method, calibration procedures and a measurement error model were developed, as well as further automated data processing were implemented [4]. The camera displayed a blur effect at high temperature gradients, which was not completely understood at the time, but which now is ascribed to a low spatial resolution.

Meanwhile, the infrastructure and expertise developed at AEDC were used to measure convective heating rates on a Space Shuttle model, under re-entry phase prevailing flow conditions, to design the orbiter thermal protection system [5]. All the aforesaid experiments were generally carried out by applying IR camera to the thin-film sensor, but this was not feasible at very high *M* values under rarefied conditions because of resultant low heat flux values. Allegre *et al.* [6] used the thin-skin sensor (see Section 2) to overcome these drawbacks.

Apart from the evaluation of heat fluxes, the characterization of flow field behaviour, with location of boundary layer transition to turbulence and of flow separation and reattachment zones, constituted a subject of great interest to aerodynamicists and efforts were devoted to acquire information on the capability of an IR camera to deal with these phenomena. The boundary layer transition over a flat plate was first examined with IR by Peake *et al.* [7] who carried out measurements on a stainless-steel plate with a Bakelite insert. In the obtained thermograms, they observed a hot front to be attributed to the different adiabatic wall temperatures which occur among laminar and turbulent flows.

A first analysis of heat transfer measurements by IR thermography with some of their applications were presented by Carlomagno and de Luca [8]. Gartenberg and Roberts [9] reported an extensive retrospective on aerodynamic research with infrared cameras.

The heat flux sensors most commonly used with IR thermography are described in the next Section 2. Thermal restoration of IR images and related processing methods are discussed in Section 3. In the last Section, some convective heat transfer measurements in complex fluid flows, performed with IR thermography by the author and his co-workers, are presented and reviewed. Such measurements span from natural convection to hypersonic regime.

## 2. HEAT FLUX SENSORS

The convective heat flux $Q_c$ (energy rate per unit area, W/m$^2$) between a fluid at temperature $T_r$ and a surface at temperature $T_w$ is governed by the *Newton law*:

$$Q_c = h(T_w - T_r) \tag{1}$$

$h$ being the *convective heat transfer coefficient* [W/(m$^2$K)] and $T_r$ a *reference temperature of the fluid*. The reference temperature depends on the stream experimental conditions. *E.g.*, for external low Mach number flows, the reference temperature practically coincides with the undisturbed stream one. Instead, for high Mach number flows [10], or for mixing of two streams at different temperatures, such as warm jets in ambient air, the correct choice is $T_r$ equal to the *adiabatic wall temperature* $T_{aw}$.

Present main interest is to measure either $Q_c$, or $h$, but it has to be noted that, as in the case of film cooling [11], IR thermography practically allows the measurement of $T_{aw}$ itself so as to determine surface adiabatic effectiveness maps.

Data are generally presented in terms of either the *Nusselt number* $Nu = hl/k_m$ (mainly for internal flows) or the *Stanton number* $St = h/(\rho c_p V)$ (external flows), where: $l$ is a characteristic length of the problem and $k_m$, $\rho$, $c_p$ and $V$ are, respectively thermal conductivity coefficient, specific heat per unit volume at constant pressure and characteristic velocity of the fluid medium.

Heat flux sensors generally consist of plane slabs, with a known thermal behaviour, where temperature has to be measured at fixed points. The energy conservation equation applied to the sensor thermal model yields the relationship which correlates measured temperatures to heat flux, or $h$. In practice, the slab can also be curved, its curvature in such a case being ignored only as long as the layer affected by the exchanged heat flux is small compared to the slab local curvature radiuses.

The sensor is supposed *ideal*, which means that its material thermo-physical properties are independent of temperature. The slab surface the flow is going over is called *front surface*, while the opposite one *back surface*. The simplest model of a sensor is the *one-dimensional* one, where the heat flux to be measured is supposed to be normal to the sensing element surface, *i.e.*, with negligible temperature gradient components parallel to the slab surfaces.

In this section, one-dimensional sensors are considered, their operation being extended to the multi-dimensional case in Section 3.

The heat flux sensors mainly used with IR thermography are:

1. *Heated-thin-foil* (HTF). This sensor consists of a *thermally thin* metallic foil, or a printed circuit board, steadily and uniformly heated mostly by Joule effect. Heating can be accomplished also in a different way, *e.g.*, by a known steady radiation input to foil. The heat flux is computed by measuring sensor surface temperature and performing an energy balance. In this sensor, due to the foil thermal thinness, either one of its surfaces can be generally viewed by IR camera. An extension to not thermally thin foils can be performed.

2. *Thin-film.* Classically, a very thin resistance thermometer (*film*) measures the surface temperature of a *thermally thick* slab to which it is bonded. The heat flux is inferred from the theory of heat conduction in a semi-infinite solid. The surface film must be very thin so as to have negligible heat capacity and thermal resistance compared to the layer affected by the exchanged heat flux. When using this sensor with IR thermography, the film does not exist and the heat exchanging surface must be necessarily viewed by camera. As such, fluids at least partially transparent in the used IR band must be investigated.

3. *Wall calorimeter*, or *thin-skin*. The slab is made *thermally thin* (so that temperature can be assumed constant across its thickness) and is used as a mere calorimeter. The heat flux is computed from the time rate of the slab temperature change, usually measured by a thermocouple. When using this sensor with IR thermography, the skin can be really thin, because there are no thermocouples, and either one of the slab surfaces can be generally viewed by the infrared camera.

Recently, heat flux sensors, based on a numerical solution of Fourier's law (typically described by an inverse heat transfer model) and surface temperature measurements, have been developed, *e.g.*, to take into account temperature dependence of sensor thermo-physical properties and/or to have slabs with high curvature but, for the sake of concision, they will not be herein described.

Thin-film and wall calorimeter sensors belong to the class of *transient techniques*, while HTF represents a *steady technique*. Transient techniques generally regard *passive heating* due to some existing temperature difference between sensor surface and flow, whereas HTF involves *active heating*.

When using IR thermography, transient techniques should be mainly applied to experimentally analyze the constant convective heat transfer coefficient boundary condition, while the heated-thin-foil steady technique is functional to the constant heat flux one. A constant temperature boundary condition can be implemented with transient techniques for large temperature differences between sensor and flow and relatively short measuring times.

## 2.1. Heated-Thin-Foil

A steady state way to measure convective heat fluxes which is very effective with IR thermography is the *heated-thin-foil* sensor. This sensor is mostly made of a thin metallic foil (*e.g.*, an AISI or constantan foil, typically tens of microns thick), or a printed circuit board (with copper tracks 5-35µm thick, mainly arranged in a Greek fret mode), which is steadily and uniformly heated by Joule effect (see sketches of Figs. **1** and **7**). Unless the foil is heated by radiation, the main limitation of this sensor is that, for practical reasons, it must have a cylindrical, or conical, geometry. Next, it is initially assumed an adiabatic back surface of HTF.

By making a simple (one-dimensional) steady state energy balance, it is found:

$$Q_j = Q_r + Q_c \tag{2}$$

where: $Q_j$ is the imposed constant Joule heating per unit sensor area, $Q_r$ is the ever present radiative heat flux to ambient and $Q_c$ is the convective heat flux to fluid.

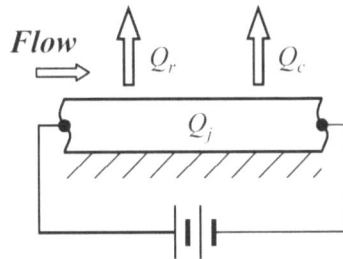

**Figure 1:** Heated-thin-foil sensor with adiabatic back surface.

Radiative heat flux is computed with what is some time called *radiosity law*:

$$Q_r = \sigma\varepsilon_t\left(T_w^4 - T_{amb}^4\right) \tag{3}$$

where: $\varepsilon_t$ is the front surface *total* hemispherical emissivity coefficient; $T_w$ and $T_{amb}$ are respectively the temperature of the sensor surface and of its ambient from Equations 1-3 it is possible to find an explicit expression for the searched $h$:

$$h = \frac{Q_j - \sigma\varepsilon_t\left(T_w^4 - T_{amb}^4\right)}{T_w - T_r} \tag{4}$$

Under the assumption that the *Biot* number, $Bi = h_t s/k$ (where $s$ and $k$ are thickness and thermal conductivity coefficient of the sensor, respectively, and $h_t$ includes convection and radiation) is small as compared to unity, temperature can be considered constant across the foil thickness. Therefore, foil surface to be measured may be either the back surface, or the front one. Tests with relatively high Biot number can be performed as well [12].

If also back surface is adiabatic (see Fig. **2**), Equation 4 must be extended by subtracting to $Q_j$ also the total heat flux from sensor to external ambient $Q_a$, *via* the back surface. Typically, this heat flux results to be the sum of radiative and natural convection heat fluxes. Radiative heat flux can again be computed by means of the radiosity law (3) while convective heat flux to external ambient may be evaluated by using standard correlations tables. However, to carefully evaluate $Q_a$, it is much better to perform *ad hoc* tests with thermally insulated front surface and which include the radiative contribution as well [13].

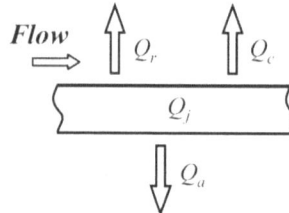

**Figure 2:** Sensor with adiabatic back surface.

## 2.2. Unsteady Techniques

Since it measures surface temperature $T_w$, IR camera output can be thought as coming from a two-dimensional array of thin-films. Yet, in transient techniques the surface measured temperature can be correlated to the heat flux by using either the semi-infinite wall thermal conduction model, (thin-film, Fig. **3**) or the wall calorimeter (thin-skin, Fig. **4**).

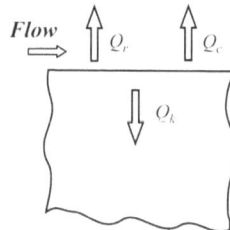

**Figure 3:** Thin-film sensor.

In the *thin-film* case, practically the heat flux sensor will be anyhow constituted by a slab of finite thickness *s*, hence the thin-film model is applicable only for somewhat small measuring times. On a quantitative basis (within less than one percent error), if $t_M$ is the measuring time, it has to be verified:

$$t_M < \frac{s^2}{4\alpha} \tag{5}$$

where $\alpha = k/\rho c$ is the slab material thermal diffusivity coefficient. Therefore, for this sensor the boundary condition on the back surface is irrelevant as long as inequality (5) holds.

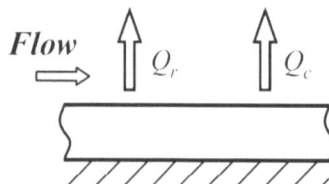

**Figure 4:** Thin-skin sensor with adiabatic back surface.

By assuming the sensor to be isothermal at initial time $t = 0$, $T_{wi}=T_w(0)$, for a variable input $Q_i$ heat flux (as shown in Fig. **3**, typically $Q_i = -Q_c - Q_r$), the classical formula to evaluate it from the measured surface temperature $T_w$ is [14]:

$$Q_i = \sqrt{\frac{\rho c k}{\pi}} \left[ \frac{\phi(t)}{\sqrt{t}} + \frac{1}{2} \int_0^t \frac{\phi(t) - \phi(\xi)}{(t-\xi)^{3/2}} \right] \tag{6}$$

where: $\phi = T_w(t) - T_{wi}$; $\rho c$ is the slab specific heat per unit volume and $k$ its thermal conductivity coefficient.

In the particular case of a step constant heat flux $Q_c$ and $Q_r = 0$, for $t>0$, the solution for the surface temperature $T_w$ as a function of time is:

$$\phi = T_w(t) - T_{wi} = -\sqrt{\frac{\alpha t}{\pi}} \left[ \frac{2Q_c}{k} \right] \tag{7}$$

Integral of Equation 6 can be numerically evaluated by using one of the algorithms accepted for aerospace applications [15]. However, such algorithms are generally sensitive to temperature measurement errors and one has to be very cautious when using them with noisy data and/or when initial time (so $T_{wi}$) is not precisely known.

Moreover, the approach based on Equation 6 needs high data sampling rate and this constraint is not often fully satisfied, in very fast transients, by an IR camera with acquisition frequency typically of the order of 100Hz.

A better approach is based on the suitable assumption that the direct problem yields a certain heat flux time variation law, where some free parameters are introduced. Then, such parameters are found so that predicted temperatures best agree with measured ones, *e.g.*, by the ordinary least squares criterion.

In the most common case of constant convective heat transfer coefficient $h$ and constant reference temperature $T_r$, the convective heat transfer rate varies linearly with wall over-temperature. Based on the above boundary condition, the solution of the wall temperature of a semi-infinite solid can be obtained by Laplace transforms as [14]:

$$T_w = T_{wi} + (T_r - T_{wi}) \left[ 1 - e^{\beta^2} erfc\beta \right] \tag{8}$$

with: $\beta = \frac{h\sqrt{t}}{\sqrt{\rho c k}}$ and $T_{wi} = T_w(t = 0)$.

In presence of radiative heat flux and under the assumption that convective and radiative contributions are uncoupled, Equation 8 can be extended to take into account the radiative correction as well [16]:

$$T_w = T_{wi} + (T_r - T_{wi}) \left[ 1 - e^{\beta^2} erfc\beta \right] - \frac{Q_r}{h} \tag{9}$$

The least squares method consists of finding $h$ and $T_{wi}$ values (which, as already said, may be not exactly known due to inaccuracies on either initial temperature and/or starting time) to minimize the function:

$$\sum_{j=1}^{n} (Y_j - T_{wj})^2$$

where: $Y_j$ is the $j$-term of the $n$ measured surface temperature values and $T_{wj}$ is the temperature predicted by Equation 9. Naturally, both temperatures must be evaluated at the same time and location.

Note that both Equations 8 and 9 are implicit (the term $Q_r$ depending on $T_w$) and valid whenever $T_{amb} = T_{wi}$.

In the case of the wall calorimeter (*thin-skin*), the sensor, practically a thin plate (Fig. **4**) of thickness $s$, is modelled as an ideal calorimeter (isothermal across its thickness) which is heated at the front surface and thermally insulated at the back one. The isothermal conditions is again accomplished as long as the Biot number $Bi = h_s s/k << 1$.

The unsteady one-dimensional energy balance gives:

$$Q_c + Q_r = -\rho c s \frac{dT_w}{dt} \tag{10}$$

Where $T_w$ is the sensor temperature. From Equations 1, 3, 10 and by recording the temperature evolution (to be measured by IR camera), it is possible to evaluate the convective heat transfer coefficient. Since temperature can be measured on either side of model, the back surface may be not insulated but, as it has been already discussed for the heated-thin-foil sensor, Equation 10 can be extended to include the total heat flux from sensor back surface to external ambient $Q_a$.

The solution of (10) for $Q_c = const$ and $Q_r = 0$ is straightforward and leads to a linear dependence of $T_w$ on $t$. Under the assumption that the convective heat transfer coefficient $h$ is constant and the convective and radiative contributions are uncoupled, Equation 10 can be solved in the form:

$$(T_w - T_r) = (T_{wi} - T_r)exp[-ht(\rho cs)] - \frac{Q_r}{h} \tag{11}$$

This relation can be used to implement a regression process to measure heat transfer coefficient $h$ as in the case of thin-film previously described.

### 2.3. Periodic Convective Heat Transfer

For both the steady and unsteady techniques, particular attention is necessary when the convective heat transfer coefficient, or heat flux, to be measured on front surface is variable with time, *e.g.*, as a simple harmonic function:

$$\begin{aligned} h &= |h| exp\,(i\omega t) \\ Q_c &= |Q| exp\,(i\omega t) \end{aligned} \tag{12}$$

where $\omega$ and $t$ are respectively the circular frequency and time.

For the HTF, characterized by a constant heat flux, the first of Equation 12 has to be considered and the developed model is still applicable whenever the condition $\omega << \alpha/s^2$ is verified.

Satisfying such condition (*i.e.*, by using very thin foils) makes the technique very attractive to measure wall heat transfer fluctuations due to turbulence [17].

In order to find the limits of operation for the simple thin-film and thin-skin sensor models, the asymptotic solution of temperature inside a slab of thickness $s$ with the second boundary condition (12) at front surface is analysed [14]. The coordinate $z$ has its origin at the front surface pointing towards the back one.

To give unitary solution to the problem, for the thin-film sensor, two boundary conditions at the back surface are considered: thermally insulated surface (*i.e.*, adiabatic,), or constant surface temperature (*i.e.*, in contact with a heat sink). Only the adiabatic boundary condition is examined for thin-skin, but solutions are presented for both front and back surfaces. $Q_r$ and $Q_a$ are neglected.

In both sensors, at a given point $z$, the *asymptotic* harmonic solution for $T$ has the form:

$$T(z,t) = [T(z)]exp\{i[\omega t + \kappa(z)]\} \tag{13}$$

where $\kappa(z)$ is the phase angle by which $T$ lags $Q$ and the temperature amplitudes $|T(z)|$ depend on the boundary conditions and on the dimensionless parameter $Fo_\omega = 2\alpha/\omega s^2$, which is a modified *Fourier Number*.

To identify the $Fo_\omega$ values that satisfy the previously developed models (semi-infinite body for thin-film, or wall calorimeter for thin-skin), temperature amplitudes are normalised with respect to the solution for $Fo_\omega \to \infty$, for the thin-skin, by obtaining:

$$T_s = \frac{k|T|\omega s}{|Q|\alpha} \tag{14}$$

or $Fo_\omega \to 0$, for the thin-film:

$$T_f = \frac{k|T|}{|Q|}\sqrt{\frac{\omega}{\alpha}} \tag{15}$$

*i.e.*, the effective harmonic solutions are compared to the previous ones for steady heat flux.

The amplitudes $T_s$ (14) at front and back surfaces of the slab and $T_f$ (15) at front surface for the two back surface boundary conditions are reported in Fig. **5** as a function of $Fo_\omega$.

Within an approximation of 2% one has:

- For the thin-skin sensor, the theoretical solution $T_s$ is valid for back surface whenever $Fo_\omega > 1$, while for front surface whenever $Fo_\omega > 4$, the difference being due to the shape of the temperature profile within the slab;

- For the thin-film sensor, the theoretical solution $T_f$ is valid whenever $Fo_\omega < 0.2$ for both back surface boundary conditions, this limit practically coinciding with what already indicated by Equation 5.

It is obviously found that, for $Fo_\omega \to 0$, the thin-skin model recovers a thin-film behaviour, while, for the thin-film with the adiabatic boundary condition, just the opposite occurs for $Fo_\omega \to \infty$.

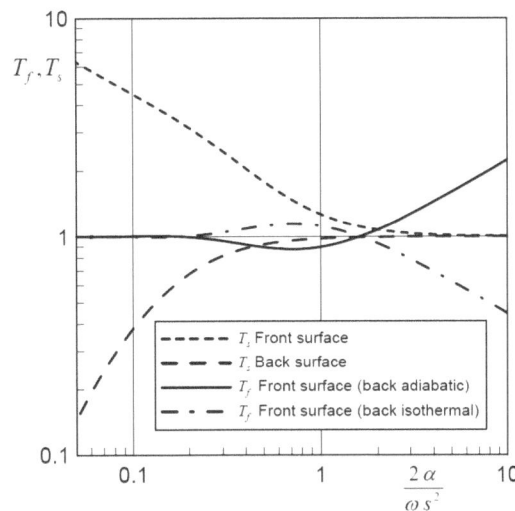

**Figure 5:** Validity of thin-film and of thin-skin sensor models for periodic boundary conditions.

## 3. THERMAL RESTORATION

The problem of restoring in a general sense the recorded thermal images is very important from both the imaging and the sensor points of view. *E.g.,* as shown by de Luca *et al.* [18, 19] and Simeonides *et al.* [20] in the study of Goertler vortices in hypersonic flow, the requirement of data restoration is stressed in heat transfer phenomena where relatively high temperature gradients are present. In fact, the possibility that accurate measurements of convective heat transfer coefficients may be performed by means of IR thermography depends on the chance that all potential error sources, linked to environment, thermal sensor and IR camera, can be removed. With regard to errors due to the environment, they can be automatically corrected by performing accurate calibration.

### 3.1. Errors Due to the Thermal Sensor

Errors due to the sensor are mainly associated with the radiation it releases towards ambient and to tangential conduction within the sensor itself. When standard transducers are used to measure wall temperature, it is possible to have a very low surface emissivity coefficient (*e.g.,* by gold plating the sensor surface) so as to neglect the radiative heat flux to ambient. Obviously, this cannot be the case when using an IR camera, since surface emissivity should be high.

Anyway, being decoupled from the conduction problem by means of radiosity law (3), the radiation problem has been already completely addressed in section 2. Thus, sensor errors to be still analyzed are only related to tangential conduction.

The assumption of zero-dimensional sensor is strictly satisfied only if the sensor temperature is constant along sensor surface (*tangential*) direction. However, with two-dimensional IR thermography, one is interested in studying complex thermo-fluid-dynamic phenomena where sensor surface temperature generally varies. Temperature variations inevitably cause conductive heat fluxes $Q_k$ in tangential direction which may constitute an important part of the exchanged heat (Fig. **6**).

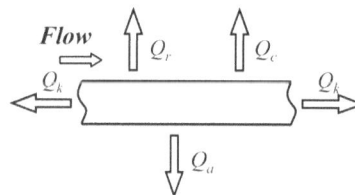

**Figure 6:** Sensor with tangential conduction.

For the *heated-thin-foil* sensor, by retaining the assumption that the sensor is thermally thin (*i.e.,* isothermal across its thickness $s$) and ideal, for isotropic foils, it is possible to evaluate the tangential conduction heat flux $Q_k$ per unit sensor *area* by means of Fourier law:

$$Q_k = -\lambda s \nabla^2 T_w \tag{16}$$

Therefore, to extend the heated-thin-foil sensor to the multi-dimensional case, one must include in the no longer one-dimensional energy balance this equivalent tangential conduction heat flux as well. Therefore, the final form of convective heat transfer coefficient becomes:

$$h = \frac{Q_j - \sigma\varepsilon(T_w^4 - T_{amb}^4) - Q_a - Q_k}{T_w - T_r} \tag{17}$$

It is essential to remark that use of IR thermography (intrinsically two-dimensional) allows, in theory, to directly evaluate the Laplacian of Equation 16 by numerical computation. However, since second derivatives are involved, this can be accomplished only after an adequate filtering of the camera signal, which is typically affected by noise [13].

As already mentioned, in many applications of the heated-thin-foil sensor, a spatially constant Joule heating can be obtained by using a printed circuit board [21, 22]). The printed circuit is generally manufactured by several adjacent thin (down to 5µm) copper tracks, closely spaced, arranged in a Greek fret mode (see Fig. 7) and bound to a fibreglass substrate. However, due to the high conductivity coefficient of pure copper, the board exhibits an anisotropic thermal conduction behaviour (along, or across, the tracks) so that it is not correct to evaluate tangential conduction by means of Equation 16.

By still retaining $T_w$ independent of coordinate $z$ normal to the board, it is therefore necessary to generalize Equation 16 to take into account this effect:

$$Q_k(x, y) = -\underline{\nabla} \cdot \left( s(x, y) \underline{\underline{\Lambda}}(x, y) \right) \tag{18}$$

where $\underline{\underline{\Lambda}}$ is the thermal conductivity tensor. To simplify Equation 18 and by neglecting the top and bottom edges of the board, it is feasible to roughly separate the effect due to copper tracks from that of fibreglass support. In particular, by choosing a Cartesian coordinate system with its axes directed along the two principal axes $x$ and $y$ of $\underline{\underline{\Lambda}}$ (see Fig. 7), one may split the effects in the two directions, normal to tracks $x$ and parallel to them $y$ [13]. As such, the total conduction heat flux can be expressed as sum of the two contributions, one along $x$, $Q_{kx}$, and the other along $y$, $Q_{ky}$:

$$Q_k = Q_{kx} + Q_{ky} \tag{19}$$

Bearing in mind the sketch of Fig. 7 and the thinness of the copper layer, it is simple to conclude that along the $y$ direction, the overall conductive heat flux is the sum of two mechanisms in parallel, one due to copper tracks and the other one to fibreglass support. So the bulk heat flux can be appraised by:

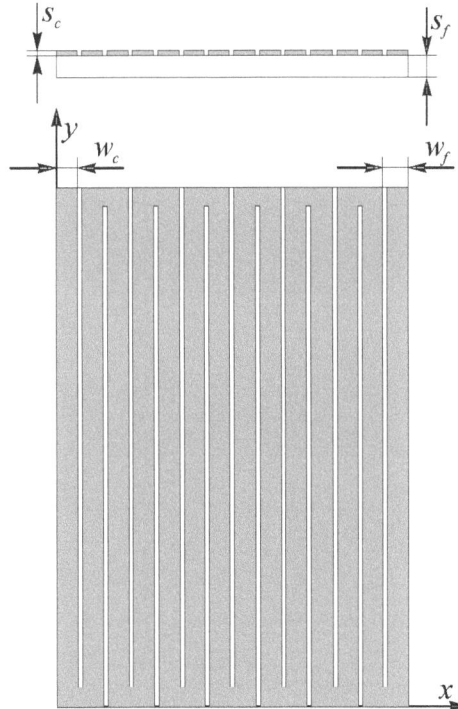

**Figure 7:** HTF made with a printed circuit board.

$$Q_{ky} = -\left( \frac{w_c(sk)_c + w_f(sk)_f}{w_f} \right) \frac{\partial^2 T_w}{\partial y^2} = -\left( \gamma^*(sk)_c + (sk)_f \right) \frac{\partial^2 T_w}{\partial y^2} = -(sk)_{ey} \frac{\partial^2 T_w}{\partial y^2} \tag{20}$$

where: $w$ indicates width; $s$ thickness; the suffixes $c$ and $f$ respectively refer to copper and to fibreglass. In Equation 20, it is introduced the *width parameter* $\gamma^*$ defined as:

$$\gamma^* = \frac{w_c}{w_f}$$

(21)

The quantity $(sk)_{ey}$ stands for the equivalent thermal conductance (per unit width) along the $y$ axis. The quantity $w_f$ represents also the Greek fret pitch.

The phenomenon is slightly more complex in the $x$ direction, normal to the copper tracks. In fact, in the copper gap only fibreglass allows conductive heat transfer while, in the track zone, both materials contribute to it. Therefore, in this case, the bulk conductive heat transfer can be estimated as due to both series and parallel processes:

$$Q_{ky} = -\left(\frac{1-\gamma^*}{(sk)_f} + \frac{\gamma^*}{(sk)_c+(sk)_f}\right)^{-1} \frac{\partial^2 T_w}{\partial x^2} = -(sk)_{ex}\frac{\partial^2 T_w}{\partial x^2}$$

(22)

where $(sk)_{ex}$ represents the equivalent conductance along the $x$ axis.

As expected, in the limits $\gamma^* \to 0$, or $\gamma^* \to 1$, Equations 20 and 22 reduce to the case of isotropic material. For the typical case of $(sk)_c/(sk)_f = 17$ ($s_c = 17.5\mu m$, $s_f = 1.7mm$). Fig. **8** shows the equivalent conductance $(sk)_e$, referred to that of fibreglass $(sk)_f$, in the direction of coppertracks and in that normal to it. For $0.25 < \gamma^* < 0.75$, the equivalent conductance in the direction orthogonal to the copper tracks, is less than one fourth of the other one. This event may be exploited to reduce tangential conduction whenever the preferred direction of the surface temperature gradient is *a priori* known. However, in order to have a uniform heating of the board, it is generally better to have $\gamma^*$ as high as possible. Values of $\gamma^* = 0.9$, with $w_f = 2mm$ and $s_c = 5\mu m$, can be easily achieved. Being the tracks gap practically constant, higher $w_f$ values allow for higher $\gamma^*$ values.

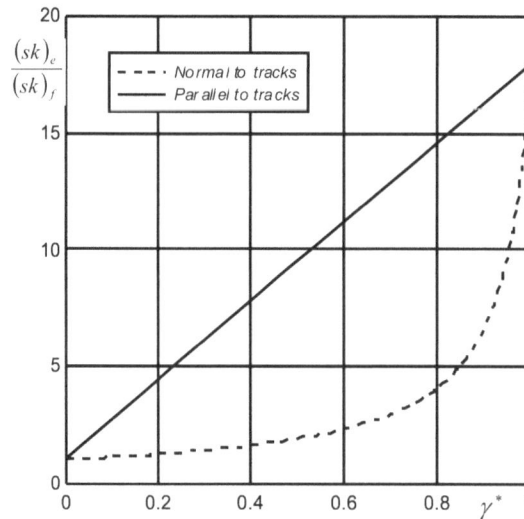

**Figure 8:** Equivalent conductance, referred to that of fibreglass for $(sk)_c/(sk)_f = 17$.

Also for both *thin-skin* and *thin-film* previously derived models, the temperature within the sensor is often supposed to be one-dimensional. Under the assumption that the sensor material is isotropic, or (as already done before) by choosing a Cartesian coordinate system with its axes directed as the two principal axes of the thermal conductivity tensor, it is possible to split conduction effects along the two tangential directions. Therefore, since the extension to any arbitrary convective heat flux is straightforward, for the sake of ease, in the following it is assumed that convective heat flux harmonically varies only along one direction parallel to the sensor surface $x$, that is:

$$Q_c(x) = Q_h\cos(\xi x) \tag{23}$$

where $Q_h$ is the heat flux amplitude and $\xi = 2\pi/L$ the wave number ($L$ being the wavelength).

The solution for the two sensors, with boundary condition (23) at the front surface and with adiabatic back surface for thin-skin, is given by de Luca *et al.* [19] in terms of difference between the surface temperature $T_w(x,t)$ at time $t$ and at the initial ($t = 0$) uniform sensor temperature $T_{wi}$: $\theta(x,t) = T_w(x,t)-T_{wi}$. For both sensors, the solution has the form:

$$\theta(x,t) = Bf(Fo_\xi)\cos(\xi x) \tag{24}$$

where $Fo_\xi=\xi^2/\alpha t$ is another modified Fourier Number.

If suffix $s$ denotes the *thin-skin* sensor and suffix $f$ the *thin-film* one, it is:

$$B_s = Q_h/(k\xi^2 s)f_s = 1 - \exp(-Fo_\xi) \tag{25}$$

$$B_f = Q_h/(k\xi^2 s)f_f = \mathrm{erf}(\sqrt{Fo_\xi}) \tag{26}$$

where: $k$ is the thermal conductivity of the sensor material and $s$ the *thin-skin* sensor thickness. Equation 24 states that, in both sensors, there is no phase difference between incident harmonic heat flux and surface temperature response. The maximum amplitudes, obtained for $Fo_\xi\rightarrow\infty$, are $B_s$ and $B_f$, respectively. For finite $Fo_\xi$ values, amplitudes are reduced by the *attenuation factors* $f_s$ and $f_f$, respectively.

To correct measured temperatures so as to take into account tangential conduction effects, it is convenient to evaluate the ratio between the temperature amplitude $B f(Fo_\xi)$ (as given by Equations 25 and 26) and that corresponding to the same value of $Q_h$ but in absence of tangential conduction (*i.e.*, that given by the one-dimensional solution). By defining this ratio as *Temperature Amplitude Transfer Function* ($F$), for the two models it results:

$$F_s \frac{1-\exp(-Fo_\xi)}{Fo_\xi} \tag{27}$$

$$F_f = \frac{\sqrt{\pi}}{2}\frac{\mathrm{erf}\sqrt{Fo_\xi}}{\sqrt{Fo_\xi}} \tag{28}$$

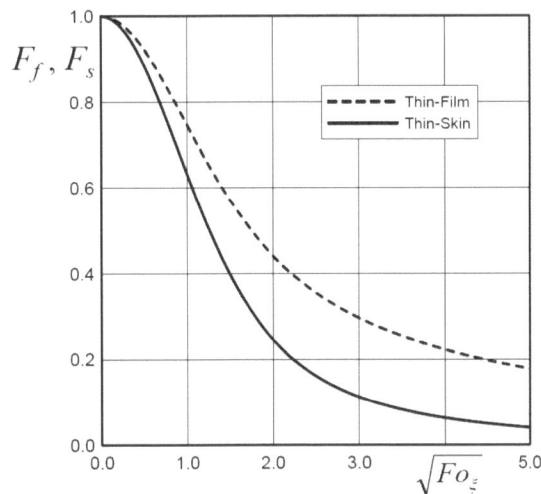

**Figure 9:** Temperature Amplitude Transfer Function for unsteady sensors.

The amplitude of each harmonic component of the measured temperature may be thus corrected and the corresponding harmonic component of the heat flux can be evaluated by using the classical one-dimensional formulae. The temperature amplitude transfer functions $F_f$ and $F_s$ are plotted as function of the Fourier number $Fo_\xi$ in Fig. **9**.

A direct comparison between the two sensors has practically no meaning because they are basically built with different thermal conductivities (high for thin-skin and low for thin-film) materials. Nonetheless, it has to be noted that for the thin-skin sensor, the tangential conduction correction can be directly computed as in the case of the heated-thin-foil with Equation 16, as a function of time.

### 3.2. Errors Due to the Camera

The relation between the original thermal object $o(x,y)$ and the recorded degraded image $r(x,y)$ is given by the two-dimensional *point spread function PSF(x,y)*, where $x$ and $y$ are the spatial coordinates. *PSF* describes the response of the measuring process to a *point source* (or *point object*). A general expression for $r(x,y)$ is:

$$r(x,y) = PSF(x,y) \otimes o(x,y) + n(x,y) \tag{29}$$

where $\otimes$ is the convolution operator, $n(x,y)$ indicating an additional noise term.

If it is assumed that the main subsystems of the whole measurement process, (thermal sensor and IR camera which includes optics, detector and electronics), are shift invariant, *PSF* may be expressed as:

$$PSF(x,y) = h_{TS}(x,y) \otimes h_{IR}(x,y) \tag{30}$$

where $h_{TS}(x,y)$ is the impulse response of thermal sensor and $h_{IR}(x,y)$, the one of IR camera. After Fourier transform, the convolution product of Equation 29, in the frequency domain $\zeta_x$, $\zeta_y$, reduces to an ordinary product:

$$R(\zeta_x,\zeta_y) = STF(\zeta_x,\zeta_y) \cdot O(\zeta_x,\zeta_y) + N(\zeta_x,\zeta_y) \tag{31}$$

where *STF* is the *System Transfer Function*, which expresses the overall degradation. In the presence of imaging degradations only, *STF* coincides with the *Optical Transfer Function,* which normalized magnitude (with unit value at zero frequency) is generally defined as the *Modulation Transfer Function* ($MTF_{IR}$) of the camera. A discussion on the $MTF_{IR}$ of IR scanning systems is reported by de Luca and Cardone [23] and of modern FPA systems by Boreman [24].

Instead, in the most general case, under the assumption that the different modulation processes are cascaded, the product of the various transfer functions yields the *STF*. E.g., for thin-film and thin-skin (see Section 3.1) the *STF* is equal to $MTF_{IR} \times F$. It has to be recalled that, for the heated-thin-foil and thin-skin sensors, the sensor modulation (heat flux losses due to tangential conduction and radiation) may be directly appraised as already indicated (see Section 3.1).

It is important to point out that $MTF_{IR}$ of an IR camera is a property of the entire system. In fact, all components (lens assembly, FPA, A/D converter, cabling, *etc.*) contribute to the final $MTF_{IR}$ of the system. Assuming that all subsystem are cascaded, the system $MTF_{IR}$ is given by: $MTF_{IR} = MTF_{Optics} \cdot MTF_{FPA}$

In Fig. **10**, the measured $MTF_{IR}$ of an IR camera, based on a 1024×1024 square pixels MWIR QWIP [25], is shown as a function of the spatial frequency $\xi$ normalized with respect to Nyquist frequency $\xi_N$ (1/2 pixel pitch). The $MTF_{Optics}$ of the spot scanner optics at $\xi_N$ is 0.2, thus the $MTF_{FPA}$ at Nyquist frequency should be around 30% and 45% along horizontal and vertical direction, respectively.

The $MTF_{IR}$ difference along horizontal and vertical directions is also present in the differently shaped *PSFs* (not shown herein). Since $MTF_{Optics}$ data do not show a large variation between horizontal and vertical directions, the difference is most probably due to the electronic readout integrated circuit $MTF_{Electronics}$.

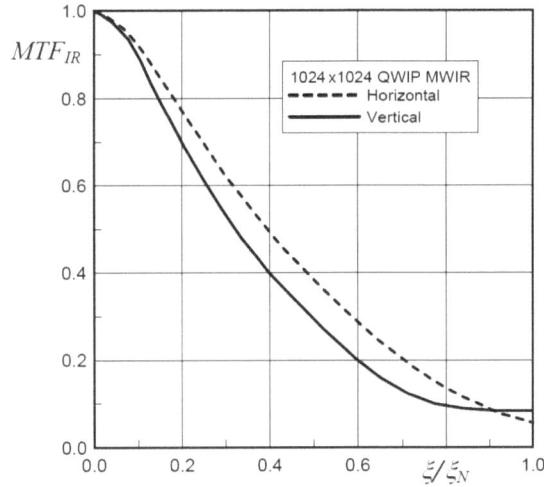

**Figure 10:** Measured $MTF_{IR}$ of an IR Camera. From Gunapala [25].

### 3.3. Image Restoration

From Equation 31, the degraded image can be restored with a restoration (inverse) filter $W(\xi_x, \xi_y)$, which is the reciprocal of *STF*, yielding the restored image $\tilde{o}(x, y)$:

$$\tilde{o}(x, y) = F^{-1}\left(W\left(\xi_x, \xi_y\right) R\left(\xi_x, \xi_y\right)\right) \tag{32}$$

$F^{-1}$ indicating the inverse Fourier transform. Restoration with the inverse filter $W(\xi_x, \xi_y)$ results in unlimited amplification of the response when modulation approaches zero. This undesired amplification can be prevented by stopping the restoration process just before the noise frequencies, or by using the Wiener filter $W_w(\xi_x, \xi_y)$ according to Jain (1989) [26]:

$$W_W\left(\xi_x, \xi_y\right) = \frac{STF^*\left(\xi_x, \xi_y\right)}{\left|STF^*\left(\xi_x, \xi_y\right)\right|^2 + \frac{\Phi_{noise}}{\Phi_{image}}} \tag{33}$$

where $STF^*(\xi_x, \xi_y)$ is the complex conjugate of the *STF* and $\Phi_{noise}$ and $\Phi_{noise}$ are the power spectral densities of noise and image, respectively. The additional ratio $\Phi_{noise}/\Phi_{image}$ (image signal to noise ratio) attenuates high noise components. The Wiener filter becomes the exact inverse filter for noise-free images.

An example of image restoration for the thin-skin sensor is presented by Bougeard [27] who studies the local heat transfer on a plate fin which had two-tube rows normal to it. Heat transfer measurements are corrected for radiation and tangential conduction effects. The fin is warmed up to $42°C$ and then cooled down by airflow with a fixed entrance temperature of $22°C$. The IR camera records temperature variations during a $6.5s$ duration (about one hundred thermal images). In order to reduce noise, Bougeard computes the integral (between initial and final time of test) of the terms of Equation 10 and evaluates the tangential conduction with Equation 16. By indicating with $e_{tot}$ the integral term corresponding to the total internal energy per unit area lost by the fin (J/m$^2$), $e_{cond}$ the integral of the conductive term, $e_{rad}$ the integral of the radiative term, and $i_{temp}$ the term corresponding to the integration of the difference of fin temperature and $T_{ref}$ (Ks), the local convection coefficient can be computed from the ratio:

$$h(x,y) = (e_{tot} - e_{cond} - e_{rad})/i_{temp} \tag{34}$$

Moreover a specific digital image restoration technique is developed to enhance the camera spatial resolution. The restoration technique uses a two-dimensional Wiener filter [28]. The $e_{tot}$ image (Fig. **11a**) shows the fin energy density lost during the experiment. High values are found at the fin leading edge because of the developing boundary layer and in front and around of the second tube row which are due to the horseshoe vortex presence.

The $e_{cond}$ image (Fig. **11b**) reveals high heat transfer by conduction at the leading edge and around the tubes because of high spatial temperature variations due to the vortex structures. On the fin remaining part, the conduction is quite small, isolated sharp variations being due to the presence of noise in infrared images. The $e_{rad}$ image (Fig. **11c**) shows that radiation heat transfer is pretty low (temperatures are relatively low), representing only a few percents of the total energy exchange which is predominant downstream of the tubes, where convection is very weak.

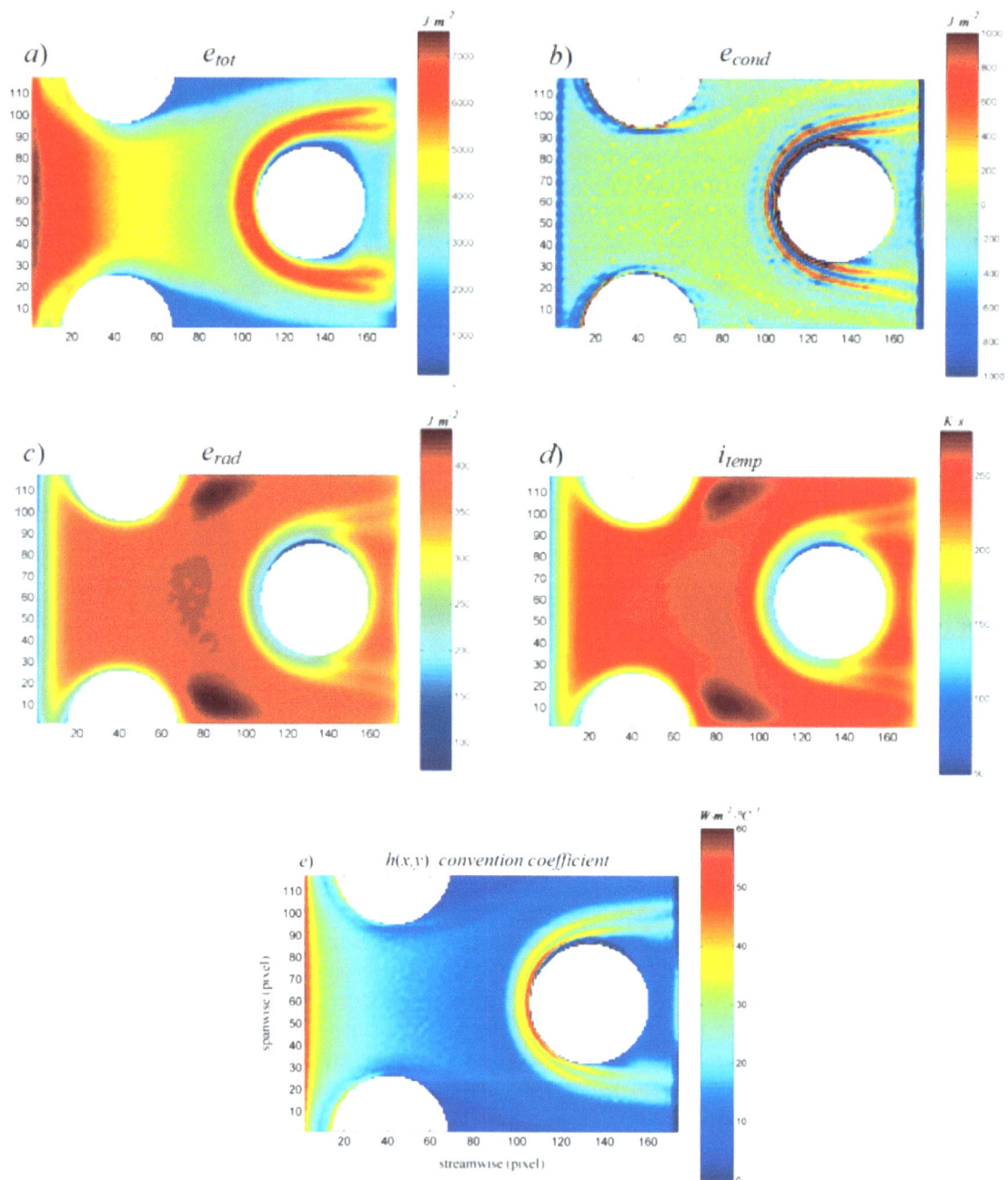

**Figure 11:** Plate fin and two-tube rows assembly (flow from left to right): *a)* total energy exchanged, *b)* energy exchanged by conduction, *c)* energy exchanged by radiation, *d)* temperature difference integration and *e)* computed convection coefficient distribution. From Bougeard [27].

The convection coefficient distribution (Fig. **11e**), computed according to (34), shows that: the first row of tubes has a very small horseshoe vortex effect; an increase of heat transfer is found at the fin leading edge; the staggering disposition of tubes decreases the wake size behind their first row.

## 4. APPLICATIONS

The applications of IR thermography to thermo-fluid-dynamics include much diversified topics which span from turbine cooling, including film cooling, to transition and separation, micro systems, rotating bodies, jets, flow instability, two-phase and hypersonic flows. In the following a few significant contributions, performed by the author and his co-workers regarding complex fluid flows, are presented and reviewed, also with the aim of pointing out a few relevant aspects.

### 4.1. Impinging Jets

Impinging jets have always received great attention from both industrial and academic points of view. Industrial interest is justified by the many applications such as to dissipate heat in electronic circuits, achieve high heat transfer coefficients at the leading edge of turbine blades, dry textiles, temper glass and heat zones which are critical for ice formation over aircraft. Research attention derives from the complex thermo-fluid-dynamics occurring in impinging jets.

Meola *et al.* [31], through measurements of adiabatic wall temperature $T_{aw}$ (not heated foil), observe instability developing for high $M$ values, such phenomena also depending strongly on the impingement distance. For $z/d$ <6, as $M$ increases, the vortex ring, which is located in the shear layer at $\approx 1.2d$ in the radial direction, strengthens up to its highest magnitude ($M \approx 0.7$) and breaks (*Widnall instability*) entailing entrainment of warmer ambient air and giving rise to the formation of structures.

Snapshots of the $T_{aw}$ map ($d = 5$mm, $z/d = 4$) for increasing Mach number is shown in Fig. **12**. Temperature colours do not change from Fig. **12c**) to Fig. **12f**). For $M = 0.3$ (Fig. **12a**), the minimum $T_{aw}$ at $1.2d$ has the shape of a completely developed and stable annulus. This annulus, which is the location of the vortex ring, for $M = 0.4$ is transformed into an unstable semicircle. In reality, the entire region outside the potential core is interested by unstable mixing phenomena. The vortex ring reinforces as Mach number increases and, for $M = 0.71$ (Fig. **12d**), breaks up in the impact with the plate, entraining warmer ambient air and giving rise to secondary minima at about $0.9d$ and $2.2d$,with a maximum between them at about $1.6d$. As $M$ further increases, such structures strengthen up and reach their highest magnitude for $M = 0.85$ (Fig. **12e**). To a further increase of $M$, the structures break up into numerous smaller structures, which tend to coalesce giving rise to a transient alternate circumferential movement.

The *recovery factor* is defined as $r_f = (T_{aw}\text{-}T_j)/(T_{oj}\text{-}T_j)$, where$T_{oj}$ and $T_j$ are the jet total and static temperatures, respectively. Carlomagno and de Luca [29] and Meola and Carlomagno [32] report tests performed with $T_{oj}$ equal to ambient temperature. In Fig. **13**, the relief map of the recovery factor $r_f$, relative to $d$=10mm, $z/d = 8$, $Re = 130,000$ and $M = 0.52$, is reported. The figure shows that the recovery factor has a peak value on the jet axis and a local minimum in the annular region at $r/d \approx 1$ ($r$ is the radial coordinate), which is attributed to the vortex ring in the shear layer surrounding the jet. A local $r_f$ maximum is located at $r/d \approx 1.8$ which is followed by another minimum at $r/d \approx 3.8$. For further $r/d$ increasing (at $r/d \approx 8$), the recovery factor eventually attains a unity value.

The relief map of the Nusselt number, for $d = 10$mm, $M = 0.11$, $Re = 28000$ and a much lower distance $z/d = 2$, is represented in Fig. **14** from Carlomagno and de Luca [29]. When the jet flow impinges on a wall relatively close to the nozzle exit, moving along the radial direction, two local minima and two local maxima values are encountered. The minima are located one at the jet axis and the other one at $r/d \approx 1.5$ (practically where also the minimum of $r_f$ is located). Instead, the inner maximum is situated at $r/d \approx 0.5$ and the outer one at $r/d \approx 2.5$. These behaviours confirm results documented in the literature.

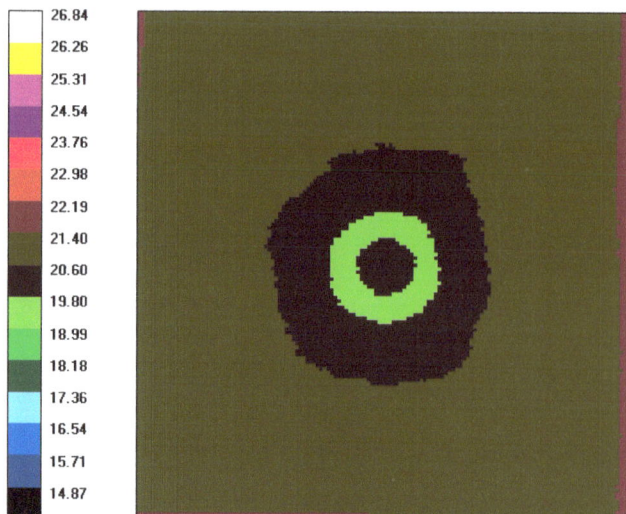

a)

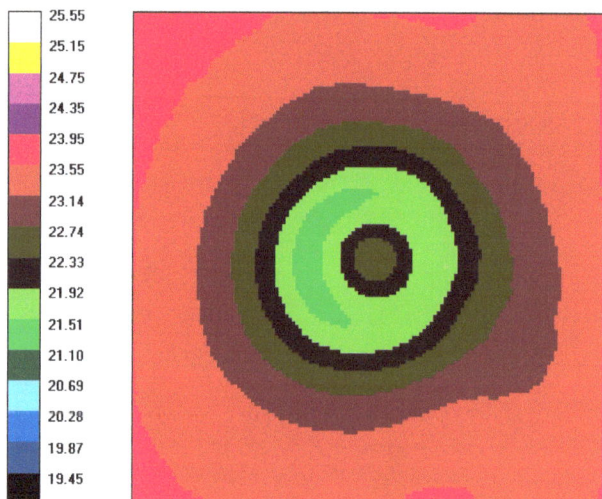

b)

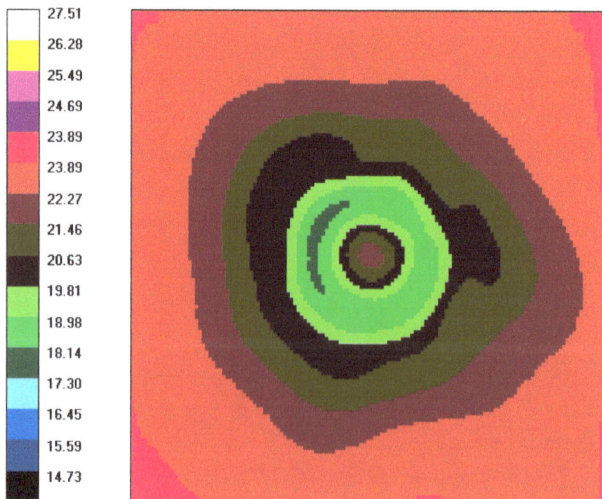

c)

*Fig. 12: cont….*

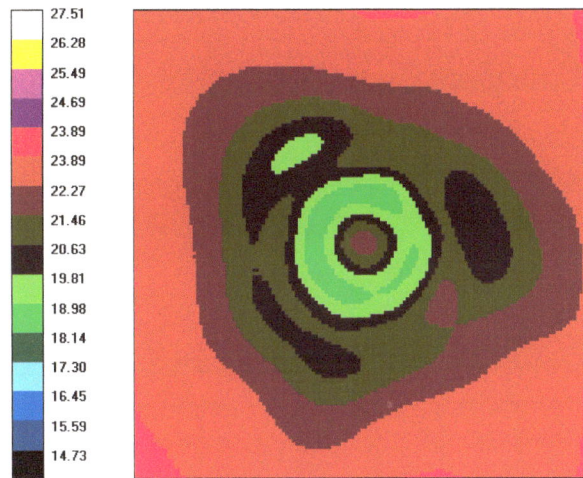

d)

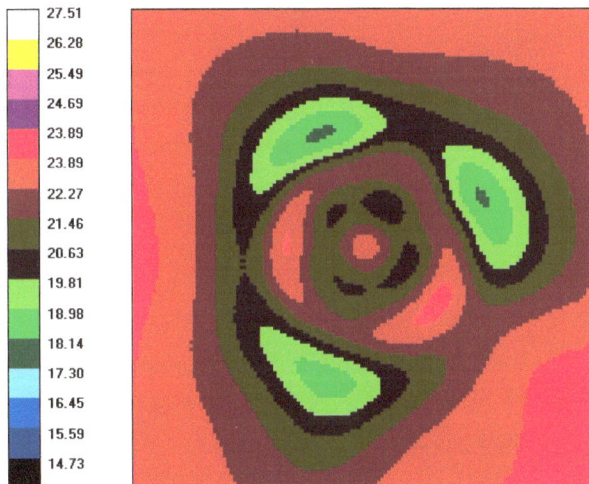

e)

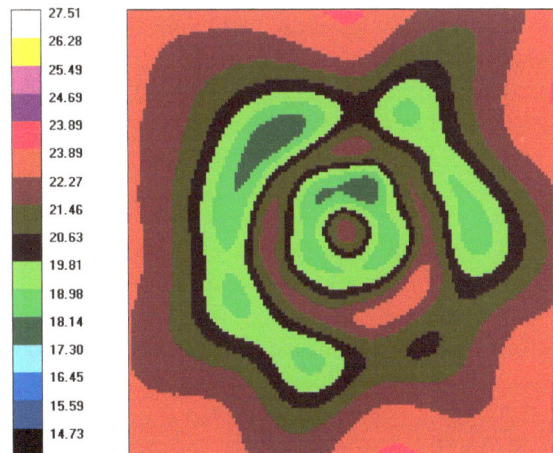

f)

**Figure 12:** Adiabatic wall temperature, taken with the Agema 880 LW, for $d$ = 5mm, $z/d$ = 4 and increasing Mach number; *a)* $M$ = 0.3; *b)* 0.4; *c)* 0.67; *d)* 0.71; *e)* 0.85; *f)* 0.91. From Meola *et al.* [31].

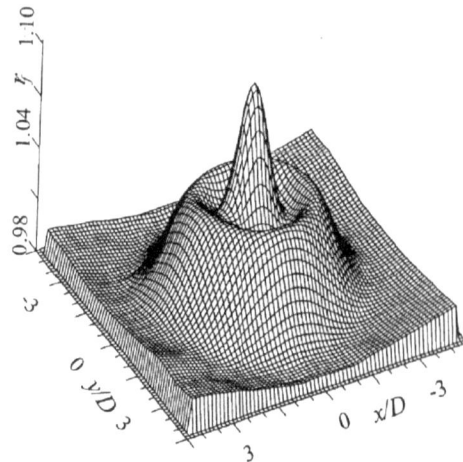

**Figure 13:** Recovery factor $r_f$ of a jet impinging on a flat plate for $d$=10mm; $z/d$ = 8; $M$ = 0.52; $Re$ = 130,000.

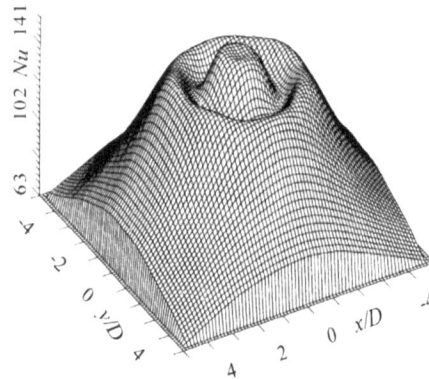

**Figure 14:** Nusselt number of a jet impinging on a flat plate for $d$=10mm; $z/d$ = 2; $M$ = 0.11; $Re$ = 28,000.

### 4.2. Transient Natural Convection

The heated-thin-foil technique for analyzing transient natural convection on a vertical plate is described by Carlomagno and de Luca [29]. A vertical stainless steel foil (245mm high, 960mm wide, 40μm thick), coated on one side with a thin layer of high emissivity ($\varepsilon$ =0.95) paint, is electrically step heated ($Q_j$ = 0-80-270W/m$^2$). By measuring apparent temperatures at both foil sides, the uncoated side emissivity is evaluated. In fact, in natural convection gas flow, the radiative heat flux is of the same order of magnitude (up to about one third) of the convective one and therefore, it must be accurately subtracted from the Joule heat input.

For a step Joule power input 0-130.5W/m$^2$, the time evolution of the foil temperature vertical profile is shown in the pseudo-thermogram of Fig. **15** where: abscissa is time (for a total of 80s), ordinate is position along the foil central vertical segment (0-245mm) and colours represent temperatures (21-42°C).

The thermogram left zone, characterized by a sequence of coloured vertical bands, points out an initial uniform time increasing of foil temperature which indicates that the foil behaves as a thin skin sensor with respect to Joule heating and a prevalent conductive and radiative heat transfer regime between foil and ambient air. Temperature evolution well agrees with theoretical calculation.

The temperature variation which later occurs in the vertical direction demonstrates the progressive onset of the natural convection flow. When steady state is reached, measured data agree within ±5% with the theoretical prediction of Sparrow and Gregg [30] for laminar flow.

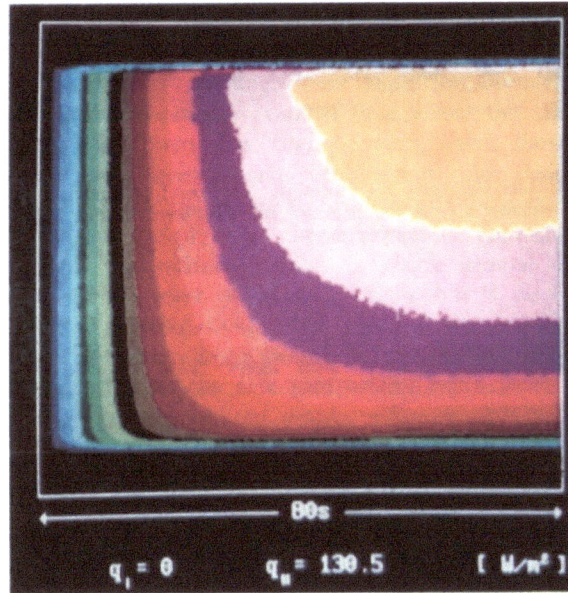

**Figure 15:** Time evolution of temperature profile on a vertical plate under transient natural convection. From Carlomagno and de Luca [29].

## 4.3. Airfoils Transition/Separation

The knowledge of the temperature distribution over an aerodynamic body is of great interest because of the contained information which can be related to heat transfer and skin friction at the wall through Reynolds analogy and to detect transition, separation and reattachment zones.

De Luca *et al.* [33] study the flow field over a Göttingen 797 airfoil (with a 180mm chord) at several angles of attack. The wing model is made of a glass-epoxy skin ($\approx$3mm thick) over polyurethane foam. In order to perform measurements by means of the heated-thin-foil sensor, the wing leeside is coated with a Joule heated thin (30$\mu$m) stainless-steel foil, 155mm wide, which starts 13mm away from the leading edge and is coated with a very thin layer of thermally black paint ($\varepsilon = 0.95$). Tests are carried out in the subsonic Pistolesi wind tunnel of Pisa University. The tunnel is a Göttingen type closed-return tunnel with circular open test section 1.1m in diameter. Results are reduced in non-dimensional form in terms of the Stanton number *St* defined in the conventional way.

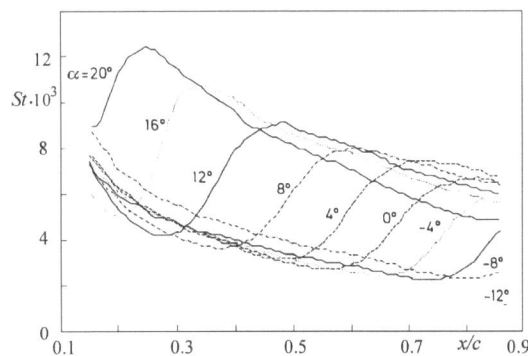

**Figure 16:** Stanton number chordwise distribution over leeside of the airfoil at several angles of attack$\alpha$. Flow from left to right. From de Luca *et al.* [33].

The chord wise *St* profile at leeside for chord based Reynolds number *Re* = 259,000 and different angles of attack $\alpha$ (in the range-12°-20°) is shown in Fig. **16**. The beginning of the plateau may be assumed to coincide with the point of laminar separation and its end with the beginning of transition in the separated shear layer. As $\alpha$ increases, the minimum *St* value moves upstream following the movement of the laminar separation point deriving from the adverse pressure gradient on the airfoil leeside. At high $\alpha$ values the plateau practically disappears from the measured zone. Data are slightly affected by tangential conduction in the model wall and by the modulation transfer function of the IR scanner that, as the authors say, has a limited spatial resolution. They were both not corrected at the time.

### 4.4. 180° Turn Channels and Ribs

To improve the performance of modern gas turbines, high temperatures are required. Often, to cool blade surface, air from the compressor goes through the hub section into the blade interior and, after flowing through serpentine passages, is discharged into the external gas. These passages are mostly made of several adjacent straight ducts, span wise aligned along the blade and connected by 180° turns which cause flow separation and reattachment.

Astarita and Cardone [13] obtain local and span wise averaged heat transfer measurements on a 180° sharp turn of a square channel using the SC3000 (Flir systems). Reynolds number *Re* based on hydraulic diameter *D* ranges from 16,000 to 60,000. Heat transfer is measured for two heating conditions (one or both channel sides) and presented in terms of the normalized Nusselt number *Nu/Nu\**, where *Nu\** is the Nusselt number predicted by Dittus and Böelter correlation for channel flows.

The normalized local Nusselt number distribution, for $D = 80$mm, Reynolds number $Re = 30,000$ and symmetric heating, is presented in Fig. **17**. Data is corrected for radiation, tangential conduction and natural convection at the sensor back side, which is the viewed surface. Flow enters the duct from the left channel (*inlet*) and exits from the right one (*outlet*).

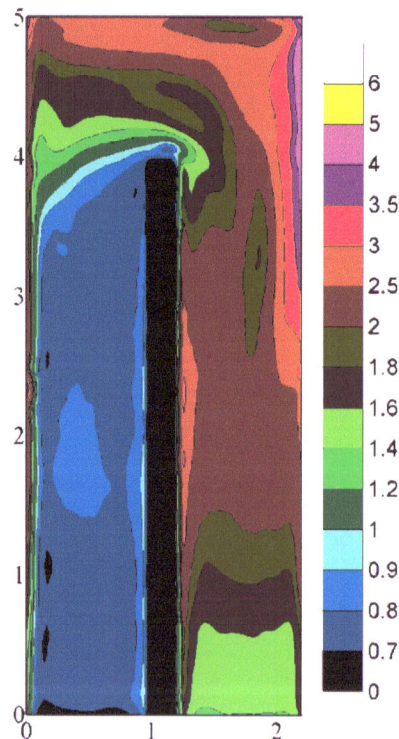

**Figure 17:** Nusselt number *Nu/Nu\** distribution for smooth channel: $Re = 30,000$, symmetric heating. Flow enters from left and exits to right.

At the inlet channel, the flow is fully developed both dynamically and thermally so the heat transfer coefficient is about that predicted by Dittus and Böelter correlation. The flow field in the turn region, and downstream of it, is highly three-dimensional and the normalised *Nu* maps show zones with low heat transfer coefficient, corresponding to recirculation bubbles, as well as zones with high heat transfer coefficient, which are due to flow reattachment and/or to an increase of the flow mean turbulent level. A first recirculation bubble is located just before the first outer corner and attached to the external wall. Two other regions with relatively low heat transfer coefficient are evident, the first nearby the partition wall tip and the second by the end wall by the outlet channel axis. In the map, it is also possible to notice three high heat transfer zones: the first placed near the end wall; the second located downward of the second outer corner and extending for about 3 diameters; the third placed about 2 diameters after the second inner corner and attached to the partition wall. They are due to the jet effect of the flow through the bend.

Rib turbulators attached to the blade heat exchanging walls enhance the convective heat transfer efficiency and completely modify the channel flow field as well as the local heat transfer coefficient distribution. With a single rib normal to the flow, the main stream first separates, generating a recirculation zone ahead of the rib, and then reattaches over the rib itself. A further separation, occurring just after the rib, creates a second recirculation zone which is followed by another reattachment at the duct wall.

For the channel described above with angled rib turbulators, Astarita *et al.* [13] and Astarita and Cardone [21] present detailed quantitative maps of the heat transfer distribution. The aluminium ribs have a square side (*e* = 8mm), are placed on the two opposite walls, at an angle between 30° and 60° with respect to the duct axis. Two rib pitches *P*, two rib arrangements (overlay or staggered), two heating conditions (from one or both channel sides) and Reynolds number *Re* varying from 16,000 to 60,000 are investigated. Data is corrected for radiation, tangential conduction and natural convection at the sensor back side.

The two-dimensional map of the local normalized Nusselt number *Nu/Nu\**, for the overlay rib arrangement, symmetric heating, rib angle 60° and *P/e*= 10 is shown in Fig. **18** for *Re* = 30,000. Ribs are clearly visible due to the higher heat transfer rate occurring over them. Indeed, the higher value of the normalized Nusselt number over the ribs is a consequence of their higher effective heat transfer surface (*i.e.*, fin effect) and of the flow reattachment there. The low *Nu* value spots on the last rib of the inlet duct and on the first rib of the outlet duct, near the partition wall, are due to a partial detachment of the ribs glued to the wall, owing to high local thermal stresses, difficult to avoid. Present author believes this has a little influence on the local nearby measurements, but it has to be highlighted that it is easily revealed by IR thermography.

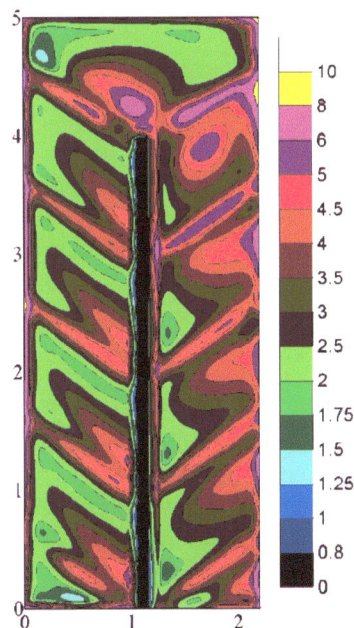

**Figure 18:** Nusselt number *Nu/Nu\** distribution for a channel with overlay ribs: *Re* = 30,000, symmetric heating, rib angle 60°, *P/e* = 10. Flow enters from left and exits to right. From Carlomagno and Cardone [53].

Before the turn, the inlet duct thermal pattern appears to be recurring (a kind of thermally fully developed flow) except for some edge effects at the duct entrance and exit, linked to the quite strong tangential conduction due to dissimilar nearby zones, there. The map also shows that the presence of the sharp turn already induces a slight modification of the shape of the contour lines before the inlet duct last rib.

The rib angle causes the formation of secondary flows under the form of two counter-rotating vortices in the channel cross section. In the inlet duct, the main flow nearby both bottom and top walls, entrapped by the ribs, is accelerated towards the side wall. The secondary flows after licking this wall merge and go back, *via* the duct centre plane, to the partition wall so as to practically generate a jet which impinges onto the latter wall. This explains the rib wise *Nu* asymmetry. In fact, the jet presence tends to increase the convective heat transfer coefficient near the partition wall with respect to that near the side wall. The map shows also that, towards the channel exit, the secondary flows which are reversed enhance the convective heat transfer coefficient near the side wall with respect to that near the partition wall.

In the inlet duct, the reattachment downstream of the ribs can be identified as the locus of the normalized Nusselt number local maxima when moving in streamwise direction. The reattachment distance increase going towards the side wall is most likely due to the interaction of the main flow with the secondary one. The separation zone after each rib is strongly influenced by the secondary flow impingement towards the partition wall. Instead, the separation zone ahead of each rib remains well visible and it is not much influenced by the secondary flow, its width remaining practically constant along the ribs.

Blades rotation strongly changes the channel fluid dynamic behaviour because there is an interaction between the following forces: Coriolis forces, pressure gradients causing main flow, centrifugal forces. For a *radially outward flow*, the Coriolis force produces a secondary flow (*i.e.*, a symmetric pair of secondary vortices) in the plane normal to main flow direction which pushes the particles in the channel centre towards the *trailing* (that follows) *wall*, then the flow continues along the side walls and finally gets to the *leading* (that goes ahead) *wall*. When the flow is reversed, *i.e.*, for a *radially inward flow*, one has to change the role played by the leading wall with that of the trailing one and *vice versa*.

One of the first attempts to measureconvective heat transfer coefficients *h* in a rotating air channel with IR thermography is performed by Cardone *et al.* [34]. The apparatus concept is a direct consequence of the used heated-thin-foil sensor. Since the foil back surface (to be viewed by the IR camera) cannot be thermally insulated, to avoid there high thermal losses by forced convection, the channel rotates in a vacuum tank (below 100Pa).

Also Gallo *et al.* [12] measure detailed *Nu/Nu\** maps nearby a 180° sharp turn of a rotating channel with the heated-thin-foil sensor and using the SC3000 (Flir systems). The experimental apparatus (Fig. **19**) consists of a Plexiglas two-pass water channel with a sharp 180° turn, mounted on a revolving platform whose rotational speed can be continuously varied in the range 0-60rpm. The channel has a square cross section 60mm on a side, its length of 1200mm ahead of the 180° turn ensuring a hydro-dynamically fully developed flow before the turn. The central partition wall, dividing two adjacent ducts, is 12mm thick. Water from a tank is pumped through an orifice meter, a rotating hydraulic coupling and, after flowing in the test channel, is discharged back into the tank.

The mass flow rate can be varied with a by-pass circuit and the tank water temperature is kept constant with a heat exchanger. A magnetic pick-up allows synchronizing IR image acquisition. The classical heated-thin-foil reducing technique is modified because the presence of water makes the Biot number not very low.

The apparatus is capable to simulate both Reynolds numbers and Rotation numbers $Ro = \varpi D/V$ (where $\varpi$ is the rotational speed of the channel) values typical of turbine blades.

In Fig. **20**, the segment-by-segment average *Nu/Nu\** profiles at the leading and the trailing channel walls, for *Re* = 20,000 and Rotation number *Ro* in the range 0.0-0.3, are reported. The length of a segment is half the channel side. In the inlet duct, the flow results fully developed for both walls being *Nu/Nu\** practically constant in the first four segments. For the leading wall (Fig. **20a**), these values are lower than those relative to the static case and decrease for *Ro* increasing. At the fifth segment, *i.e.,* about one hydraulic diameter before the turn, the Nusselt number starts to increase and reaches a maximum value in segment seven. In fact, in the turn zone, the radial velocity component suddenly decreases and changes sign with consequent decrease and inversion of the Coriolis force. This inversion causes flow separation by the trailing side and an abrupt reattachment toward the leading side, with strong increase of *Nu/Nu\**. For all rotation numbers, profiles in the turn region and in the outlet duct, exhibit analogous trends with values of the normalised Nusselt number increasing for increasing the Rotation number.

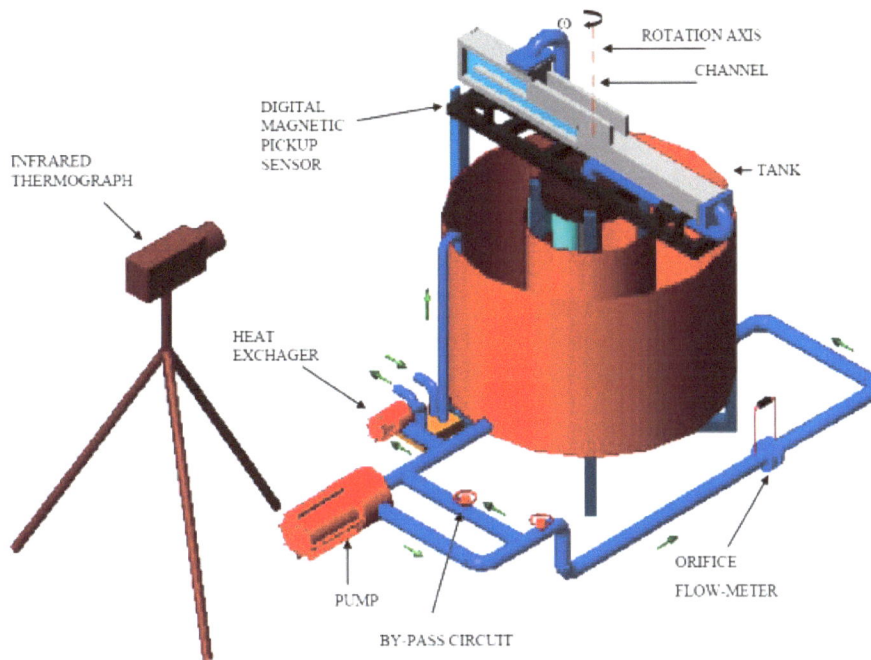

**Figure 19:** Apparatus for rotating channel heat transfer measurements. From Gallo *et al.* [12].

At the trailing wall (Fig. **20b**), the *Nu/Nu\** values are higher than those for the static case and slightly increase as *Ro* increases. In the first corner and in the first half of the second corner (segment 7, 8 and 9), the profiles relative to all Rotation numbers present the same trend of the curve pertinent to the static case but with higher *Nu/Nu\** values. For rotation number increasing, in the outlet duct, it is possible to note that the maximum of the convective heat transfer coefficient (located downstream of the leading wall case) increases and tends to move upstream, *i.e.*, towards the tip of the partition wall.

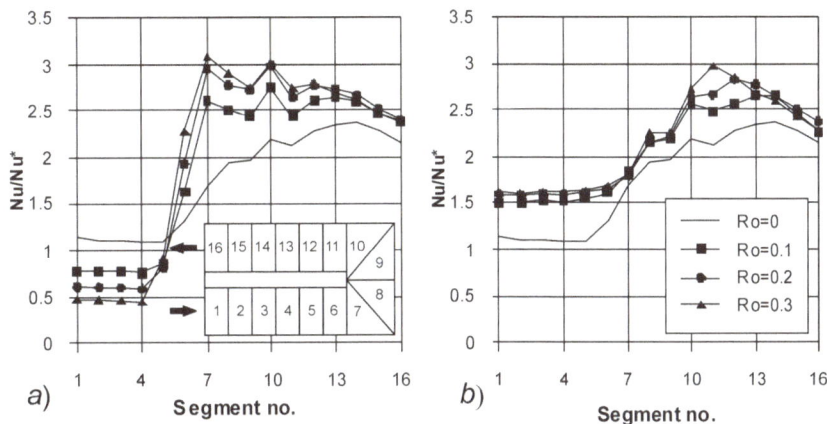

**Figure 20:** Normalized Nusselt number profiles: *a*) leading wall, b) trailing wall. From Gallo *et al.* [12].

## 4.5. Rotating Surfaces

Rotating devices are quite relevant in several applications and the rotating disk represents the simplest configuration. As a matter of fact, flywheels, turbine disks to which blades are attached, disk brakes and even modern high speed CD-ROMs are all examples of practical application of such a model. Often, the fluid resistance due to rotation is irrelevant but there are a number of cases where the disk thermal behaviour may be important. A rotating disk gives rise to a stream that impinges on it which, freely speaking, could be considered as a jet flow.

Astarita and Cardone [35] perform convective heat transfer measurements on a heated rotating disk, with a relatively small centred jet impinging on it using the Agema 900. The disk section consists of a printed circuit board glued on a 20mm thick polystyrene foam slab contained in a 450mm diameter steel cup, which rotates in the range $\omega$ = 100-4400rpm. An optical transducer precisely monitors the angular speed fluctuations which are below 1%. Air, passing through a heat exchanger and a rotameter, goes through a nozzle to produce the jet. The heat exchanger ensures a jet bulk temperature practically equal ($\pm 0.1°C$) to the temperature of the ambient air the jet mixes with. Three nozzles with exit diameter $d$ from 4 to 8mm are in turn used during the tests, the nozzle exit to disk distance varying between $3d$ and $75d$. The jet always impinges perpendicularly at the centre of the disk.

This topic allows to report two cases where the intrinsic two-dimensionality of IR camera is reduced to a single dimension only, during tests performed in the above described apparatus. The first one is to detect the spiral vortices, *attached* to the surface, which occur in the transitional regime of a simply rotating disk (without jet) and cause $h$ azimuthal variations. Since for the tested conditions, during a frame acquisition time, the disk would rotate about two turns producing a blurred image, the vortices attached to the surface are detected by the line-scan option of the used Agema 900 IR camera [36].

While the disk is rotating at thermal steady state, the temperature profile along one radius is acquired as a function of time. Then, the temperature map of the disk surface is reconstructed by taking into account its rotation, *i.e.,* by azimuthally displacing the acquired radial profile sequence so as to generate the thermal map attached to the disk surface.

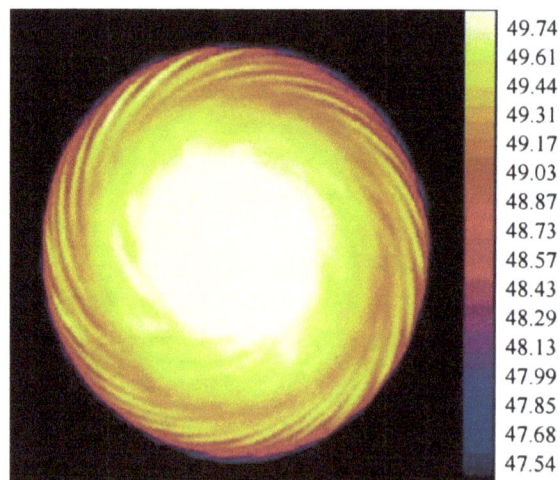

| 49.74 |
| 49.61 |
| 49.44 |
| 49.31 |
| 49.17 |
| 49.03 |
| 48.87 |
| 48.73 |
| 48.57 |
| 48.43 |
| 48.29 |
| 48.13 |
| 47.99 |
| 47.85 |
| 47.68 |
| 47.54 |

**Figure 21:** Reconstruction of spiral vortices on a rotating disk. From Astarita *et al.* [36].

Fig. **21** shows this map, obtained by azimuthally displacing about 15,000 profiles, where the footprint of the vortices is clearly visible. The disk diameter is equal to the side of the externally black square since the temperature range to detect the vortices is very low.

Before treating the disk with the jet, it is important to address the relevance of the adiabatic wall temperature $T_{aw}$ in high speed flows; this is illustrated with thermograms obtained in the normal way (full frame acquisition) in the same apparatus. The upper half of Fig. **22** [37] shows the thermogram of the disk rotating at 4390rpm with a Joule heat flux $Q_j = 871\text{W/m}^2$. A small region ($\approx$16% of the disk surface) at the disk centre, with a constant wall temperature $T_w$, and where the flow is laminar, is clearly evident. In fact, since the rotating disk laminar solution of Millsaps and Pohlhausen [38] predicts that the $h$ coefficient has to be constant over the surface, so, for a constant heat flux boundary condition and constant $T_{aw}$ distribution, the wall temperature $T_w$ must be constant. Actually, in the low subsonic regime, $T_{aw}$ coincides with fluid temperature and near disk centre speed is low.

Outside the laminar zone, the temperature $T_w$ decreases, first rapidly in the transitional regime and then, more slowly in the turbulent one. Immediately after (besides some edge effects at the disk periphery) the temperature trend is reversed as $T_w$ begins to slowly rise while, according to the turbulent correlation law [39], the convective heat transfer coefficient should increase ($T_w$ decrease) for increasing local radius. This behaviour is explained by examining the half bottom of Fig. **22**, which shows the adiabatic wall temperature $T_{aw}$ map (simply obtained with $Q_j = 0$) at the same disk angular speed.

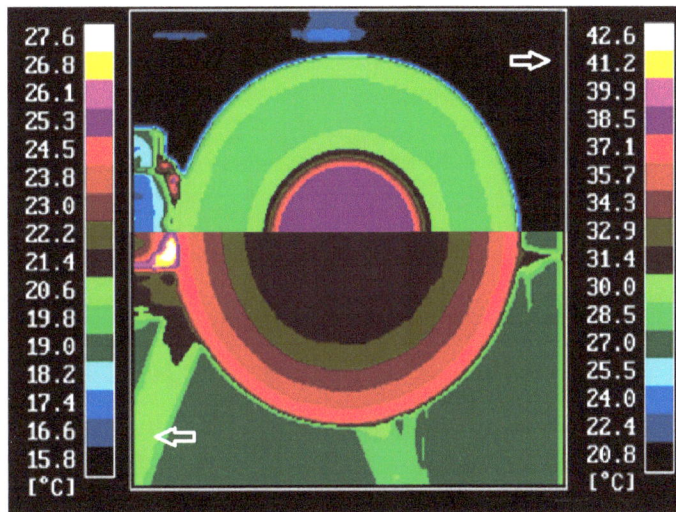

**Figure 22:** 450mm disk rotating at 4390rpm. Upper half $T_w$ (right scale): $Q_j = 871\text{W/m}^2$. Lower half $T_{aw}$ (left scale): $Q_j = 0\text{W/m}^2$. From Carlomagno and Cardone [53].

In the map, $T_{aw}$ is practically constant (and equal to $T_{amb}$) only within the circle of diameter about 60% of the disk diameter; afterwards $T_{aw}$ shows a significant increase (about 4°C over $T_{amb}$ near the disk edge). Since for the tested conditions, $T_w$-$T_{aw}$ is of the same order of magnitude of $T_{aw}$-$T_{amb}$, the $T_w$ behaviour is readily explained.

In order to double the camera spatial resolution, the azimuthal reconstruction described before is again used by Astarita and Cardone [35] in the tests of a rotating disk with the centred jet, where the acquired thermal profiles are averaged in time. For each test, about 16,000 lines are used for both map reconstruction and profile average.

In order to reduce the number of governing parameters in absence of theoretical analysis, it is essential to find a rational way for accounting the relative importance of the jet influence with respect to the disk rotation. On the assumption that heat transfer coefficient depends on the momentum flow rate, a reasonable dimensionless parameter is the ratio between the two momentum rates (one due to jet, the other one to disk rotation).

By assuming a jet width proportional to the nozzle-to-disk distance and laminar flow over the disk, such a parameter can be put in the form:

$$\chi = \left(\frac{dV}{v}\right)(\omega z^2)^{\frac{3}{4}} \tag{35}$$

where $v$ is the fluid kinematic viscosity coefficient, $z$ is nozzle-to-disk distance (that should be high as compared to $d$ so to have jet width proportional to $z$), other quantities being already defined.The convective heat transfer coefficient at the jet stagnation point $h_o$ is examined.

By considering only disk rotation, the flow is always laminar at the disk centre and in the case of a jet impinging on it, the $h_o$ departure from the single rotation value should be a function of $\chi$ only.

Therefore, the dimensionless quantity $h_0/k_m\sqrt{v/\omega}$ is plotted in Fig. **23** as a function of $\chi$, where 160 tests, performed by randomly varying disk angular speed, nozzle diameter, jet initial flow rate and nozzle-to-disk distance, are reported. Since the similitude parameter holds only for high values of the $z/D$ ratio, even if randomly chosen, $z$ always fulfils the condition $z/D > 14$. A first attempt is made to correlate all measured data by an equation of the type:

$$\frac{h_0}{k_m}\sqrt{\frac{v}{\omega}} = 0 - 33 + b\chi^c \tag{36}$$

where the constants $b$ and $c$ turn out to be, $b = 1.342$ and $c = 0.688$ and the constant 0.33 derives from the laminar solution for the simply rotating disk (no jet, $\chi = 0$) of Millsaps and Pohlhausen [38] for $Pr = 0.71$.

However, correlation (36), which is indicated by the broken line in Fig. **23**, deviates from the experimental data especially at low $\chi$ values. In fact, the experimental data shows a different behaviour for small and large values of $\chi$. Really, for $\chi > 1$, data appears to be well correlated, in the log-log plane, by a straight line, while the same is not true for smaller $\chi$ values, where a linear regression shows a more satisfactory behaviour.

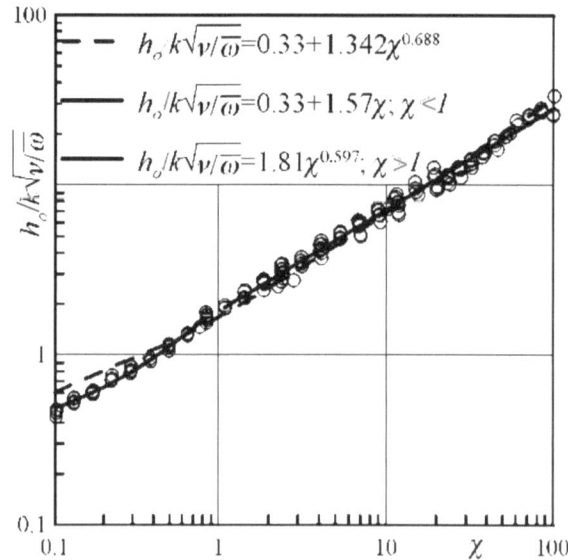

**Figure 23:** Convective heat transfer coefficient at the jet flow stagnation point.

Thus, the equations of the two correlation curves, also shown in Fig. **23** with continuous lines, are:

$$\frac{h_0}{k}\sqrt{\frac{v}{\omega}} = 0.33 + 1.57\chi \text{ for } \chi < 1 (\rho^2 = 0.988) \tag{37}$$

$$\frac{h_0}{k}\sqrt{\frac{v}{\omega}}\ 1.81\chi^{0.597} \text{ for } \chi > 1 (\rho^2 = 0.989) \tag{38}$$

The two correlations (37) and (38) are respectively obtained by using 42 and 122 data points.

### 4.6. Hypersonic Flows

The shock-wave/boundary-layer interaction (SWBLI) is a widely studied problem of hypersonic flows, its importance stemming from applied problems, such as efficient design of control surfaces, high-speed air inlets or thermal protection systems of re-entry vehicles.

De Luca *et al.* [16] analyze SWBLI in a two-dimensional hypersonic wedge flow over a flat plate/ramp configuration. Flow is only geometrically two dimensional because, due to formation of Goertler type vortices, some spanwise heat flux periodic variations over the ramp at reattaching region are observed. Measurements are performed on a flat plate followed by a ramp in a blow down wind tunnel at Mach number $M = 7.14$ and unit Reynolds number $Re_u = 8.6 \times 10^6$/m. The influence of leading-edge shape (bluntness and geometry), flat plate length, and ramp angle on heat transfer at reattachment and on periodic heat transfer variations are analyzed.

The mean Stanton number $St$ profiles, measured over a flat plate (90mmx90mm) followed by a flap (with angles $\alpha$ of $10°$, $15°$, and $20°$) are shown in Fig. **24**. The increase of the flap angle moves the location of the (laminar) separation upstream. The Stanton number over the ramp attains its highest values for the highest ramp angle of $20°$, where the very clear $St$ peak denotes the presence of turbulent reattachment.

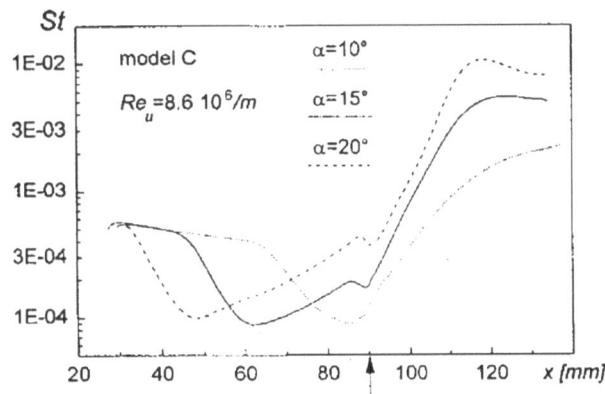

**Figure 24:** Streamwise $St$ profiles for a flat plate-ramp configuration and three ramp angles $\alpha$: $M = 7.14$; $Re_u = 8.6 \times 10^6$/m; arrow indicates hinge line position. From de Luca *et al.* [16].

It has to be stressed that Stanton number distributions over the ramp of Fig. **24** refer to spanwise averaged values since, in all of the tested flow conditions, the heat transfer coefficient exhibits a spanwise periodic variation to be ascribed to Goertler type vortices developing in the reattaching flow region. The onset of such three-dimensional variations strongly depends on leading-edge shape and/or its non uniformities.

Cardone [40] tests SWBLI in the High-Enthalpy Arc-heated Tunnel (HEAT) of Centrospazio which produces $M = 6$ flows with a specific total enthalpy up to 2.5MJ/kg, low to medium unit Reynolds number range ($10^4 \div 10^6$/m) and a nozzle exit (to the test chamber) diameter of 60mm. The tunnel operates in a pulsed, quasi-steady mode, with running time ranging up to 200ms. The model consists of a flat plate followed by a compression ramp with spanwise dimension of 100mm. The hinge line is positioned at 50mm from the leading edge and the ramp angle is $15°$ and MACOR® is chosen as model material for its relatively low thermal conductivity, as required for using the thin-film sensor.

A thermal map recorded about 80ms after starting of the wind tunnel is reported in Fig. **25** (flow direction is from left to right). The temperature distribution is almost two-dimensional only near the model leading edge. Moving downwind, the continuous decrease of wall temperature shows the boundary layer development. By the hinge line, is clearly visible a region where the temperature attains a minimum which is due to the presence of a flow separation region. Continuing to move along the symmetry axis, after the hinge line, the temperature reaches a maximum due to flow reattachment on the ramp. Fig. **25** is mainly reported to show how IR thermography allows for easy detection of non uniformities in the flow, which, in the present case, are due to the expansion fan starting from the lips of the nozzle exiting in the low pressure test chamber.

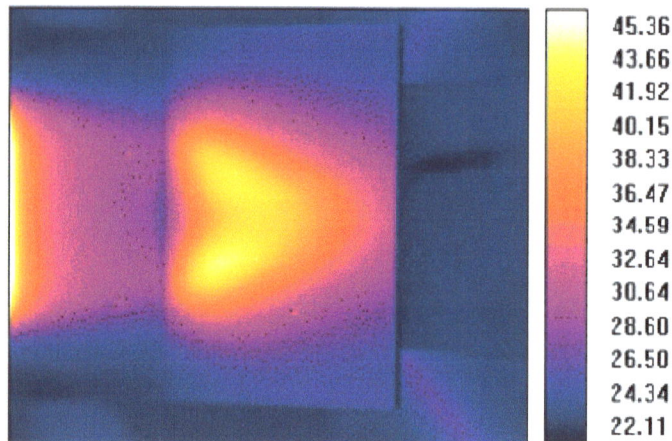

**Figure 25:** Thermogram of a flat plate/ramp in hypersonic flow: $M = 6$. Flow from left to right. From Cardone (2007) [40].

## 5. CONCLUSIONS

The capability of infrared thermography for convective heat transfer measurements was thoroughly analyzed with regard to its various relevant aspects. Since convective heat fluxes measurements must be performed with sensors where proper temperatures have to be measured, an IR camera appears advantageous from several points of view compared to standard transducers. In fact, the IR camera being a fully two-dimensional transducer (up to 1M pixels), allows for accurate measurements of surface temperature maps even in the presence of relatively high spatial gradients and, therefore, it provides also an easier evaluation of errors due to radiation and tangential conduction. Besides, thermography is non-intrusive (*e.g.,* allowing to get rid of conduction through thermocouple or RTD wires), it has high sensitivity (down to 20mK) and low response time (down to 20μs), which makes it attractive also for rather fast transients. As such, IR thermography can be effectively employed to measure convective heat fluxes with either steady or unsteady techniques [41-53].

In this chapter, after a short introduction recalling the main historical steps of IR thermography in thermo-fluid-dynamics, the most commonly used steady and unsteady heat flux sensor models were described and discussed, by addressing in particular the problem of thermal restoration. Finally, some applications to complex fluid flows, ranging from natural convection to hypersonic regime, performed by the author and his co-workers were presented and reviewed. In conclusion, IR thermography represents a very powerful optical tool that can be beneficially exploited in complex fluid flows to measure wall convective heat fluxes. As such, it gives a wide range of experimental opportunities to fluid-dynamic and heat transfer researchers.

## DISCLOSURE

Part of information included in this chapter has been previously published in "Experiments in Fluids Volume 49, Number 6, 1187-121".

## REFERENCES

[1]     Thomann H, Frisk B. Measurement of heat transfer with an infrared camera. Int J Heat Mass Transfer 1968; 11: 819–26.
[2]     Compton DL. Use of an infrared-imaging camera to obtain convective heating distributions. AIAA J 1972; 10: 1130-2.

[3]    Bynum DS, Hube FK, Key CM, Dyer PM. Measurement and mapping of aerodynamic heating in VKF tunnel B with an infrared camera. Arnold Engineering Development Center AEDC-TR-76-54, Arnold Air Force Station TN 1976.

[4]    Noble JA, Boylan DE. Heat-transfer measurements on a 5 sharp cone using infrared scanning and on-board discrete sensor technique. Arnold Engineering development Center AEDC-TR-78-V51, Arnold Air Force Station TN 1978.

[5]    Stalling DW, Carver DB. Infrared and phase-change paint measurements of heat transfer on the space shuttle orbiter. Arnold Engineering Development Center, AEDC-TSR-78-V13, Arnold Air Force Station, TN 1978.

[6]    Allegre J, Dubreuilh XH, Raffin M. Measurements of aerodynamic heat rates by infrared thermographic technique at rarefied flow conditions. Rarefied Gas Dynamic: Physical Phenomena. Progress in Astronautics and Aeronautics, AIAA 1988; 117: 157-67.

[7]    Peake DJ, Bowker AJ, Lockyear SJ *et al.* Non intrusive detection of transition region using an infrared camera. AGARD-CP-224, 1977.

[8]    Carlomagno GM, de Luca L. Infrared thermography in heat transfer In: Yang WJ, editor. Handbook of flow visualization. London, Hemisphere 1989; Chapter 32: 531-53.

[9]    Gartenberg E, Roberts AS. Twenty-five years of aerodynamic research with infrared imaging. J Aircraft 1992; 29(2): 161-71AGARD-CP-224.

[10]   Shapiro AH. The dynamics and thermodynamics of compressible fluid flow, Vol. II. New York: Ronald Press, 1954.

[11]   Dhungel A, Lu YP, Phillips W *et al.* Film Cooling From a Row of Holes SupplementedWithAntivortex Holes Journal of Turbomachinery-Transactions of the ASME 2009; 131(2): Art. No. 021007.

[12]   Gallo M, Astarita T, Carlomagno GM. Heat transfer measurements in a rotating two-pass square channel. QIRT J 2007; 4 (1): 41-62.

[13]   Astarita T, Cardone G. Thermofluidynamic analysis of the flow in a sharp 180 degrees turn channel. Experimental Thermal and Fluid Science 2000; 20 (3-4): 188-200.

[14]   Carslaw HS, Jaeger JC. Conduction of heat in solids, 2nd ed. Oxford Science Publications, 1959.

[15]   Cook WJ, Felderman EJ. Reduction of data from thin-film heat-transfer gages: a concise numerical technique. AIAA J 1966; 4: 561–2.

[16]   de Luca L, Cardone G, Aymer de la Chevalerie D *et al.* Viscous interaction phenomena in hypersonic wedge flow. AIAA J 1995; 33 (12): 2293-8.

[17]   Hetsroni G, Kowalewski TA, Hu B *et al.* Tracking of coherent thermal structures on a heated wall by means of infrared thermography. Experiments in Fluids 2001; 30 (3): 286-94.

[18]   de Luca L, Cardone G and Carlomagno GM. Image Restoration in Thermo-fluid-dynamic Applications of IR Digital Imagery. Infrared Technology and Applications SPIE 1990; 1320: 448-57

[19]   de Luca L, Cardone G, Aymer de la Chevalerie D *et al.* Goertler instability of a hypersonic boundary-layer. Experiments in Fluids 1993; 16 (1): 10-6.

[20]   Simeonides G, Vermeulen JP, Boerrigter HL *et al.* Quantitative heat-transfer measurements in hypersonic wind tunnels by means of infrared thermography. IEEE Trans Aerosp Electron Syst 1993; 29(3): 878-93.

[21]   Astarita T, Cardone G. Convective heat transfer in a square channel with angled ribs on two opposite walls. Experiments in Fluids 2003; 34(5): 625-34.

[22]   Fénot M, Dorignac E, Vullierme, JJ. An experimental study on hot round jets impinging a concave surface. Int J Heat Fluid Flow 2008; 29 (4): 945-56.

[23]   de Luca L, Cardone G. Modulation transfer-function cascade model for a sampled IR imaging-system. Applied Optics 1991; 30(13): 1659-64.

[24]   Boreman GD. Modulation Transfer Function in Optical and Electro-Optical Systems. SPIE Press, 2001.

[25]   Gunapala S D, Bandara SV, Liu JK *et al.* 1024 × 1024 pixel mid-wavelength and long-wavelength infrared QWIP focal plane arrays for imaging applications. Semicond Sci Technol 2005; 20: 473–80.

[26]   Jain AK. Fundamentals of digital image processing, Prentice-Hall International, Inc, Englewood Cliffs, NJ 07632, 1989.

[27]   Bougeard D. Infrared thermography investigation of local heat transfer in a plate fin and two-tube rows assembly. Int J Heat and Fluid Flow 2007; 28 (5): 988-1002.

[28]   Bougeard D, Vermeulen JP, Baudoin B. Spatial resolution enhancement of an IR system by image restoration techniques. EUROTHERM Seminar 42, Quantitative Infrared Thermography QIRT 94 Sorrento, Italy, Editions EuropennesThermique et Industrie 1994; 3-8

[29]   Carlomagno GM, de Luca L. Infrared thermography for flow visualization and heat transfer measurements. Proc. Workshop "Stato dell'arte del rilevamento con camere termiche nella banda 8-15 micron" Firenze 1991.

[30]   Sparrow EM, Gregg JI. Laminar free convection from a vertical plate with uniform surface heat flux. Trans. ASME 1956; 78: 435-40.

[31]   Meola C, de Luca L, Carlomagno GM. Azimuthal instability in an impinging jet-adiabatic wall temperature distribution. Experiments in Fluids 1995; 18(5): 303-10.

[32]   Meola C, Carlomagno GM. Recent advances in the use of infrared thermography. Measurement Sci Technol 2004; 15(9): 27-58.

[33]   de Luca L, Carlomagno GM, Buresti G. Boundary-layer diagnostics by means of an infrared scanning radiometer. Experiments in Fluids 1990; 9(3): 121-8.

[34]   Cardone G, Astarita T, Carlomagno GM. Wall heat transfer in static and rotating 180 degrees turn channels by quantitative infrared thermography. Revue Generale de Thermique 1998; 37 (8): 644-52.

[35]   Astarita T, Cardone G. Convective heat transfer on a rotating disk with a centred impinging round jet. Int J Heat Mass Transfer, 2008; 51(7-8): 1562-72.

[36]   Astarita T, Cardone G, Carlomagno GM. Spiral vortices detection on a rotating disk proc. 23rd cong. int. Council Aeronautical Sciences, paper n. ICAS2002-3.6.4, Toronto, 2002.

[37]   Cardone G, Astarita T, Carlomagno GM. Infrared Heat Transfer Measurements on a Rotating Disk. Optical Diagnostics in Engineering 1996; 1(2): 1-7.

[38]   Millsaps K, Pohlhausen K. Heat transfer by laminar flow from a rotating plate. J Aeronaut Sci 1952; 19: 120-6.

[39]   Cardone G, Astarita T, Carlomagno GM. Heat Transfer Measurements on a Rotating Disk. Int J Rotat Mach 1997; 3: 1-9.

[40]   Cardone G. IR heat transfer measurements in hypersonic plasma flows. QIRT J 2007; 4(2): 233-51.

[41]   de Luca L, Cardone G, Carlomagno GM *et al.* Flow visualization and heat transfer measurement in a hypersonic wind tunnel. Experimental Heat Transfer 1992; 5(1): 65-78.

[42]   de Luca L, Guglieri G, Cardone G, *et al.* Experimental-analysis of surface flow on a delta-wing by infrared thermography. AIAA J 1995; 33(8): 1510-2.

[43]   Meola C, de Luca L, Carlomagno GM. Influence of shear layer dynamics on impingement heat transfer. Experimental Thermal and Fluid Science 1996; 13(1): 29-37.

[44]   Aymer de la Chevalerie D, Fonteneau, A, de Luca L, Cardone G. Goertler-type vortices in hypersonic flows: the ramp problem. Experimental Thermal and Fluid Science 1997; 15 (2): 69-81.

[45]   Cardone G, Buresti G, Carlomagno GM. Heat transfer to air from a yawed circular cylinder. Atlas of Visualization III, Nakayama Y. and Tanida Y. eds, 1997; Ch. 10, 153-68.

[46]   Astarita T, Cardone G, Carlomagno GM, Meola C. A survey on infrared thermography for convective heat transfer measurements. Optics and Laser Technology 2000; 32(7-8): 593-610.

[47]   Carlomagno GM, Astarita T, Cardone G. Convective heat transfer and infrared thermography. Annals of the New York Academy of Sciences 2002; 972: 177-86.

[48]   Astarita T, Cardone G, Carlomagno GM. Convective heat transfer in ribbed channels with a 180 degrees turn. Experiments in Fluids 2002; 33(1): 90-100.

[49]   Carlomagno GM, Nese FG, Cardone G *et al.* Thermo-fluid-dynamics of a complex fluid flow. Infrar Phys Technol 2004; 46(1-2): 31-9.

[50]   Astarita T, Cardone G, Carlomagno GM. Infrared thermography: An optical method in heat transfer and fluid flow visualisation. Optics and Lasers in Engineering 2006; 44 (3-4): 261-81.

[51]   Carlomagno GM, Colours in a complex fluid flow. Optics and Laser Technology 2006, 38(4-6): 230-42.

[52]   Carlomagno GM, Heat flux sensors and infrared thermography. J Visualization 2007; 10(1): 11-6.

[53]   Carlomagno GM, Cardone G. Infrared Thermography for Convective Heat Transfer Measurements. Experiments in Fluids 2010; 49: 1187-218.

# CHAPTER 7

## Infrared Imaging to Combustion Systems

### Christophe Allouis[*] and Rocco Pagliara

*Istituto di Ricerche sulla combustione – CNR, P.le V. Tecchio 80 – 80125 Napoli, Italy*

**Abstract:** This chapter covers a wide range of applications of infrared imaging systems to combustion, from both specific and practical scientific points of view. Recent new applications are also presented in the last part. The availability of new sophisticated and flexible IR sensors makes possible the applicability of IR devices, as complementary tools, to existing techniques for combustion characterization. In fact, an IR camera allows visualization of phenomena that cannot be caught by the UV-Visible cameras. Different aspects of InfraRed imaging systems utilization are presented here from the choice of the sensor to the specific application. Particular attention is focused on the use of Fast IR Imaging of flames applied in two different types of gas turbine burners and in two traditional industrial burners. We illustrate how useful is the IR imaging device for temporally and spatially characterizing the combustion fluctuations. Finally, we discuss the undergoing developments of the application of IR-based sensors for burner control.

**Keywords:** Flow visualization, humming characterization, instability characterization, combustion control.

## 1. INTRODUCTION

Combustion is the widest source for energy production in the World. The improvement of combustion efficiency is still under study to minimize the use of mineral resources and to reduce the emission of by-products which are released in the atmosphere.

Combustion and optical diagnostics have been connected since Swan's observations in 1857 of the $C_2$ emission bands in a candle flame. Flames became a perfect source for optical emission spectroscopy and this was early reviewed by Gaydon in 1957 [1].

The fundamental understanding needed for the improvement of combustion devices and processes requires a strong collaboration between scientists of different research fields. In fact, each model of practical combustion device requires knowledge of different disciplines such as chemistry, fluid dynamics and heat transfer. Experimental tests are also needed for the validation of numerical combustion models. Laser-based diagnostic techniques currently offer direct interrogation of the combustion process. These diagnostic measurements provide excellent results for understanding of combustion mechanisms. They stimulate our ideas on how to apply this knowledge to the control and optimization of practical combustion systems.

Combustion systems constitute a severe, high temperature environment, which often entails complex time-dependent chemistry in a turbulent medium. In addition, practical combustion often occurs in two-phase flow and/or at elevated pressure. The experimental determination of the basic parameters requires significant efforts since temperature, species concentration, flow velocities, and potential temporal and spatial fluctuations of these quantities must be measured, often simultaneously. Limited optical accesses, window deterioration constraints in measurement time, vibrations and other effects may pose secondary problems when coupling a laser technique, a visualization techniques and/or a laser based instrument for measurements in practical environment.

It is evident that not one single optical technique or measurement approach will provide all the necessary quantities to characterize a complex practical combustor. A review of all the possible techniques was given by Kohse-Höinghaus and Jeffries [2]. In this chapter we will present the specific use of Infrared imaging to combustion processes.

---

*Address correspondence to Christophe Allouis: Istituto di Ricerche sulla combustione – CNR, P.le V. Tecchio 80 – 80125 Napoli, Italy, Tel: +390817683256, Fax: +390815936936; E-mail: allouis@irc.cnr.it

**Carosena Meola (Ed)**

## 2. BACKGROUND

The multidisciplinary nature of the combustion diagnostics has been breeding ground for development of a wide and varied literature. The fundamental physics underlying diagnostic techniques, which are used for spatially resolved gas-phase measurements was reviewed by Eckbreth [3]. Another interesting review concerned with laser spectroscopy was done by Demtröder [4]. General design considerations for optical diagnostics in combustion research, including the choice of the laser sources, the selection of the optical components and the working wavelengths, and the detection devices are also available in referecne [3]. Moreover, for a better understanding of the Infrared detection technology, the work by Henini and Razeghi is recommended [5].

In this chapter, we only discuss the two Dimensional application of the infrared imaging technique applied to combustion problems. For further information about mono dimensional infrared techniques for combustion systems, we address the readers to referecne [2].

## 3. SENSORS OVERVIEW

Infrared detectors have seen a remarkable surge in interest over the past decades. This is thanks in part to the successful development of high performance devices. The natural progression of the infrared detectors is a multispectral, cooled and/or uncooled, detector, which can, by itself, address many requirements in: fluid-dynamics, infrared astronomy, medical diagnostics and material science. As in nature, a good system should be flexible, power efficient, light weight, high sensitive and easy to use. Most infrared detectors have a limited photocarrier lifetime and peak detection wavelength that is fixed by the band gap of the material. Without changing the chemical composition of the material, patterning on an atomic scale can allow an increase in carrier lifetime and tuning of the peak detection wavelength. While we cannot expect to match the sophistication of natural systems, we can be inspired by them to build new sensors.

Most of the applications involving infrared cameras in combustion systems have been performed using detectors working in different infrared bands. The common bands are 2-5μm (mid IR), 8-12μm or 14μm (long wave IR), and in the near IR 0.8-1.5μm [6-8]. The choice of the spectral sensitivity of the sensor is crucial and strongly connected to the application and to the parameters under investigation during the combustion process. In the case of waste incinerator applications the cameras operated in the mid-infrared range 2.5 – 5.6μm as presented in Refs. [6] and [8] to control the combustion quality. Manca and Rovaglio preferred to limit the camera sensitivity to a specific wavelength (3.9μm) to measure temperature in a combustion chamber (Thermovision 450 from AGEMA) [7]. In other applications, the broadband short infrared (1.5-5μm) InSb (Indium Gallium Arsenide) sensor was used for flame stability study (PHOENIX camera by FLIR Systems) [9]. Finally, techniques of *in situ* visualization of the CO and $CO_2$ species used an InSb camera (SBF134 by Focal Plane-Santa Barbara) [10-13].

## 4. IR LASER-INDUCED FLUORESCENCE IMAGING

Laser-Induced Fluorescence (LIF) is one of the most important diagnostics techniques for combustion characterization. It provides spatial chemical species characterization and it allows taking specific images of flame structures [2]. Recent development of infrared lasers and fast time-gated infrared cameras has enabled point wise and two-dimensional images (PLIF) using vibrational transitions of important combustion species that do not have suitable electronic transition for convenient LIF. Many combustion species, including important combustion products and fuels, *e.g.*, CO, $CO_2$, $H_2O$, $CH_4$, do not have accessible single-photon electronic spectra and hence have not been amenable to PLIF imaging. Kirby and Hanson [10-13] excited such species using vibrational overtone transitions and detect infrared emissions from vibrational transition. Fig. **1** represents the excitation, energy transfer, and IR LIF from CO. Here the light is near 2.3 μm from a high –pulse energy tunable IR optical parametric oscillator. It excites the ground state of CO molecules into the excited level $v = 2$.

Molecules can undergo fast vibrational transfer with near resonant nitrogen or resonant CO colliders. Some of the excited molecules radiate single quanta transitions 2 →1 and 1 →0 near 4.7μm, which is detected using a fast time-gated InSb camera. A scheme of a typical IR PLIF experimental set up is presented in Fig. **2**.

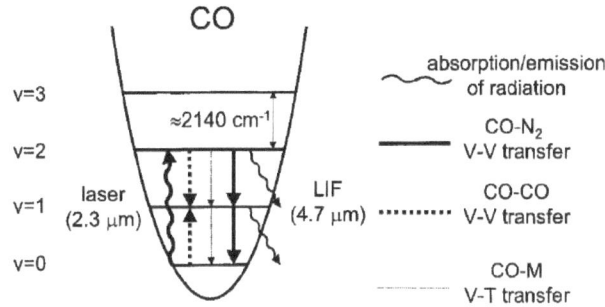

**Figure 1:** Energy-level diagram for the CO-air system. Fluorescence yield is determined by the competition between the characteristic times of (1) fluorescence at 4.7μm, (2) V-V transfer with CO and $N_2$, and (3) integration time of the camera.

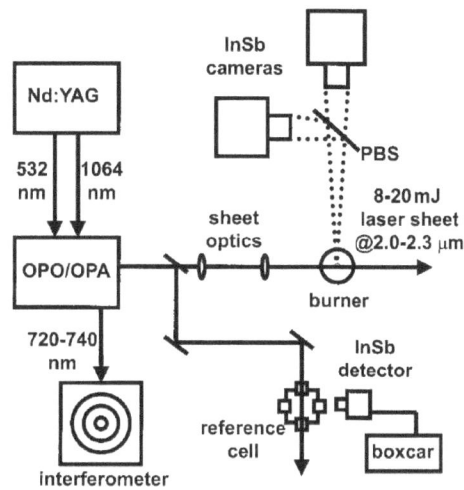

**Figure 2:** Typical IR PLIF experimental set up (PBS: pellicle beam splitter) [13].

An example result of IR PLIF of carbon monoxide is shown in Fig. **3**, demonstrating the quality of the image that is achievable with this technique.

The combination of new InSb cameras and laser development with novel excitation and detection wavelength strategies to avoid IR flame emissions may lead to a combustion diagnostic tool that will supplement existing PLIF imaging techniques.

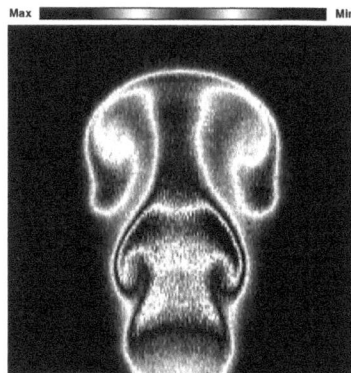

**Figure 3:** Sample IR PLIF visualization of a room-temperature, 6-mm-diameter forced $CO_2$ jet mixing with ambient air. A single 8-mJ pulse at 2.0μm pumps the $(20^01)_{II} \leftarrow 4 (00^00)$ transition in $CO_2$, and an InSb camera collects fluorescence at 4.3 μm. (Courtesy by Brian J. Kirby and Ronald K. Hanson [13]).

## 5. THERMOGRAPHY FOR COMBUSTION CONTROL

### 5.1. Combustion Control for Waste Incinerators

The basis for combustion control in municipal waste incinerators with grate firings consists in the coordination of the primary combustion with the secondary combustion in the furnace [14]. Thus, the control of fuel bed reactions plays an important role in the optimization of refuse incinerators. The in homogeneity of refuse as a fuel represents the main problem of the process control. It requires an immediate and sensitive adaptation of the combustion parameters *versus* the calorific value. The Conventional Combustion Control System (CCS) aims to reduce emissions and to optimize combustion parameters. These systems suffer either from being only indirectly correlated to the quality of combustion or due to the measurement techniques, from showing only a relative slow response time *versus* rapid process changes.

The fuel bed in a furnace is a classical example of a "cool" target in a hot combustion enclosure. The two main problems in measuring pyrometric temperature are the background reflections and the sight path emissions [15, 16]. Both effects give a contribution to the total measured radiance signal; thus, a correction is needed to obtain temperature measurements. Strong interferences may be caused by the flame overlying the fuel bed. The hot flame gases do not emit continuous spectra, but radiate in discrete bands [17] as represented in Fig. **4**.

Their influence can be suppressed by using appropriate band pass filters blocking then the gas radiations. Careful selection of transmitted wavelengths is critical to avoid interferences by gas species and to obtain images of the underlying fuel bed. Suitable spectral windows generally suit in the near and middle infrared IR at about 1.7, 2.2, 3.4 and 3.9μm. A somewhat different treatment is necessary to estimate the interference by fumes and soot particles. Their spectra are continuous and, due to the small particles size, are characterized by strong variation with wavelength. The radiation theory predicts a decrease of the emission coefficient of small particles when increasing the wavelength [17].

Since the flame atmosphere is significantly hotter than the fuel bed, it is interesting to estimate the ratio of blackbody spectral radiance at the fuel bed and flame temperature. As neither the particulate clouds nor the fuel bed represent true black bodies, the influence of emissivity has to be taken into account. For typical values of particle cloud density ($1g/m^3$) and cloud dimension (1m), which are encountered in practical combustion systems, the average spectral emissivity of hot particle clouds is approximately $\varepsilon_1 = 0.1$ [17], whereas a typical value for refuse materials is approximately $\varepsilon_2 = 0.8$. Radiation from the hotter flame exceeds that from the fuel bed by far in the near IR, whereas at longer wavelengths it is the fuel bed contribution that dominates [6].

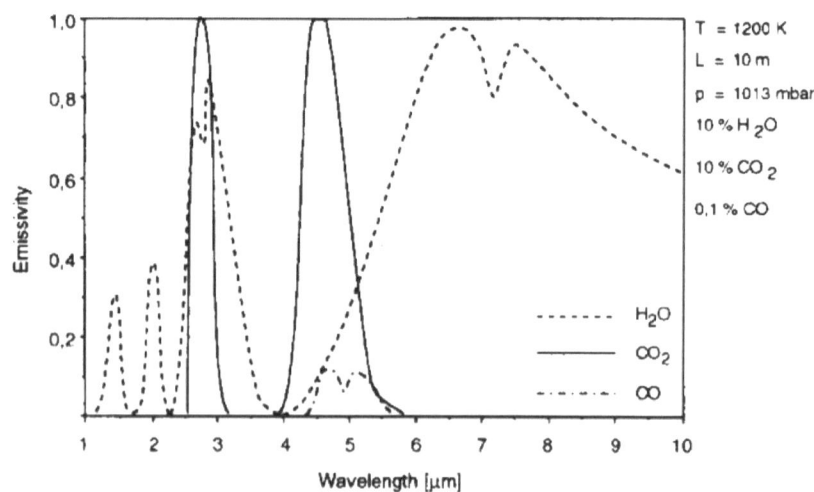

**Figure 4:** Spectral emission coefficients for the main radiative species within a furnace [17].

The radiation conditions within a furnace generally impose a double requirement on the operating wavelength of the imaging system: the wavelength must be in a suitable spectral window between gas radiation bands and as long as possible. This uniquely specifies the wavelength interval around 3.9µm, because out of the three suggested spectral windows between the gas bands it is the only one that is less influenced by particle radiation [6, 7]. Moreover, the spectral region beyond 10µm was not considered for reasons of system practicability.

Consequently, the adopted Thermographic camera was designed to operate in a narrow region at $\lambda = 3.9$µm. The infrared camera was connected to a personal computer to grab and process the gray scale images taken by the camera. The acquisition system is described in Fig. **5** [6, 7].

In practical systems, the information is extracted from the IR images and is used to optimize the control of the various plant parameters. The fast response time of the thermography system allows fine tuning of the total under fire air flow. The distribution of under fire air can be readily adjusted to variations in local demand caused by non uniform feeding or variations in calorific value that can be detected thermographically.

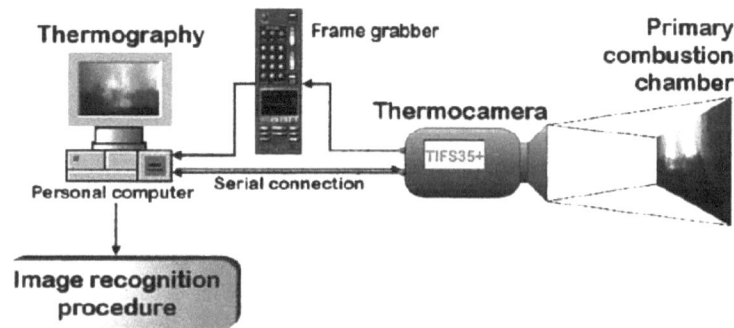

**Figure 5:** Typical image acquisition system installed for combustion control [7].

## 5.2. Combustion Control for Burner Stability

The environmental aspects become always more important for the development of energetic and propulsive technology; in particular, the new regulations require a better control of nitrogen oxides, particulate and inorganic compounds emitted by combustion sources. Fuel injection and depletion processes notably influence the formation of the pollutants in the combustion chamber and consequently their emissions.

Many researches are focused on new concepts for ultra-low emissions combustors, with developments in fuel preparation and wall cooling techniques in the case of gas turbine combustors. Moreover, coal came back as an important and challenging fuel for energy production. New burner technologies have to be explored to obtain clean coal combustion in the future power plants. Concerning Gas Turbine, a possible technological solution for the reduction of pollution could be the use of lean mixtures, which are premixed before the fuel/air mixture enters the combustion chamber. Such a solution is the LPP (Lean Premixed Prevaporized) for liquid and LP (Lean Premixed) for gas based technologies. Nevertheless, these new emergent technologies are affected by many problems that must be solved in order to make them better reliable.

Numerous experimental studies and numerical models were developed to understand the behaviour of gas turbines; in particular, there are current researches to understand fuel/air interactions in the premixing duct, upstream of the combustion chamber [18-21]. Indeed, the characterization of the fuel mixing with air is essential for the optimization and the choice of the injection technology. The latter is important since it influences the homogeneity of the fuel/air mixture. Unfortunately, developments in premixed combustion are generally accompanied by an increase in the occurrence of oscillating combustion [22, 23]. Unstable combustion refers to self-sustained combustion oscillations at/or near the acoustic frequency of the combustion chamber, which are the result of the closed-loop coupling between unsteady heat release and pressure fluctuations. The heat release fluctuations which produce pressure fluctuations is well known and well understood; however, the mechanisms whereby pressure fluctuations result in heat release fluctuations are not.

Rayleigh [24] postulated that, for the pressure oscillations to be amplified, the heat release and pressure fluctuations must be in phase. The exact mechanism of unstable combustion is not yet completely understood. Moreover, in order to validate the different numerical models of combustion instabilities, real time measurements are needed giving thus the possibility for a better description of the phenomena. Acoustic frequencies can be easily measured by fast transducers, while the oscillations of the equivalence ratio can be obtained performing real time measurements [25]. Other time averaged techniques (LIF, PLIF, Chemiluminescence, *etc.*) give an idea of the combustion process [25]. Usually, UV-Visible spectroscopy is used to obtain information on the flame structures. New available technology allows to shift the 2-dimensions studies towards the Infra Red region [26].

The objective of this chapter is to present the capability of the fast infrared imaging in individuating the flame and fluid dynamic fluctuations and the eventual correlations between them. To reach this goal, different experiments were carried out: a LPP burner as reference system and industrial conventional LP burner with pilot flame. Fast transducers and Fast Infra Red Imaging (FAIRI) were also compared. Finally, the capability of the FAIRI technique was applied to an industrial burnerfuelled by natural gas and coal to study flame stability.

### 5.2.1. Experimental Procedure

The experiments were performed using firstly a 100 kW Lean Premixed Prevaporized burner fuelled with Jet-A1 fuel and using a test rig of AnsaldoEnergia equipped with a 3MW atmospheric gas turbine burner. A sketch of the LPP burner was presented in a previous paper [9]. The swirl number of the co-rotating blades is fixed at $S = 0.5$. The LPP burner was inserted in a large furnace in order to avoid acoustic interferences. Air was supplied by pressurized tanks filled by two compressors.

At the Ansaldo Caldaie Combustion research center, the 3 MW test rig was used to operate the full scale turbine combustor in atmospheric conditions. It is a cylindrical refractory lined combustion chamber. Particular attention have to be paid to the acoustic characteristics of the system in the experimental set up design. The rig is fully instrumented and allows for all the operative conditions to be measured and continuously registered. Fast data logging coming from pressure piezo meters was used to describe the acoustic behaviour inside the combustion chamber and an FFT calculation have been performed. It is possible to say that the pilot flame fuel flow rate represents 10% of the total fuel mass flow. During our experiments it was fuelled with methane. Normal conditions together with humming conditions were studied.

For the LLP burner, the fuel was pressurized by nitrogen in a tank before atomization in order to avoid fluctuations due to pumping. Atomization pressure was kept constant at 7bar. Different Air/Fuel ratios (AFR), ranging between 22-30, were used for the measurements. The Kerosene flow rate was kept constant at 0.5 USGal/h. The spray was generated by a hollow cone nozzle with a 60° fixed angle. The combustion air was preheated at 230°C and regulated by a mass flow controller. The experiments were performed at atmospheric pressure for both burner types. A Fast InfraRed PHOENIX Camera from FLIR Systems, equipped with a detector working in the 1.5-5µm wavelength, frame rate up to 38kHz and with spatial resolution of 320x256 pixels was used to analyze the flame fluctuations of the natural flame emission. For the LP burner fast transducers were connected to data acquisition board and Fast FFT were performed through Labview program.

A typical blue flame for stoechiometric premixed conditions on LPP burner was obtained at AFR= 25 [9].The X-Z axis system is presented in Fig. **6**. The origin corresponds to the centre of the exit plane of the mixer (Z=0, X=0). For the Ansaldo burner, analysis points will be indicated in the next paragraph.

Further experiments on industrial burner were performed in a 50MWth plant. This test rig boiler can welcome different burner types fuelled with different fuels (coal, heavy oils, *etc.*). The 50 MWt combustion chamber has an horizontal shape with a section of 5.5 x 4.5m² and a length of 12.5m; an overview of the plant is presented in Fig. **7**. The experimental boiler has several windows on its left side wall that allow an optimized view of the flame root.

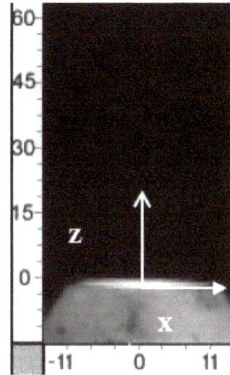

**Figure 6:** Axis system expressed in mm.

In the experiments two kinds of burner were used: one fuelled with natural gas alone and in blend with hydrogen, and another one fuelled with pulverized coal.

**Figure 7:** 50 MWtAnsaldo boiler overview.

### *5.2.2. Results and Discussion*

### *5.2.2.1. Results on LPP burner*

The frequency analyses of complete flame (up to 100mm above the burner) for the different AFR was performed using a recent available technology such as the high speed infrared imaging system described in the previous section 5.2.1 with sensor spectral response in the wavelength band ranging between 1.5-5μm. This technique is much more sensitive than a visible CCD camera and, it is less photon demanding [9]. The image acquisitions were performed with a window size of 128x18 pixels. The frequency analysis of the images acquired at different flame position was performed by calculating the power spectrum of the maximum intensity found in the rectangle surface indicated on the images presented in Table **1**. Namely, the program found the maximum intensity in the rectangle and then calculated the power spectrum of the signal intensity.

**Table 1:** Fixed flame positions considered for IR analysis.

| Position | Analyzed zone | X (mm) | Z (mm) |
|----------|---------------|--------|--------|
| 1 | | 0-10 | 0-5 |
| 2 | | 0-10 | 5-10 |
| 3 | | 0-15 | 12-18 |
| 4 | | 0-15 | 22-26 |
| 5 | | 0-15 | 28-33 |

For a more simple representation, the thermograms of the flames are horizontal but the real flame develops along the vertical direction. Considering results obtained in past investigations [9], it was decided to limit the frequency analyses to the areas inside the white vertical rulers on Table 1. The following study will concern the AFR=25. The results of the frequency analysis under isothermal conditions is presented in Fig. 8 for the five positions presented in Table 1. The results are presented in arbitrary unit (a.u.).

The result of the position #4 is very close to that for position 3, so, for clarity reasons, it is not showed in Fig. 8. We clearly observe for all the positions a fundamental frequency at ~620Hz, while its harmonics: 1240Hz, 1850Hz and 2480Hz are present only for the first four positions. The intensity of the harmonics decreases when increasing height above the burner. Moreover, a fit curve is presented in the range 600-3000Hz with a slope of -5/3. We can note a slope change around 600Hz probably due to different phenomena. In the low frequency zone (< 600Hz) the slope is-1/4 that indicates a low rate of turbulence of the system [25]. This slope could be representative of the low turbulence due to the precession visible during the experiments. Successively, frequency analysis was performed under combustion and the results are presented in Fig. 9.

We can observe an intense peak frequency at ~480Hz and some lower intense harmonics (960Hz and 1920Hz) for the first two positions, namely close to the burner exit. We can also observe minor harmonics at 120, 240 and 360Hz. These frequencies are similar to those found by Vauchelles [26] and Cabot [27]. They correlated frequencies up to 530Hz to combustion phenomena and found that the 130Hz frequency is characteristic of a rich combustion regime. In fact, we observe a peak of the 130Hz frequency and its harmonics in the zone close to the burner exit, namely for position 1 and 2. The combustion-biased frequency peaks become weaker when going downstream the flame. Another interesting point is the slope change after 800Hz *versus* the position in the flame. The first two position plots show a slope of $F^{-1}$, while downstream of the flame the slope can reach $F^{-2}$ as shown in Fig. 9. After the integrated study of the flame fluctuations for the AFR=25, the interest was focused on the first part of the flame varying the Air Fuel ratio in order to study the fluid dynamic influence on flame fluctuations. The considered spatial positions are presented in Table 2. The frequency analysis was performed using the power spectrum of the maximum intensity included in the black rectangle surface indicated on the images presented in Table 2. The results of the frequency analyses for the positions 1, 2 and 3 are presented in Fig. 10.

We can notice that respect to the broad study for AFR=25, a narrower window size increases the precision of the spectrum and it is possible to more precisely individuate the frequency peaks. We can also observe for all the positions higher peak intensity for frequencies around 2kHz. For positions 1 and 3 we clearly identify the lower frequency harmonics.

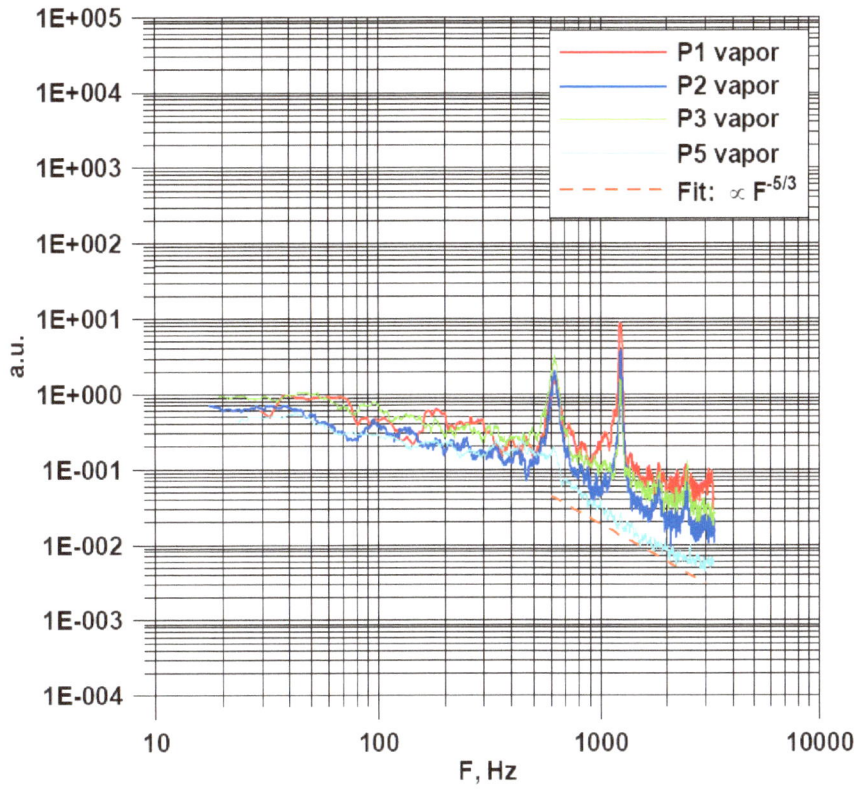

**Figure 8:** Frequency analysis of IR images under isothermal conditions (AFR= 25).

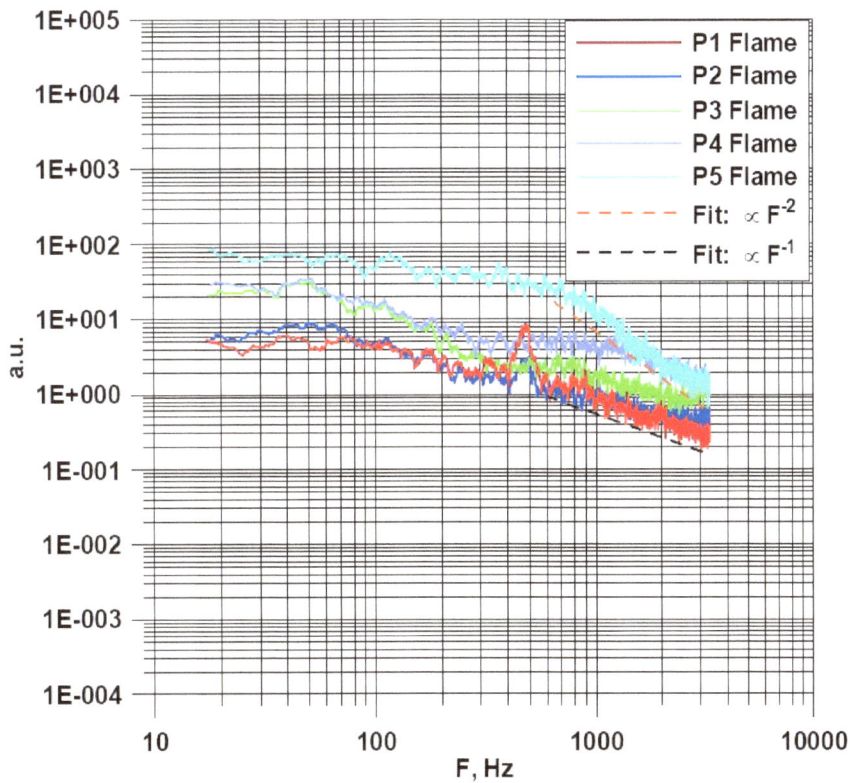

**Figure 9:** Frequency analysis of IR images for the different flame positions (AFR= 25).

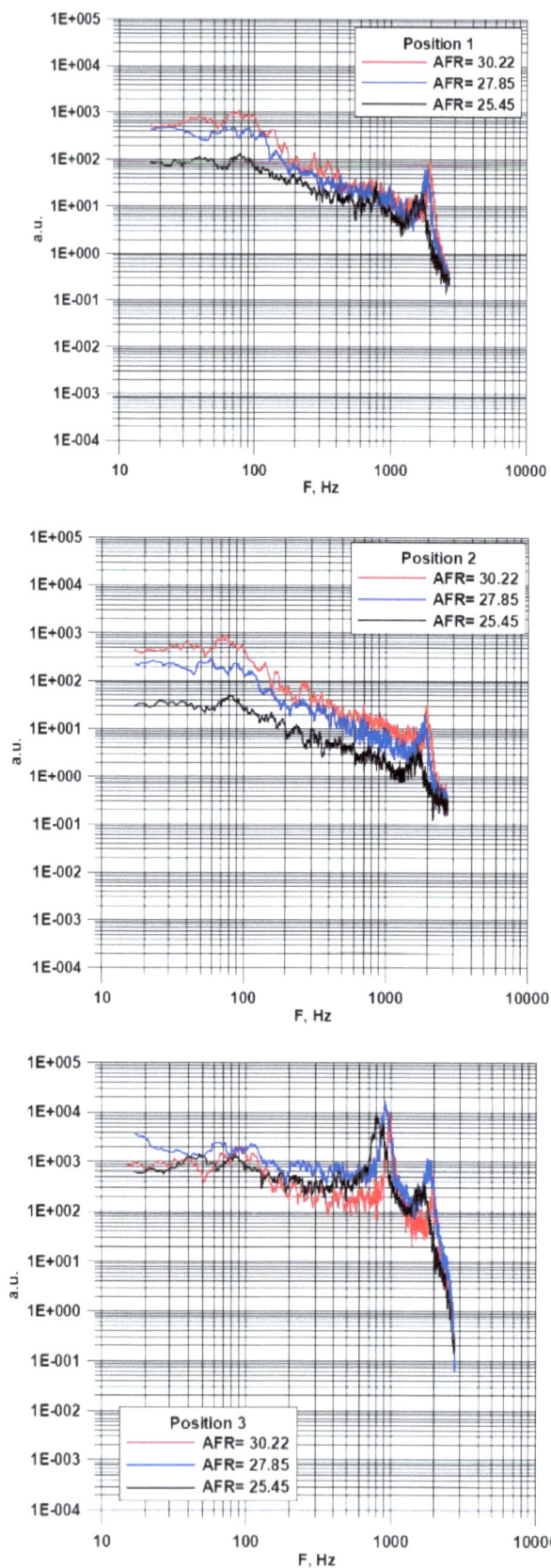

**Figure 10:** Frequency analysis for position 1, 2 and 3.

We hypothesized that the frequency around 2kHz is due to fluid dynamic conditions since the frequency peak increases while increasing the air flow rate. On another hand, the presence of a second intense peak for positions 1 and 3 with a typical frequency around 900Hz was also assigned to a fluid dynamic process since this frequency shifts up when increasing the air flow rate. This phenomenon could represent a recirculation zone at the centre of the flame. For all the positions we always observe the low frequencies biased towards combustion as explained before. The frequencies kept the same value even though the Air Fuel Ratio changed. Another interesting feature of these plots is that the slope changes *versus* the position in the flame. If we consider the positions 1 and 2 we have a slope of $F^{-5/3}$ for all the AFR over the whole frequency range, while for the position 3 we have a slope of $F^{-1/3}$ up to 1kHz and $F^{-14/3}$ above 1kHz. This indicates a strong turbulence change in the zone 3.

### *5.2.2.2. Results on LP Burner*

Due to the lack of optical accesses, IR Imaging was performed only together with fast transducers recording. Measurements were carried out by exploiting a window situated in front of the burner at a distance of about 2 m from the burner throat. The three investigated positions are presented in Table **3**.

**Table 2:** Spatial position of the IR analysis of the different flames.

| Position | Flame zone analyzed | X (mm) | Z (mm) |
|---|---|---|---|
| 1 | | 0 | 5 |
| 2 | | 15 | 8 |
| 3 | | 0 | 10 |

**Table 3:** Spatial position of the IR analysis of the LP burner.

| Position | Flame zone analyzed in front of the burner |
|---|---|
| 1 | |
| 3 | |
| 5 | |

These three positions are typical area of investigation for burner optimization. Pressure oscillations and infrared images were simultaneously recorded. The position 1 represents the pilot flame, while position 2 and 3 represent respectively the secondary and the tertiary airs.

Result of FAIRI analysis is presented in Fig. **11** for normal combustion conditions and under humming conditions.

We observe that flame intensity, represented by the signal intensity background, is low under normal conditions, while humming generates more intense flames. A typical frequency is found around 50-60Hz under normal combustion, while humming generates frequency around 25Hz and 125Hz. Moreover, the turbulence intensity of LP burner is much lower than that for the case of LPP burner. The humming phenomenon was confirmed by the fast transducers records analysis presented in Fig. **12**.

A good agreement was found between traducer results and the Fast infrared analysis. Finally, the IR analyses allowed to spatially locate the humming phenomenon in positions #1 and #3, while transducers give an overall result of the presence or not of the phenomenon.

### 5.2.2.3. Results on Traditional Burners

Two burners were studied in terms of flame stability: one fuelled with natural gas (NGB), and the other fuelled with coal (CB). The objective of this study was to control the flame stability while burning different fuels (methane, methane+$H_2$ and coal) and introducing water steam and re-circulated exhaust gases to lower emissions. The operative conditions are explained in the figure labels (Fig. **14**).

In Fig. **13** are presented normal and filtered IR views of the NGB. We can observe in the image on the left side the red rectangle that represents the zone of interest for flame stability.

The frequency analysis of the maximum intensity present inside the red rectangle was performed for different combustion conditions. The measurements were carried out for the maximum burner load with methane, and blend 50%methane-50%hydrogen, steam injection and exhaust gas recirculation. The results are presented in Fig. **14**.

We can observe in this figure that the presence of hydrogen did not perturb the flame stability either in terms of frequency, or in terms of curve slope, namely in terms of the turbulence scale. Moreover, the addition of water steam and recirculation exhausts leads to a slope reduction, and thus to the turbulence intensity reduction.

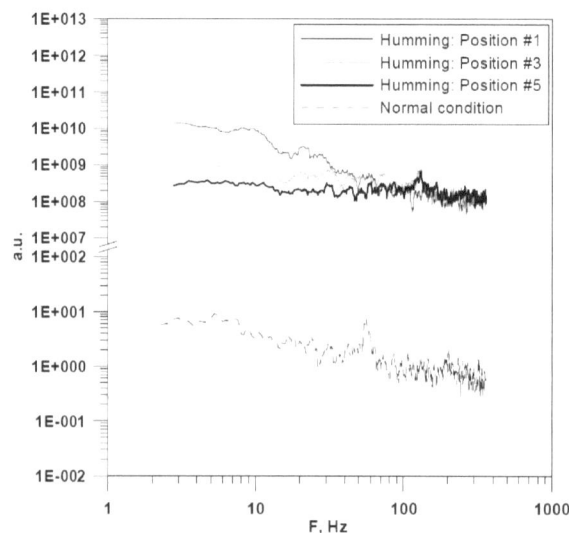

**Figure 11:** Frequency analysis of IR images under normal and humming conditions.

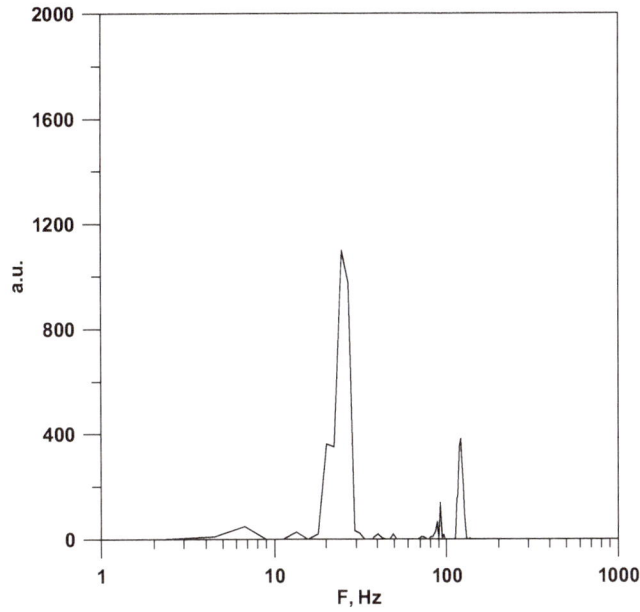

**Figure 12:** Frequency analysis of transducers under humming conditions.

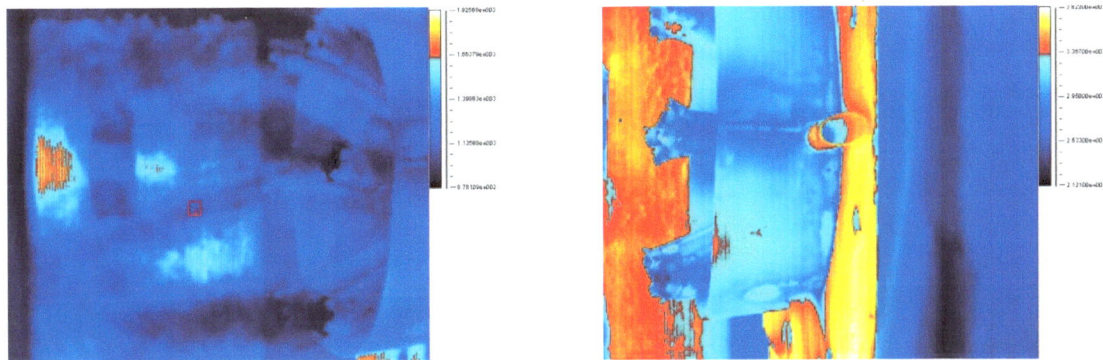

**Figure 13:** Normal IR view of the flame (left) and 3.9μm filtered view of the flame (right).

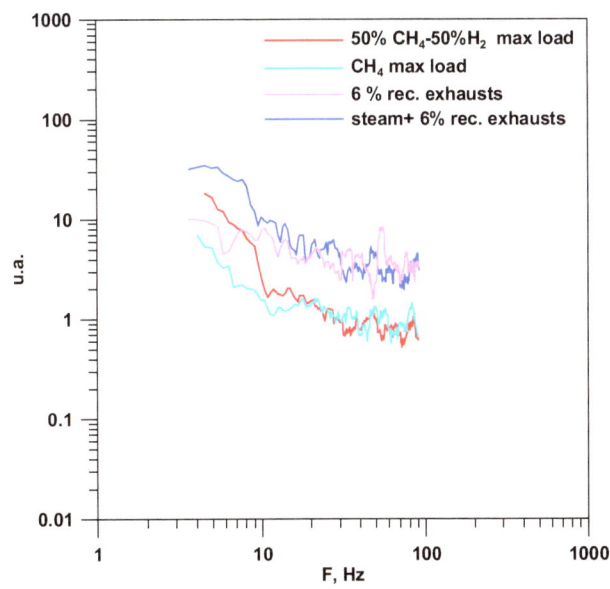

**Figure 14:** Frequency analysis under different combustion conditions.

In this combustion application a very special design of the burner was adopted in order to have broad operative flexibility regarding fuel characteristics. A fuel/exhaust gas dilution was made internally, which lead to a strong reduction of the flame front intensity. In fact, the presence of Hydrogen lead to a reduction of the flame front intensity. Further external dilution has an effect in the NOx emission reduction.

Further experiments were performed on the coal burner (CB) by changing the fluid dynamic conditions, namely primary and secondary air blades. The operative conditions are presented in Table **4**. The test was made using a pulverized sub bituminous South African coal.

The indicative burner position is tagged by greencircle in Fig. **15**. The black rectangle on the right part represents the refractory wall seen in the window.

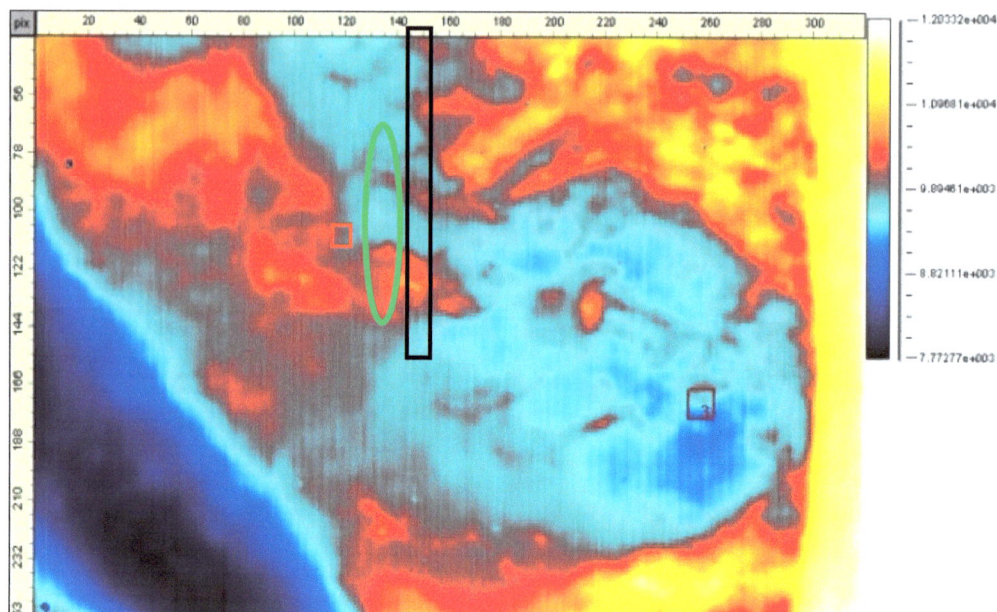

**Figure 15:** IR visualization of the flame and presentation of the zone of interest.

In order to obtain information about the flame stability, a frequency analysis was performed. The zone of interest was chosen close to the burner exit; it is represented by the red rectangle in Fig. **15**. In fact, the flame intensity close to the burner strongly affects the behaviour of the flame and the unburnt emission. The temporal evolution of the maximum intensity measured in the rectangle was analyzed. Then the FFT of the signal was preformed and the results are presented in Fig. **16**.

**Table 4:** Different conditions of burner set up.

| Condition Number | Time | Primary Air Swirl | Secondary Air Swirl |
|---|---|---|---|
| 1 | 15h39 | 25° | 35° |
| 2 | 16h07 | 25° | 45° |
| 3 | 16h45 | 35° | 45° |
| 4 | 17h23 | 45° | 45° |
| 5 | 17h47 | 40° | 45° |

We can observe the presence of a frequency peak at 40 Hz in the case of primary air set at 35°, while for PA at 40° the frequency shifts down at 20Hz (Fig. **16**). For all the conditions, the slopes of the FFT plots are similar, thus indicating a similar level of turbulence. The absolute signal intensity depends on the flame emission and consequently on the temperature. So, we can hypothesize that in the case of a PA swirl of 40°, the temperature is lower than in the other cases. Moreover, the high sensitivity of the infrared camera allows individuation of high and low temperature/emissivity scatterers, namely hot gases ($CO_2$) and small particles (coal, particulate).

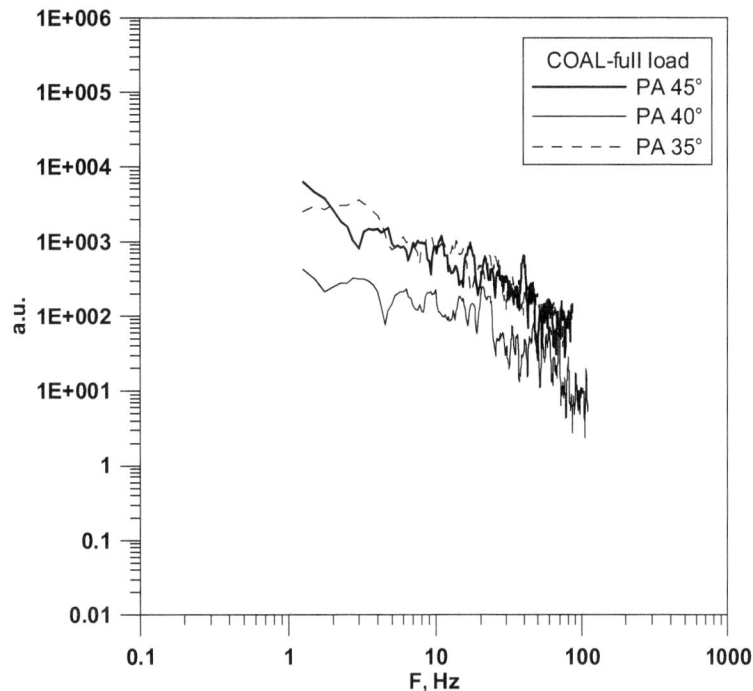

**Figure 16:** Frequency analysis for coal burner.

## 6. CONCLUSIONS

This chapter offered a large panorama of infrared camera applications to combustion, from both specific scientific and practical points of view. Recent new applications were also presented in the last part paragraphs. The tool based on FAst IR Imaging (FAIRI) of flames was applied on two different types of gas turbine burners and on two traditional industrial burners varying the fluid dynamic conditions. IR imaging allowed for characterization of temporal and spatial combustion fluctuations. The extension of this technique to a practical Lean Premixed gas turbine gave good results for both frequency analysis of fluctuations, and location of the humming phenomenon. Moreover, regarding industrial boiler combustion the FAIRI allowed identification of the proper location of the flame root and of its better performance in stability in both gas and coal combustion. Further development and investigations are undergoing for a wider application of IR-based sensors for burner control.

## REFERENCES

[1]     Gaydon AG. The Spectroscopy of Flames, Chapman and Hall, London, UK, 1957.
[2]     Kohse-Höinghaus K Jeffries JB. Applied Combustion Diagnostics, Kohse-Höinghaus and Jeffries, Combustion: An International Series, Taylor & Francis, 2001.
[3]     Eckbreth AC. Laser Diagnostics for Combustion Temperature and Species, 2nd Ed., Gordon & Breach, UK, 1996.
[4]     Demtröder W. Laser Spectroscopy, 2nd Ed., Springer-Verlag, Berlin, 1981.
[5]     Henini M, Razeghi M. Handbook of Infrared Detection Technologies, Elsevier, 2002.
[6]     Schuler F, Rampp F, Martin J, Wolfrum J. TACCOS – A Thermography –assisted Combustion Control system for Waste Incinerators. Combustion and Flame 1994; 99: 431-9.
[7]     Manca D, Rovaglio M. Infrared Thermographic Image Processing for the Operation and Control of Heterogeneous Combustion Chambers. Combustion and Flame 2002; 130: 277-97.
[8]     Madsen OH. New Technologies for Waste to Energy Plants, 4th International Symposium on Waste Treatment Technologies, Sheffield, UK, 29 June – 2 July 2003.

[9]     Allouis C., Beretta F., Amoresano A., "Experimental Study of Lean Premixed Prevaporized Combustion Fluctuations in a Gas Turbine Burner". Combustion Science and Technology 2008; 180 (5): 900-9.

[10]    Kirby BJ, Hanson RK. Planar laser-induced fluorescence imaging of carbon monoxide using vibrational (infrared) transitions. Applied Physics B—Lasers and Optics 1999; 69: 505-7.

[11]    Kirby BJ, Hanson RK. Imaging of CO and CO2 using infrared planar laser-induced fluorescence. Proceedings of the Combustion Institute 2000; 28: 253-9.

[12]    Kirby BJ, Hanson RK. CO2 imaging with saturated planar laser-induced vibrational fluorescence. Applied Optics 2001; 40: 6136-44.

[13]    Kirby BJ, Hanson RK. Linear excitation schemes for IR PLIF imaging of CO and CO2. Applied Optics 2002; 41: 1190-201.

[14]    Seeker WR. Waste Combustion. Proceedings of the Combustion Institute 1990; 23: 867-85.

[15]    Ridley U, Beynon TGR. Infra-red temperature measurement of bright metal strip using multiple reflection in a roll-strip wedge to enhance emissivity. Measurement 1989; 7: 171-6.

[16]    Ariessohn PC, James RK. Imaging of hot infrared emitting surfaces obscured by particulate fume and hot gases. US 4539588 (1985).

[17]    Ludwig CB, Malkmus W, Reardon JE, Thomson JAL. Handbook of Infrared Radiation from Combustion Gases. NASA SP-3080, Washington D.C., 1973.

[18]    Cowell LH, Smith KO. Development of a Liquid-Fueled, Lean-Premixed Gas Turbine Combustor. Journal of Engineering for Gas Turbines and Power 1993; 115: 554-62.

[19]    Gradinger TB, Inauen A, Bombach R, Kappeli B, Hubschmid W, Boulouchos K. Experiments on swirl stabilized non-premixed natural gas flames in a model gasturbine combustor. Combustion and Flame 2001; 3: 124-49.

[20]    Brandt M., Rachner M, Schmitz G. An Experimental and Numerical Study of Kerosine Spray Evaporation in a Premix Duct for Gas Turbine Combustors at High Pressure, Combustion Science and Technology, 1998; 138: 313.

[21]    Snyder TS, Rosfjord TJ, McVey JB, Hu AS, Schlein BC. Emission and Performance of a Lean-Premixed Gas Fuel Injection System for Aeroderivative Gas Turbine Engines, Journal of Engineering for Gas Turbines and Power 1996; 118: 38-45.

[22]    Dowling AP, Hubbard S. Instability in Lean Premixed Combustors. Proc Instn Mech Part A: 2000; 214: 317.

[23]    Venkataraman KK, Preston LH, Simons DW, Lee BJ, Lee JG. Santavicca DA. "Mechanism of Combustion Instability in a Lean Premixed Dump Combustor". Journal and Propulsion and Power 1999; 15(6): 909-18.

[24]    Rayleigh JWS. The Theory of Sound, Macmillan, New York, 1994.

[25]    Lieuwen TC, Yang V. Combustion Instabilities in Gas Turbine Engines. Progress in Astronautics and Aeronautics 2005; vol. 210.

[26]    Vauchelles D, Etude de la stabilité et des emissions polluantes des flammes turbulentes de prémélange pauvre à haute pression appliquées aux turbines à gaz, Ph. D. thesis, University of Rouen – CORIA, France, 2004.

[27]    Cabot G, Vauchelles D, Taupin B, Boukhalfa A, Experimental Study of Stability, Structure and CH* Chemiluminescence in a Pressurized Lean Premixed Methane Turbulent Flame. Experimental Thermal and Fluid Science 2004; 28: 683-90.

# CHAPTER 8

## Pulse and Lock-In NDT Infrared *Active-Source* Techniques

### Ralph A. Rotolante[*]

*Vicon Infrared 98 Baldwin Dr. C-8 Boxborough, MA 01719 USA*

**Abstract:** Infrared Pulse and Lock-In techniques are in use for NDT applications worldwide for Defect Imaging, Stress Imaging, Solar-Cell Shunt Imaging, Solar-Cell Carrier-Density/Lifetime Imaging, and Semiconductor Photo-and Electro-Luminescence imaging. A variety of active-sources, various infrared camera types and a selection of analytical tools are available for monitoring the responses of the materials to the active-sources. This chapter presents an overview of this rich choice of NDT tools used in academia, laboratories and factories for studies of material types ranging from composites, metals, ceramics, polymers and concrete, to semiconductors, solar cells, bones, leather and cheese, and projects strong future growth for the technology.

**Keywords:** NDT, infrared, thermography, lock-in, transient, flash, stress, delamination, cracks, shunts, photoluminescence, electroluminescence, LIT, vibrothermography, defects.

## 1. INTRODUCTION

Infrared Pulse and Lock-In techniques are in use today for NDT applications worldwide for Defect Imaging, Stress Imaging, Solar-Cell Shunt Imaging, Solar-Cell Carrier-Density/Lifetime Imaging, and Semiconductor Photo-and Electro-Luminescence imaging.

A variety of active-sources, various infrared camera types, with a variety of lenses and a selection of analytical tools are available for monitoring the responses of the materials to the active-sources.

This chapter presents an overview of this rich choice of NDT tools used in academia, laboratories and factories for studies of material types ranging from composites, metals, ceramics, polymers and concrete, to semiconductors, solar cells, bones, leather and cheese.

## 2. HISTORICAL BACKGROUND

Infrared *Active-Source* techniques are a sub-set of general infrared thermography.

Infrared thermography dates from the 1800s and the discovery of thermal radiation by Sir William Herschel. The fundamentals of infrared science were laid by pioneers, also in the 1800s, Gustav Kirchhoff, Slovak Joseph Stefan, Ludwig Boltzmann, and Scot James Maxwell with his overarching theory of electromagnetic waves.

By 1900, Max Plank described the spectral distribution of radiation from a blackbody which is the theoretical foundation for today's multi-billion dollar infrared sensor business. Of course, as in much of science, the practical breakthroughs occurred during periods of world conflict—crude thermopile imagers in WWI [1], Pb-salt imagers in WWII [2], and modern high-performance cameras, based upon HgCdTe and InSb detectors, during the Vietnam War [3]. These heavy, expensive imagers used single detectors or linear arrays (up to 180 pixels long) and complex scanning mirrors to develop images [4].

A review of the history of general thermography which used these imagers is given by Meola and Carlomagno [5]. This review includes thermo-fluid dynamics, starting with the 1968 success of Thomann

---

**\*Address correspondence to Ralph A. Rotolante:** Vicon Infrared 98 Baldwin Dr. C-8 Boxborough, MA 01719 USA, E-mail: ralph@viconinfrared.com

and Frisk who successfully imaged the temperature distribution of a part at Mach 7 in a wind tunnel; and continues on to NDE applications, traced by Vavilov [6], as the techniques moved from military to civilian uses, again starting seriously in the 1960s. In the 1960s and 1970s commercial companies such as Inframetrics and Agema began marketing cameras and the thermal NDT business began in earnest.

During the 1980s, Balageas [7] laid the analytical foundation for the imaging of subsurface defects, with a focus on coatings and laminated materials. The early applications mainly used passive thermography.

Late in the decade and into the early 1990s, with the advent of two-dimensional Focal Plane Arrays (FPAs) and thus, simpler cameras, prices dropped dramatically, capability soared, and commercial applications began to flourish worldwide.

Then when powerful, affordable computers combined with the FPA cameras, *active-sources* could readily be used with complex analytical tools. For example, a 640x480 pixels lock-in system, operating at 30Hz frame rate, requires about $10^7$ Fourier Transforms per second to produce the final image. A comprehensive explanation of image processing by Fourier Transforms is given by Owens [8].

Commercial companies began to market active source solutions, the first using mirror-scan cameras, such as Ometron's SPATE (Stress Photonics in North America), and Agema'sThermovision 900 (now FLIR) with the IRLockin option. Then came to market advanced FPA systems from Cedip in 1989 (now FLIR) and Thermal Wave Imaging in 1992, and Automation Technology's IrNDT (MoviTHERM in North America). Now the stage was set for the dramatic growth we see today in this powerful NDT tool—*active-source thermography*.

## 3. CLASSIFICATION

In the literature, there is some confusion in the names of active-source techniques, as one example, some refer to mechanical stress, such as a shake table as Vibrothermography, while others mean ultrasound to be the source.

Additionally, the active-source excitation sources are no longer just heat, but include laser stimulation for photoluminescence and electrical stimulation for electroluminescence imaging.

Thus, it is proper to classify the techniques. In this paper, the techniques are first named by the waveform: *Pulsed*, which includes Flash (ms duration pulses), and Transient (s duration pulses); and *Lock-In* (repetitive sinusoidal or rectangular pulses), as shown in Table **1**.

In Table **1**, we see that the three techniques are analogous to a "blink," and a "glance," and a "stare, and, just as in everyday life, each has its own advantages and disadvantages.

A further breakdown relates to the method of excitation, as shown in Tables **2** and **3** for pulsed and lock-in, respectively.

What is common to all these techniques is the use of infrared cameras. The most commonly used are shown in Fig. **1**.

MW, with an InSb FPA, requires a Stirling-cycle cooler and 80K operation. It offers superior thermal and spatial resolution and is the most common camera used today for active-source thermography. It is the camera of choice for challenging applications such as deep defects in composites or metals, or microscopic shunts in solar cells.

The uncooled, microbolometer-type offers a more affordable solution, costing one-half the amount of the MW. As its FPA operates at room temperature, it does not require a cooler, and so offers light weight and small size for applications such as portable systems with modest performance.

NIR cameras are also about one-half the cost of MW, and are used for electro-and photo-luminescence and paint and glass applications.

**Table 1:** Classification of IR active source techniques.

| Excitation Waveform | Category | Source Duration |
|---|---|---|
| PULSE | Flash | Millisecond |
| | Transient | Seconds to Minutes |
| LOCK-IN | Repetitive | Sinusoidal or Rectangular 0.01Hz to 20Hz Seconds to Hours |

**Table 2:** Pulsed excitation sources.

| Technique | Excitation Source | Typical Application |
|---|---|---|
| Pulsed Flash | Xenon Flash Lamp | Corrosion under paint |
| Pulsed Transient | Halogen Lamp | Delamination in large structures |
| | Laser | Weld integrity |
| | Heat Gun | Water porosity in building materials |

**Table 3:** Lock-in excitation sources.

| Technique | Excitation Source | Typical Application |
|---|---|---|
| LOCK-IN | Halogen Lamp Heat | Voids in multi-ply composites |
| | Laser Heat | Cracks in solar cells |
| | Ultrasound Heat | Cracks in metals |
| | Eddie Current Heat | Voids in complex geometry metal |
| | Mechanical Heat | Fatigue limits in automobile parts |
| | Electrical Heat | Solar cell Shunts-Power Losses |
| | Optical Bias | Solar cell Shunts-Power Losses |
| | Laser Carrier Pumping | Semiconductor material quality |
| | Laser Photoluminescence | Semiconductor band-gap defect levels |
| | Electrical Electroluminescence | Semiconductor processing defects |

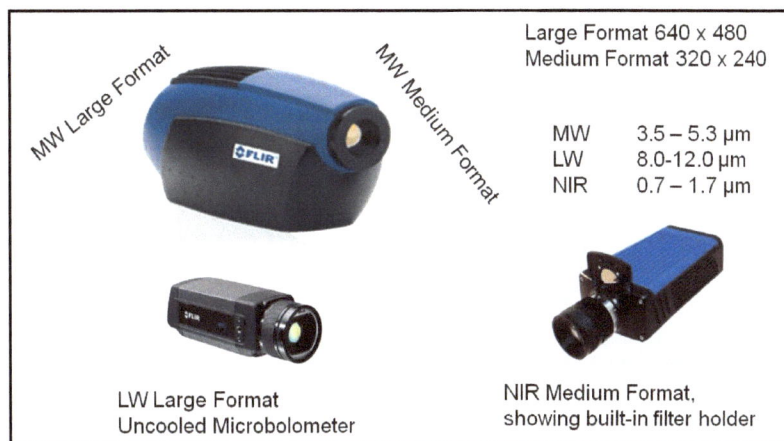

Large Format 640 x 480
Medium Format 320 x 240

MW Large Format

MW Medium Format

MW    3.5 – 5.3 μm
LW    8.0-12.0 μm
NIR   0.7 – 1.7 μm

LW Large Format
Uncooled Microbolometer

NIR Medium Format, showing built-in filter holder

**Figure 1:** Typical infrared cameras used for NDT inspections.

Long-Wave HgCdTe cameras with high sensitivity and fast response times are also available. These cooled cameras are twice the cost of a MW and are used in specialty applications requiring very high-speed or cold background temperatures.

For all cameras, wide angle lenses are available for large targets, such as bridges or large solar panels. At the other extreme, magnifying lenses (up to 5μm per pixel resolution) are available for microscopic defects such as semiconductor faults or coating failures.

## 4. PULSED THERMOGRAPHY—FLASH AND TRANSIENT--PRINCIPLES OF OPERATION

### 4.1. Pulsed Equipment

In pulsed thermography, the stimulus is applied with a Xenon Flash Lamp for a Flash pulse, or with a Halogen Lamp in the Transient case. In both cases the equipment setup is according to that shown in Fig. **2**.

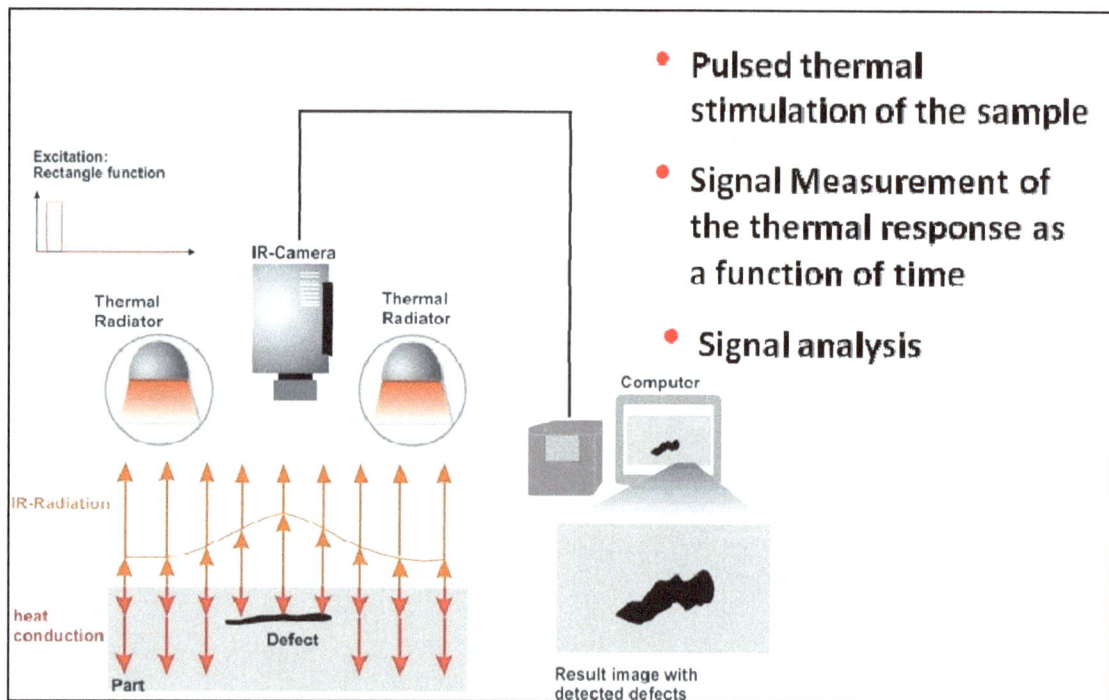

**Figure 2:** Equipment setup for Pulse Thermography—Flash or Transient.

### 4.2. Pulsed Mathematical Background

The Heat Conduction Equation, with the usual simplifying assumptions (isotropic; infinite x and y; semi-infinite z), has the solution:

$$T(z,t) = \frac{Q}{e\sqrt{\omega}} e^{-j\frac{\pi}{4}} e^{-\frac{z}{\mu}} e^{-j\frac{z}{\mu}} \tag{1}$$

$$e = \sqrt{K\rho c_p} \qquad \text{thermal effusivity}$$

$$\omega = 2\pi f \qquad \text{angular frequency}$$

$$\mu = \sqrt{\frac{2K}{\rho c_p \omega}} \qquad \text{thermal penetration depth}$$

$Q$ is the input energy at the sample surface, $T$ = temperature, $t$ = time, $\rho$ = density, $c_p$ = specific heat capacity, and $K$ is the thermal conductivity.

This gives a thermal propagation time $t^*$ to the depth $z$ of a subsurface defect:

$$t^* \sim \frac{z^2}{K}. \tag{2}$$

$K$ is the thermal diffusivity, $K/\rho c_p$.

So, $t^*$ for 2mm of aluminum is about 40ms and $t^*$ for 2mm of graphite epoxy is about 30s. This means Flash is better for materials of high thermal diffusivity, *e.g.*, metals. Materials with a low thermal diffusivity, *e.g.*, composites, have a long thermal propagation time, which limits Flash to the detection of shallow defects.

The solution of the Heat Conduction Equation at the surface, which is what the camera sees, is:

$$T = \frac{Q}{2e\sqrt{\pi t}} \tag{3}$$

An instructive visual representation of the signal is shown in Fig. **3**.

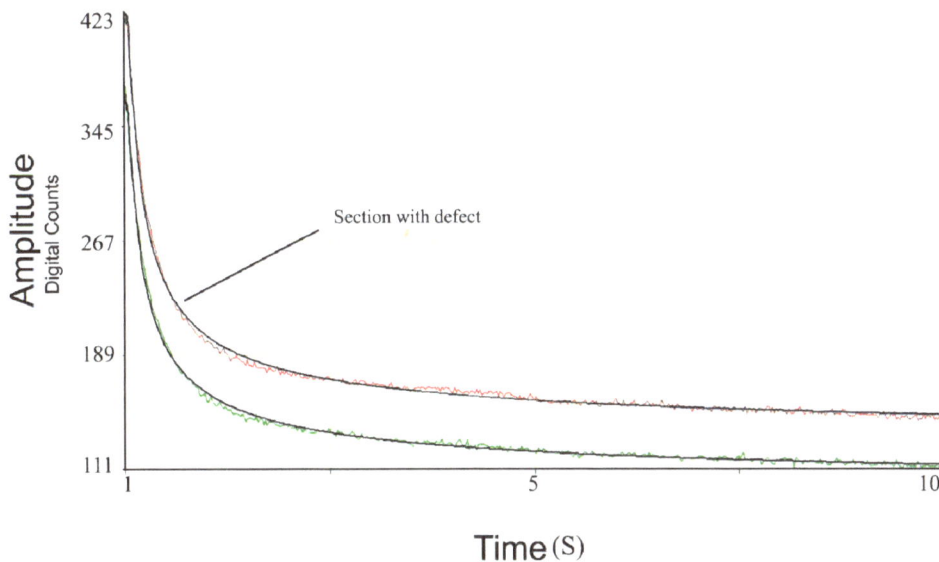

**Figure 3:** Typical temperature signal of one pixel at the surface of a part after applying a heat pulse.

In Fig. **3**, note the reconstructed profiles in black from an analytic approximation function. With the recorded infrared image sequence the information is available for the evaluation of the temperature signal in each individual pixel of the image. Each pixel can be regarded as a discrete measuring point on the surface of test object, giving a high spatial resolution with the infrared camera.

Profile approximation functions have the form of a polynomial of variable order where different mathematical approaches can be used to fit curves. A good treatment of these principles is given by Maldague [9], where he uses examples such as *polyfit* in Matlab, and then uses a quadratic fit to show good agreement with an example temperature profile.

This approach effectively removes the noise completely for image evaluations.

Commercial versions of thermal analysis software offer consideration of the analysis start point, while the temperature is rising; and allow different degrees of polynomial fit; and offer choices such as first-and second-derivative of the curve; and a display of the coefficients calculated.

Also, a Pulse-Phase calculation can be done which performs a Fourier Transform of the recorded signal and gives an amplitude and phase spectrum as a function of the frequency.

The recorded temperature information is transformed into the frequency domain. In each pixel the measured temporal evolution of the temperature is Fourier transformed for all images of the recorded sequence:

$$F_n(x,y) = \sum_{k=0}^{N-1} T(k)e^{\frac{j2\pi kn}{N}} = \text{Re}(F_n) + j\text{Im}(F_n) \qquad (4)$$

with:

$F_n$    Fourier transform for frequency $n$

$n$    increment offrequency, n = 1, 2…N/2

$N$    number of images in the sequence

$x, y$  index of image pixel

$Re$    real part

$Im$    imaginary part

A calculation of the phase information indicates the transit time of the waves until they reach a defect:

$$\Phi_n(x,y) = arctan\left(\frac{\text{Im }(F_n(x,y))}{\text{Re }(F_n(x,y))}\right) \qquad (5)$$

A display of the phase information is presented as an image. Thus, even for pulse stimulation a phase image can be obtained.

The Pulsed Phase Evaluation Method has two distinct and powerful advantages:

- All temperature data from the recorded image sequence are used for the evaluation. The result is displayed as one image. Compared to common evaluation methods the Pulsed Phase Evaluation gives a significantly improved sensitivity, a much better contrast for defects in the result image and a much greater depth range.

- The method allows the calculation of depth-resolved evaluations. It calculates the phase information for the complete frequency spectrum with very short computing times. This gives a sequence of result images for all frequencies. The corresponding frequency of a result image indicates the depth range.

It is helpful to be able to use several evaluation techniques to analyze a sequence, and then chose the best for a given test.

The key features of the Flash and Transient methods can now be summarized.

The main advantages of the Flash are short measuring times and the ability to perform depth-resolved inspections, providing excellent performance for inspection of thin layers and for the detection of shallow defects.

The main disadvantages of Flash are: A limited inspection area due to the energy of the flash lamps, a depth range limited to shallow defects and a high-thermal loading on the sample.

The main advantages of Transient are: (1) Applicable for large-area measurements, (2) Affordable halogen lamp heat-sources and (3) Low thermal load to the inspected component.

**USE TRANSIENT**

...if the sample gets thick!

Glass reinforced plastic ½ in thick

...if the area gets big!

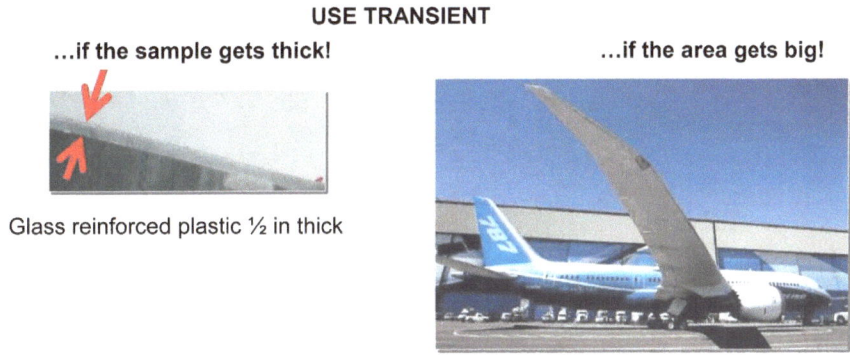

**Figure 4:** Transient is best for composite samples that are large area and/or thick.

A rough rule-of-thumb for composite defect inspection is up to 3mm for Flash (see Shepard [10]), up to 6mm for Transient (and up to 12mm for Lock-In, which is discussed below). Of course, many material and configuration variables are present, so each application must be evaluated individually. Fig. **4** summarizes a common rule-of-thumb.

## 4.3. Pulsed Thermography—Examples

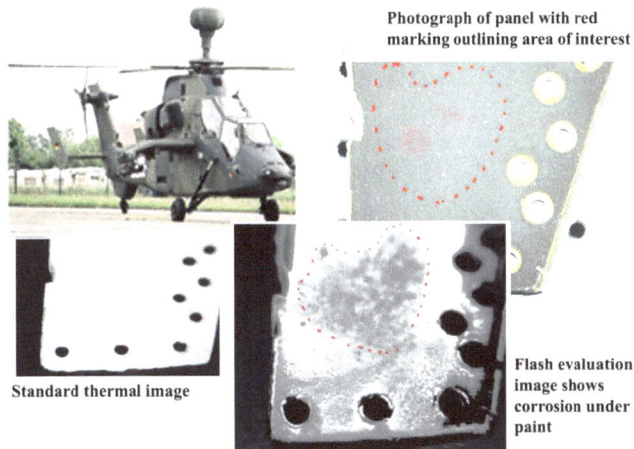

Photograph of panel with red marking outlining area of interest

Standard thermal image

Flash evaluation image shows corrosion under paint

**Figure 5:** Flash Thermography is well suited for very thin samples.

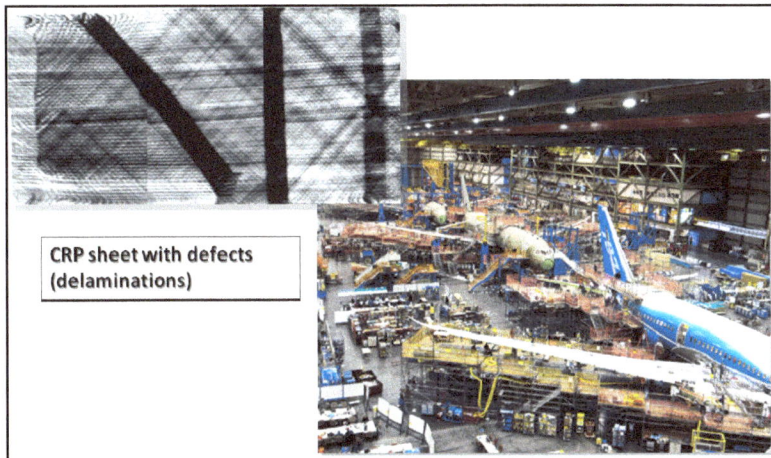

CRP sheet with defects (delaminations)

**Figure 6:** Delamination in an aircraft hull, Carbon-fiber Reinforced Polymer structure. Image taken with MoviTHERM/Automation Technology IrNDT commercial system with a FLIR 320x240 pixels cooled MW camera. A Transient Mode, 1st derivative evaluation shows the dark stripe delaminations as well as fiber structure.

**Figure 7:** Bond voids (about 15mm deep) in a rotor blade Glass-Fiber Reinforced Polymer structure. Blade section is about 4ft long, and was tested with a Transient Pulse of 15 seconds, 30 second decay, and evaluated in the Pulse-Phase Mode. A phase image is shown which assists in the problems of reflections, dirty surfaces and emissivity differences.

### 4.3.1. Flash Thermography Example

Flash thermography with a high-power Xenon lamp, is a powerful tool for thin samples, shallow defects, such as corrosion under paint. A dramatic example is shown in Fig. **5**, a helicopter panel, 8mm thick Al, with a special military-class paint. A single 6kJ Xenon lamp was used with a 2ms pulse. The each pixel (640x480) of the MW thermal decay curve was fit with a polynomial function, as described in Section 4.2, using the IrNDT software of Automation Technology (MoviTHERM). Note that the standard thermal image and the visible image show little effect, while the Flash result dramatically shows the extensive corrosion. This is a situation where Transient or Lock-in clearly would not work.

### 4.3.2. Transient Thermography Examples

Transient Thermography becomes important for large structures, as shown in the aircraft factory image of Fig. **6** and the rotor blade example of Fig. **7**. Depending upon the minimum defect size to be investigated, sections of an airplane fuselage or an AWACS dome as large as 4ft x 8ft can be imaged in a single shot. Typical set-ups will have two to four Halogen lamps (1kW each), used with a 10s pulse and a 20s cool-down.

## 5. LOCK-IN THERMOGRAPHY

### 5.1. Historical Background

An historical background of Lock-in thermography can be found in the Meola and Carlomagno review article [1]. There is cited the early works of Wu *et al.* [11, 12] on ceramic coatings and on aircraft structures with flash lamps and ultrasound excitation, and Giorleo and Meola [13] on bonded joints with adhesive and weld.

More recent pioneering work in 2001 was done by Bremond and Potet [14] on imaging dissipated energy and fatigue measurements.

New applications for lock-in thermography continue to be found, such as recent work imaging electromagnetic waves as described by Levesque *et al.* [15].

### 5.2. Principles of Operation

### 5.2.1. Lock-In Equipment

In lock-in thermography, the dynamic stimulus can be applied with a wide variety of sources. For composite inspection this includes Halogen lamps, Ultrasound, and Mechanical stimulation. In most cases the equipment setup is according to that shown in Fig. **8**.

An important variation is the illumination of the sample from the backside [16]. This is particularly useful for samples with thin inclusions and for samples with complex geometry on the front surface.

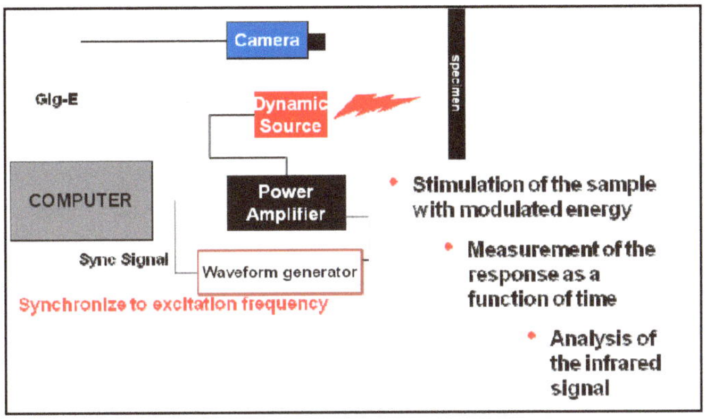

**Figure 8:** Schematic of equipment setup for Lock-In measurements.

### 5.2.2. Lock-In Mathematical Background

Mathematically, in the lock-in case, the recorded temperature information is transformed into the frequency domain. In each pixel the measured temporal evolution of the temperature is Fourier transformed for all images of the recorded sequence. The Phase and Amplitude information are derived and presented as an image.

Advantages of the lock-in method include:

- The summation causes a noise filtering of the data which enhances the contrast in the resulting images.

- Depth range for the phase information is twice that of the Pulse Mode, $\sim 2\mu$ with:

$$\mu = \sqrt{\frac{2K}{\rho\, c_p\, \omega}} \tag{6}$$

- Lock-in allows detection of thermal waves with a sensitivity of 100 to 1,000 greater than the best thermal camera – down to $\mu$-Kelvin range.

- The phase image is insensitive to external effects, such as sunlight, reflections, dirt and emissivity differences – problems common to conventional thermography.

- The phase information is insensitive to uneven distribution of the applied heat.

- Allows for large areas to be examined within a few minutes from a distance, non-contact.

- Un-cooled IR-camera is normally good enough.

- Affordable heat source (*e.g.,* halogen lamps).

- Visualization of deeper defects than with pulse thermography.

Fig. **9** shows the measuring principle, using CFRP material with a sinusoidal excitation. The curves are the typical temperature signals at the surface. Curves are shown for pixels with and without delamination. The time scale is in units of *frames* of the camera (usually 30 frames/s). The temperate value is read in digital counts, every pixel, every frame.

Note that important signal information is evident in both amplitude and phase. Now imagine one period of the heat-driving sine wave superimposed on Fig. **9**. A Fourier Transform of a given pixel image will give magnitude and phase information with respect to the input, synchronized driving signal. This is all done in the frequency domain, rather than the time domain shown in the figure.

The resulting frequency domain images contain the data we seek. Several examples follow. Be cautioned that we live and think daily in the time domain, so one should not try to over-analyze the images. In some cases the amplitude image is superior, in many the phase images are best, for the reasons stated above.

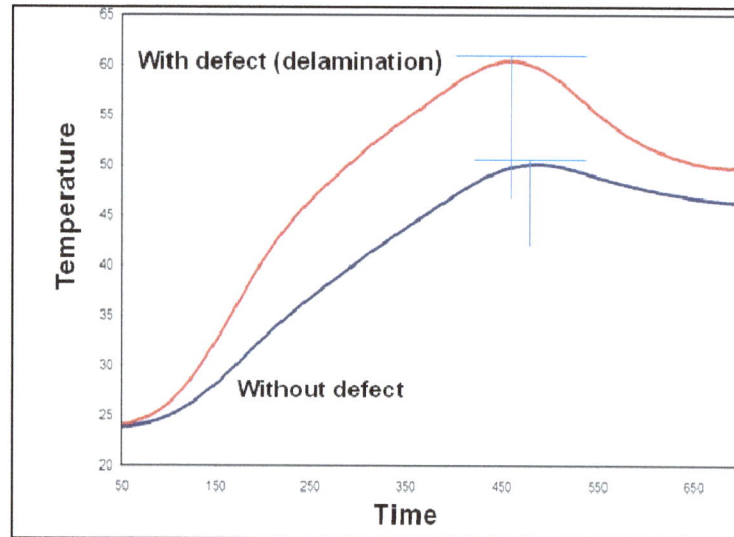

**Figure 9:** Example of the lock-in measurement principle.

### 5.2.3. Lock-In Thermography—Examples

#### Honeycomb Structures

The most common excitation source for lock-in is the affordable Halogen lamp. Fig. **10** shows an example of voids in a new composite honeycomb. Other defects like delaminations or inclusions also can be detected reliably with lock-in.

The challenging measurement of Fig. **10** was done at a frequency of 0.05Hz, and two periods—a forty second measurement, acceptable for many applications. The panel is 1 inch thick with defects on back side.

#### Aircraft Structures

Lock-in systems are available for field/depot use in the aircraft industry. The design and operation of both military and commercial aircraft have stringent requirements in order to insure structural integrity [17]. Yet, even with these requirements there have been numerous structural failures in both military and commercial aircraft due to operating the aircraft beyond their fatigue life [18].

An example is the Boeing-737; the most widely used commercial jet in the world, the first being rolled out in January of 1967. Traditional Inspection procedures require an estimated 1,500 man-hours for a complete inspection of each 737 [19]. The hull of an aircraft is manufactured using a variety of different materials and fasteners such as aluminum and rivets, carbon fiber reinforced polymers (CFRP), honeycomb structures and other composite materials. Material fatigue occurs due to constant temperature changes, pressurization and depressurization, vibration and high wind loads. Typical damage modes include: rivet failure, disbonding, impact damage, cracking and delamination. Of special interest are condensate and water inclusion in composite material as these can cause cracks at high altitudes as a result of water expansion when turning into ice.

Fig. **11** demonstrates that lock-in thermography can be used to quickly and effectively inspect large sections of the fuselage of an aircraft [18]. The area to be inspected can include several square meters at a time and the process may only take a few minutes to complete thus dramatically reducing the total inspection time needed per aircraft to about 100 hours.

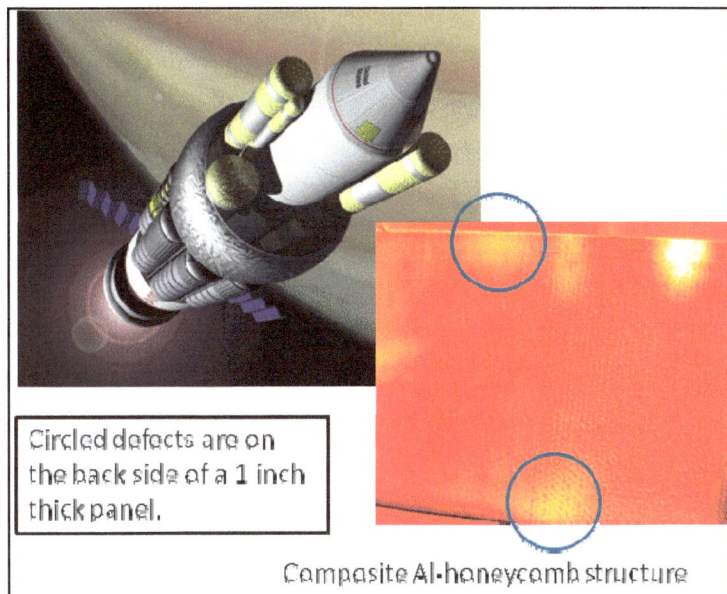

Circled defects are on the back side of a 1 inch thick panel.

Composite Al-honeycomb structure

**Figure 10:** Voids in composite Al-honeycomb structure.

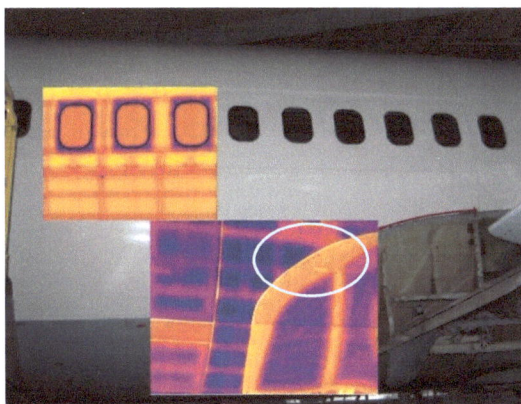

**Figure 11:** IR lock-in phase image of Boeing-737 fuselage with large area disbonding showing in lower image.

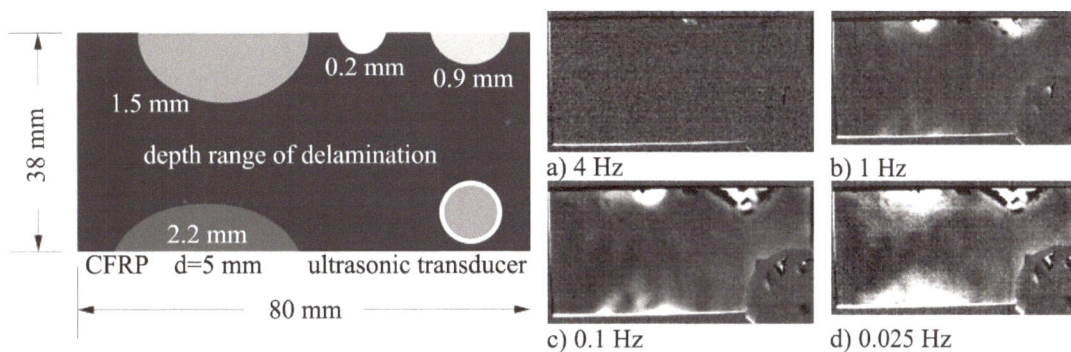

**Figure 12:** Ultrasound phase images of CFRP sample with delaminations (Zweschper [20]).

## Lock-In Ultrasound Source

An important source of heat is ultrasound. Typical settings are 100W at 20KHz with a 200ms burst frequency for synchronization. A disadvantage of the ultrasound technique is that it can be destructive—care is required in the temperature rise.

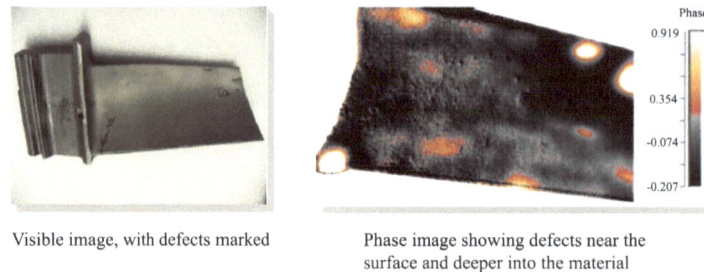

Visible image, with defects marked          Phase image showing defects near the
                                             surface and deeper into the material

**Figure 13:** Infrared lock-in phase image with ultrasound as the active-source.

Fig. **12** shows a test sample of CFRP with delaminations at different depths. The defects at 2.2mm are clearly visible.

The use of ultrasonic stimulation allows deep penetration into metal or composite structures, as shown in Fig. **13**, while retaining the large-area image advantage of thermography. In a few seconds an image is obtained that would take an hour using traditional ultrasonic means.

## Lock-in Thermal Stress Analysis

A powerful tool for laboratory and factory is the use of mechanical excitation for heat generation through the thermoelastic effect, shown in Fig. **14**.

Applications of this technique include measurement of fatigue limits, imaging of stress patterns, crack propagation studies, and imaging of vibration patterns, as shown with turbine blades in Fig. **14**.

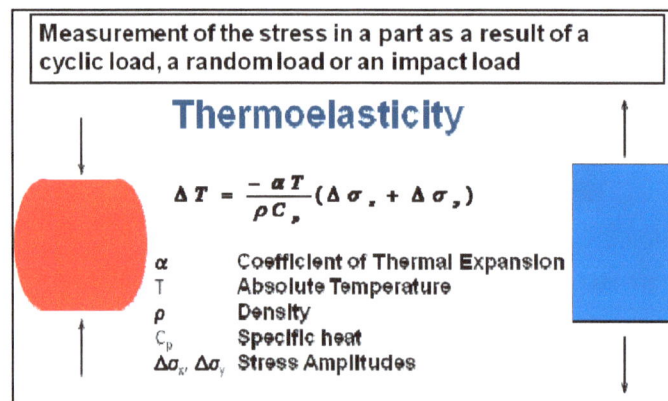

**Figure 14:** Compression and expansion give temperature changes proportional to the stress.

The stress images of Fig. **15** are equivalent to having 320 x 240 = 77,000 strain gages on the sample. Engineers often compare results such as these to FEA calculations and images.

Polymers present a challenge for stress imaging thermography because they are often orientation dependent, as shown in Fig. **16**. The different thermal characteristics of the fibers are evident in the figure. Features of 1mm size are evident. A standard microscope lens could be used with this FLIR MW camera to examine detail as small as 5μm.

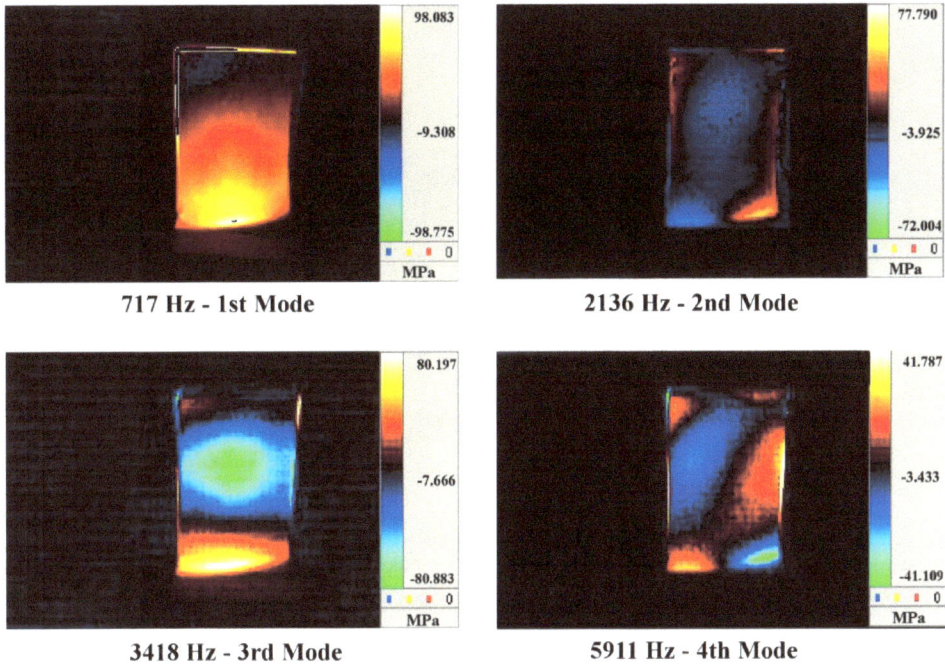

717 Hz - 1st Mode

2136 Hz - 2nd Mode

3418 Hz - 3rd Mode

5911 Hz - 4th Mode

## Tests performed on shake table

**Figure 15:** Vibration mode on a turbine blade tested on a shake table.

## Thermo-mechanics on Polymers

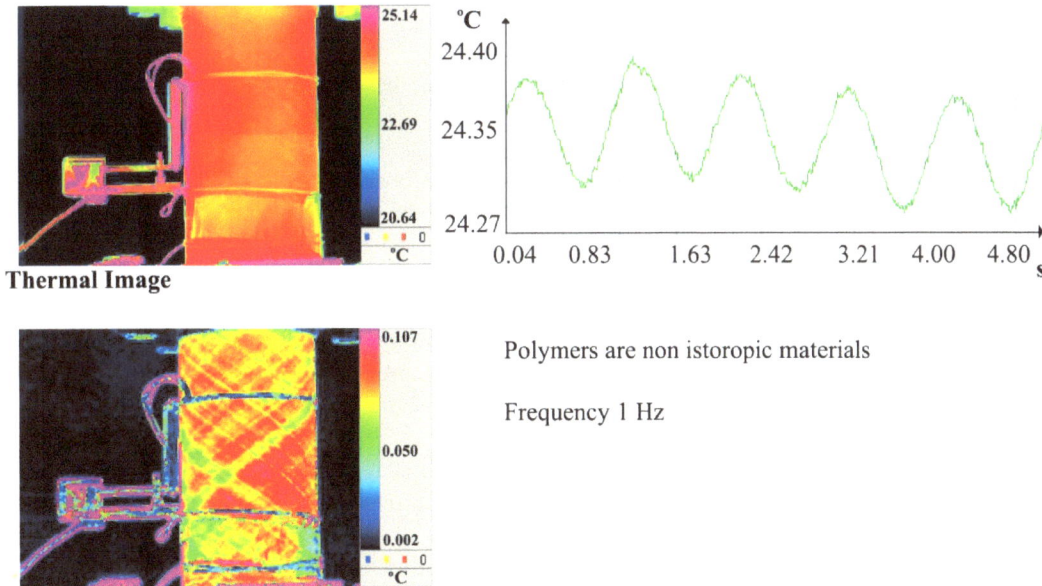

**Thermal Image**

Polymers are non istoropic materials

Frequency 1 Hz

**Figure 16:** Stress patterns on a polymer sample.

### *Lock-in Imaging of Dissipated Energy*

Microscopic cracks occur where the dynamic mechanical energy is being dissipated. A small, non-adiabatic signal is observed in the lock-in harmonics. It is observed [13] that the maximum stress point can be different from the point of maximum energy dissipation, as in the case of shear.

Fig. **17** shows an example in automotive steel. The limit of fatigue shown in the graph is given by the point where the lock-in dissipated energy phenomena starts.The total test took 2 hours, as compared to days or even weeks using traditional methods.

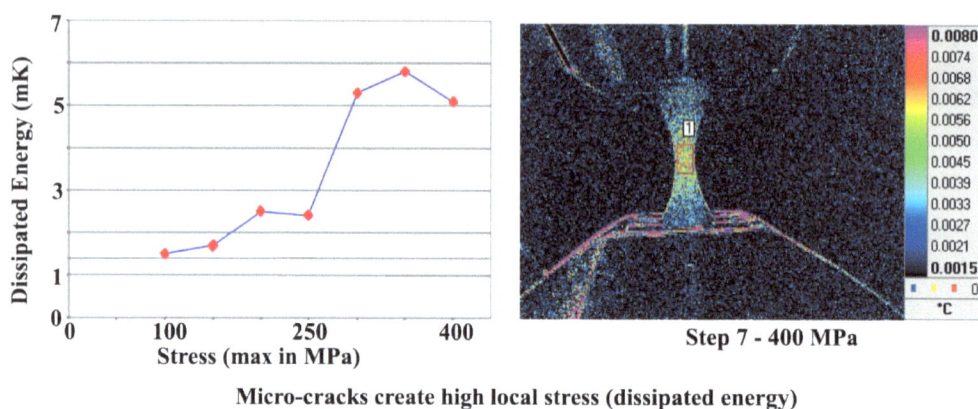

**Micro-cracks create high local stress (dissipated energy)**

**Figure 17:** Dissipated energy is observed as a harmonic signal of the lock-in frequency. The signal values are plotted as a function of the measured stress.

## Lock-In Semiconductor and Solar-Cell Imaging

The use of lock-in thermography for diagnostics of electronic components, including solar-cells, was pioneered by Breitenstein. His monograph [21] outlines the basics of the techniques.

An overview of the various lock-in techniques is given by Tarin [22].

In Fig. **18** a solar cell or other component is stimulated with a pulsed bias current. The local heating due to defects is readily observed in a lock-in image.

A Medium Wavelength (MW) camera is typically used in these measurements, although a low-cost microbolometer can observe many of the same defects.

In Fig. **19** is shown an image of shunts in a solar cell while observing the entire wafer, and then while observing with a magnifying lens. Note the standard thermal image shows no effect. This test is referred to as Dark Lock-In Thermography (DLIT).

DLIT is the most widely used active-source technique to be used with solar cells. A reverse or forward electrical bias is applied as in Fig. **18**, and the localized heating due to shunts is observed with an MW camera. Usually a large format camera (640 x 480) is used because its superior spatial resolution allows the identification of the shunt position—edge, bias line, grain boundary, *etc.*

Breitenstein has identified at least seven different kinds of shunts in solar cells [23], the two main categories being resistive or diode-like, readily differentiated in forward *vs.* reverse bias DLIT.

Optical bias, rather than electrical bias, can also be used to stimulate shunts in a solar cell. Different kinds of shunts are seen with electrical *vs.* optical stimulation, referred to as Illuminated Lock-In Thermography (ILIT). Laser illumination (usually around 810nm) or lamps are commonly used for ILIT. Because ILIT shunts are those seen in actual operation, they are important indicators of power loss. ILIT is sometimes combined with electrical bias for further differentiation of shunt types.

Also, a MW camera can be used to image carrier density, minority-carrier lifetime in semiconductor materials. Lifetime is reduced when defects are present, so lifetime maps can show the semiconductor quality, all along the fabrication process.

A map of carrier lifetime is shown in Fig. **20**. In this case, a black-body infrared source (as simple as a black plate) is placed behind the wafer and the wafer transmission of MW radiation is observed from the front side. As carriers are pumped into the conduction band with a modulated laser source (about 7Hz), the transmission of the wafer is altered. In the mid-to far-infrared region, the transmission of the semiconductor is a linear function of the carrier concentration. Minority carrier lifetime is dependent upon the defect and impurity levels in the material, and controls the decay of the carriers in the band [24].

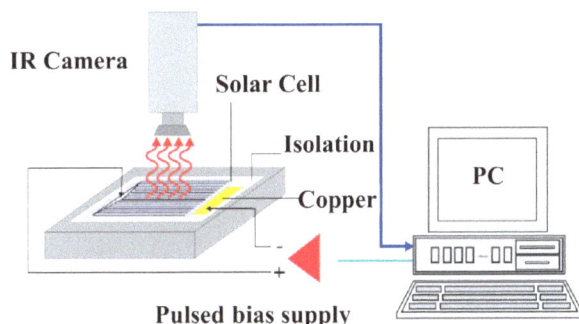

**Figure 18:** Typical set-up for measuring shunts in electronics components.

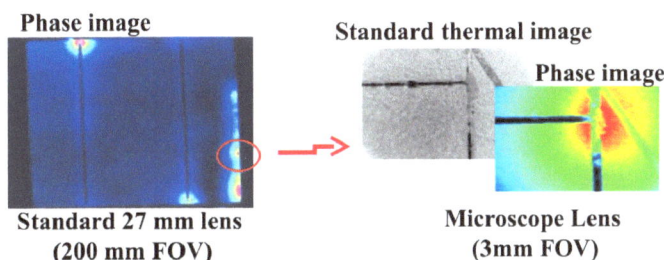

**Figure 19:** Lock-in phase images of a solar cell under rectangular bias—4V, 0.3A at 2Hz, 10 second test time. A FLIR MW, large format camera was used with a MoviTHERMIrNDT lock-in system.

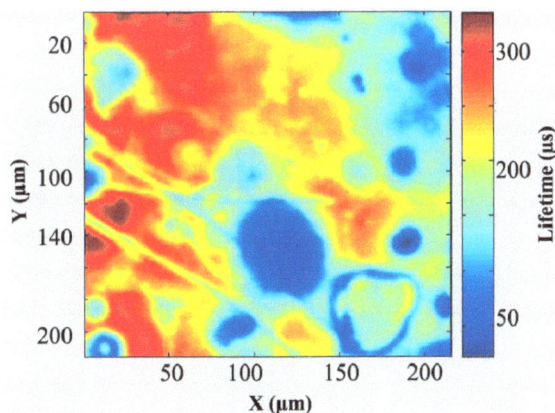

**Figure 20:** A wafer has been purposely degraded with iron spots and then the lifetime map formed using an 808µm laser source [25].

Another valuable tool for semiconductor studies is a Near IR (NIR) camera (0.9 to 1.7µm). Lock-in with an electrical bias or an optical bias produces electroluminescence (EL) and photoluminescence (PL), respectively. Both techniques give lifetime and defect-level information, the advantage of PL being the ability to image wafers before contacts are applied.

An example of the power of PL is shown in Fig. **21** [26]. Carriers are excited with a modulated laser, as above, but in this case the NIR camera observes the radiative emission of the semiconductor, due to recombination of the carriers [27]. Areas on the sample with longer lifetimes will have higher steady state density of free carriers [28].

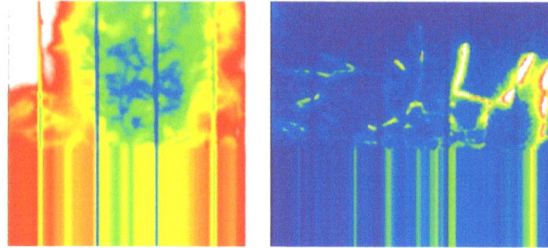

**Figure 21:**-Photoluminescence with and without a camera-lens filter. Left image shows the 0.9 to 1.7µm result. Right image shows the 1.32 to 1.7µm result using a band-pass filter.

The figure image on the left includes band-to-band plus defect emissions, while the image on the right shows luminescence from a defect level, only (probably 0.8eV). PL and EL images must be obtained on a dark housing, which complicates a test structure.

## 6. CONCLUSIONS

It is evident from the discussion above that Pulsed Flash, Pulsed Transient and Lock-In Thermography are important tools for research and for quality assurance in most areas of manufacturing. The particular technique used must be determined on a case-by-case basis. A starting point for the choice is shown in Fig. **22**—do you need a Blink…a Glance…or a Stare.

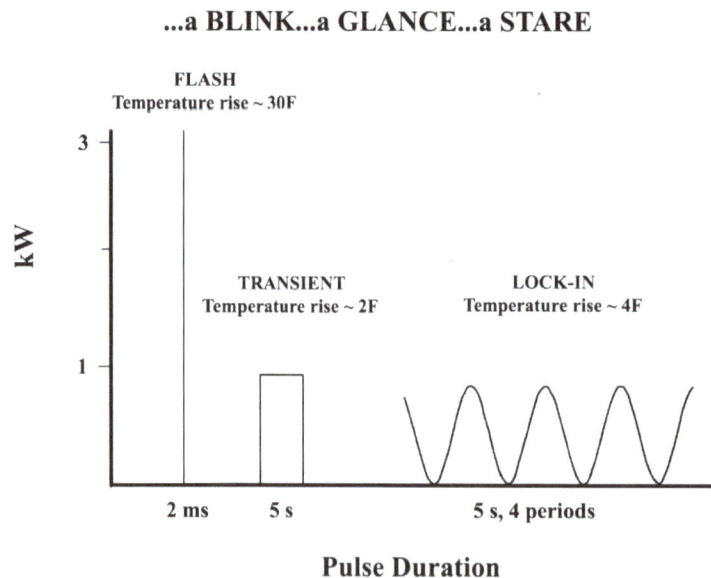

**Figure 22:** Time, power and temperature regimes for Pulsed Flash, Pulsed Transient and Lock-In.

Pulse Flash is shown to be especially useful for thin samples, and excels at observing corrosion under paint, for example.

Pulse transient is a fast, and with uncooled microbolometer cameras, an affordable, lightweight solution for automation, factory and field applications, mainly for defects in materials such as composites, polymers and leather.

Infrared Lock-In NDT is the most sensitive and versatile of the techniques. It finds use in challenging composites, metals, ceramics, polymers, plant materials, foods, semiconductors and even microwave patterns. From metal fatigue to bone strength to minority-carrier lifetime mapping, the lock-in technique continues to improve and grow in applications and importance.

Education, equipment and software advances will continue to propel *active-source NDT* to new applications and to further, continuing growth in the established techniques.

In education, powerful learning opportunities are offered by venues such as the Thermosense Conferences and resources, and its association with SPIE; Quantitative Infrared Thermography (QIRT) Conferences and resources; Inframation Conferences (sponsored by FLIR), offering hands-on clinics in addition to conference activities; and web-based resources such as NDT.org (http://www.ndt.org/).

Advances in excitation sources, including low-cost laser diodes offers new NDT opportunities, such as in welding studies. Advances in computers allow smaller, more powerful systems. New automation tools, such offered by National Instruments, allow production line applications, sometimes in conjunction with new robotics capabilities.

Advances in infrared camera technology are a continuing force which expands the uses of active-source NDT. The recent proliferation of lower-cost microbolometer cameras opens up new opportunities in the marketplace, such as hand-held systems for field use. The advent of FPA –based NIR cameras expands the capability to new materials and phenomena, such as semiconductor manufacturing. Ever-larger format cameras including 1024x1024 InSb offer new opportunities, especially in research.

As shown in this chapter, infrared *active-source* techniques have proven to be a powerful and growing tool for NDT applications covering a broad range of academic, research, factory and field applications.

## REFERENCES

[1]    Miller JL. Principles of Infrared Technology, New York Van Nostrand Reinhold, 1992.
[2]    Kondas DA. Introduction to Lead Salt Infrared Detectors, ARFSD-TR-92024, U.S. Army Armament Research, Development and Engineering Center; 1993.
[3]    Kruse PW. Uncooled Thermal Imaging Arrays, Systems, and Applications. SPIE Press, (2001).
[4]    Rotolante RA. Changes mark the military IR-detector market". Laser Focus World 1991; 93.
[5]    Meola C, Carlomango GM. Recent advances in the use of infrared thermography. Measurement Science Technology 2004; 15:27-58.
[6]    Vavilov VP. Thermal non destructive testing: short history and state-of art, in Proc. Quantitative Infrared Thermography, QIRT 92 EETI editions Paris, France 1992; 179-93.
[7]    Balageas DL. Krapez, JC, Cielo P.Pulsed photothermalmodeling of layered materials, J Appl Physics 1986; 59: 348-57.
[8]    Ownes R. Fourier Transform Theory. http://homepages.inf.ed.ac.uk/rbf/CVonline/LOCAL_COPIES/OWENS /LECT4/node2.html (Accessed January 2011).
[9]    Maldague X. Theory and Practice of Infrared Technology for Nondestructive Testing, John Wiley & Sons, (2001).
[10]   Shepard S. Understanding Flash Thermography. Materials Evaluation May 2006; 454-64.
[11]   Wu D, Rantala J, Karpen W, *et al.* Applications of lockin-thermography methods, QNDE: Review of Progress in Quantitative Nondestructive Evaluation 1996; 15: 511-9.
[12]   Wu D, Zweschper T, Salerno A, Busse G. Lock-in thermography for nondestructive evaluation of aerospace structures, NDT.net 3, (1998).
[13]   Giorleo G, Meola C. Non destructive evaluation of defects in composite materials by means of infrared thermography, 25th AIAS Naz. Cong on Mat Eng 1996;1: 431-8.
[14]   Bremond P, Potet P. Lock-In Thermography: A tool to analyze and locate thermo-mechanical mechanisms in materials and structures, Thermosense XXIII April 2001; 4360-76.

[15] Levesque P, Bremond P, Lasserre J-L, Paupert A, Balageas DL. Compared improvement by time, space and frequency data processing of the performances of IR camera. Application to electromagnetism, 16th World Conference on NDT, (2004).

[16] Meola C. Nondestructive Evaluation of Materials With Rear Heating Lock-In Thermography, Sensors Journal, IEEE, 2007; 7: 1388-9.

[17] Tarin M, Kasper A. Fuselage inspection of Boeing-737 using lock-in thermography, SPIE Proceedings March 2008; Vol. 6939: 693919.1-693919.10

[18] Gandhi MV, Thompson BS. Smart Materials and Structures, Chapman and Hall, London 1992.

[19] Laird J. Lufthansa Technik keeps the fleet flying, Aerospace Engineering and Manufacturing, SAE International Magazines, PA, USA, April 2008; 12-4.

[20] Zweschper T, Dillenz A, Scherling D, Riegert G, Busse G. Ultrasound excited thermography using frequency modulated elastic waves, Insight 2001; 43: 173-9.

[21] Breitenstein O, Langenkamp M. Lock-in Thermography Basics and Use for Functional Diagnostics of Electronic Components, Springer Series in Advanced Microelectronics, Springer-Verlag Berlin Heidelberg, 2010; Vol. 10.

[22] Tarin M, Overstreet R. Lock-in Thermography Enables Solar Cell Development, Photovoltaics World Magazine Nov 2009; Vol. 2009, Issue 5.

[23] Breitenstein O. Rakotoniaina JP, Al Rifai MH, Werner M. Shunt Types in Crystalline Silicon Solar Cells, Progress in Photovoltaics: Research and Applications 2004; 12: 529-38.

[24] Linnros J. Carrier Lifetime Measurements Using Free Carrier Absorption Transients. I. Principle and Injection Dependence, J. Appl. Phys. 1998; 84: 275-83.

[25] Berman GM, Johnson SW, Call NJ, Ahrenkiel RK, Rotolante RA. A Comparison of Transient and Imaging Techniques for Measuring Minority-Carrier Lifetime, 18th Silicon Solar Cell Workshop, Vail, CO, 2008; Proceedings NREL/BK-520-45745; 151-4.

[26] Johnston SW. Silicon Solar Cells Imaging Using Different Cameras, 20th Workshop on Crystalline Silicon Solar Cells & Modules, Aug. 2010; Breckenridge, CO, proceedings not yet published.

[27] Trupke T, Bardos RA, Schubert MC, Warta W. Photoluminescence imaging of silicon wafers. Allied Physics Letters 2006; 89: 044107-1--04107-3.

[28] Berman GM, Call N, Ahrenkiel RK, Johnson SW. Evaluation of Four Imaging Techniques for the Electrical Characterization of Solar Cells, Material Research Society Fall Meeting 2008; Proceedings NREL/CP-520-44607; 1-6.

# Concluding Remarks to Part II – Section 3

Section 3 of Part II was devoted to the employment of infrared thermography in industrial engineering. Two main topics were addressed, which are concerned with thermo-fluid-dynamics phenomena (chapters six and seven) and nondestructive evaluation of materials (chapter eight).

An overview on the applications in thermo-fluid-dynamics starting from the first attempts performed in the middle 60s with the AGA Thermovision 680 to the most recent measurements of convective heat transfer with the SC3000 (the camera models were described in Part I) was given in Chapter 6. A description of the heat flux sensors, mostly used with infrared thermography, was provided in section 2 of this chapter; the problem of image restoration was also addressed. In this chapter were also illustrated some examples of applications to fluid dynamics, ranging from natural convection to hypersonic regime, and involving transition, separation and reattachment phenomena, jets, flow instability, rotating devices, *etc.* which bear witness for the capability of infrared thermography to deal with complex fluid flows.

A particular application to the visualization of combustion phenomena was described in Chapter 7. A fast InfraRed PHOENIX Camera equipped with a detector working in the 1.5-μm wavelength, and of frame rate up to 38kHz was used, which is a particular sophisticated camera model for specific applications. It is worth noting that suppliers of infrared cameras are generally open to customer requirements. The authors of Chapter 7 focused on the usefulness of infrared thermography for characterization of temporal and spatial combustion fluctuations. They also highlighted the high sensitivity of the infrared camera, which allowed for individuation of high and low temperature/emissivity scatterers, namely hot gases ($CO_2$) and small particles (coal, particulate).

Infrared thermography is also a powerful technique for non destructive testing of materials; this topic was addressed in Chapter 8. After a short historical overview, a description of the main techniques, pulse and lock-in, with a panorama of stimulation sources and types of detected defects was given; the most important relationships were also described. The description was accompanied by examples of applications in the aerospace field. Indeed, infrared thermography is particularly useful in the inspection of advanced composites which are ever widely employed in the fabrication of aerospace structures. As remarked in the conclusions to this chapter, as the technological progress in both excitation sources and infrared cameras go forward, the inspection of ever novel materials and structures becomes possible.

Infrared thermography, as non destructive evaluation technique, is mostly useful also in the architecture and civil engineering field. This topic is covered in the following last section 4.

# CHAPTER 9

# State of the Art and Perspective of Infrared Thermography Applied to Building Science

## Ermanno Grinzato[*]

*Construction Technology Institute (ITC)-National Research Council (CNR), Corso Stati Uniti, 4, 35127 Padova, Italy*

**Abstract:** IR thermography is more and more extensively used; amongst its applications, civil engineering is one of primary importance. In fact, performance indicators for new and existing buildings are derived from the temperature pattern. This is true, for both the envelope and plant investigations. The preventive maintenance and energy savings are addressed with qualitative or quantitative techniques. Non-destructive testing and evaluation of the repair and decay status is a hot issue. In this chapter, some examples of structural analysis aided by IR thermography, are presented and discussed. In particular, it is evidenced that help to the comfort monitoring given by the distributed temperature map measured by an infrared device is impressive. Using a novel method, it is possible to "see" the main environmental quantities such as air temperature, relative humidity and velocity, obtained from thermographic readings. Also, the well established application of thermography in building science for the detection of moisture is addressed. The discussion includes both the classical and innovative techniques, which are now available for dealing with usual or more complex problems. As the characterization of building materials by thermal properties is impressively improved by thermographic techniques, it is also demonstrated. The dedicated procedures, which have been developed since a long time ago for heritage building, are also reported and discussed herein.

**Keywords:** IR thermography, heritage buildings, NDE, moisture detection, thermal transmittance, HVAC preventive monitoring, structural analysis, thermal inertia, thermal conductivity, thermal diffusivity.

## 1. INTRODUCTION

The primary issue of this chapter is to illustrate how infrared (IR) thermography ($\Theta\gamma$) could help classical and new tasks, dealing with building science as non-destructive and non-contact measuring method. The use of $\Theta\gamma$ is widely spreading, due to the sharp cost reduction of the equipment and impressive features enhancement. Nevertheless, its full exploitation could be more complex than the vendors say.

IR thermography has been applied for more than 40 years to the buildings monitoring in qualitative or quantitative ways [1, 2]. Generally speaking, there are many possibilities to transform the temperature mapping of the surface into the investigated parameter. Up to now, $\Theta\gamma$ has been successfully applied by the authors for many purposes, as the identification of hidden structure or the wall bonding knowledge, detection of cracks and detachment of frescoes, moisture mapping and the analysis of Heating-Ventilating-Air Conditioning systems. Finally, the study on the internal microclimate has been also addressed. In the following, these issues will be presented in some details.

### 1.1. A Few Words about IR Thermography

IR thermography, if a correct procedure is followed, allows measuring the surface temperature of the building at a glance. Hence, thermal image of the inspected object represents a huge amount of information suitable for direct or more sophisticated analysis. Sometimes, the qualitative information is processed by human brain, offering a global view of the structure, although, a semi-automatic procedure is needed to achieve temperature data. The temperature is a physical quantity very useful to know how thermal energy is exchanged between the building and the environment. Furthermore, using indirectly the temperature map it is possible to monitor many other aspects of paramount importance, as the presence of material defects or the moisture distribution.

---

**\*Address correspondence to Ermanno Grinzato:** Building Technology Institute (ITC)-National Research Council (CNR), Corso Stati Uniti, 4, 35127 Padova, Italy; E-mail: ermanno.grinzato@itc.cnr.it

The heat flux crossing the envelope is due to a temperature gradient. Such a temperature distribution detected on the surface by the thermal camera is very useful to discover many hidden conditions related to the building performance and maintenance. If the heat flux is generated by natural boundary conditions, the inspection is called passive. Often, the natural heat fluxes and temperatures vary slowly, because of the high thermal inertia of building materials. In the case of a quasi-stationary thermal process a single thermogram is processed, or the average of a few thermograms just to reduce the random noise. A complete scanning of wide surfaces using the natural heat source is easier from the point of view of the needed equipment, but attention must be paid to geographical aspects and topology of the building.

Sometime, especially when interiors are analyzed, artificial heat sources, *e.g.,* lamps or air streams are adopted to generate the needed heat flux, for the thermographic inspection. If such is the case, a sequence of thermal images has to be analyzed in the time domain. Updated thermographic devices work in real time and can record the sequence at easy. Unfortunately, this feature is not available for entry-level IR cameras. Time dependent processing is known as dynamic thermography. For this purpose, there are several techniques well established that will be approached in this chapter.

## 1.2. Pro and Cons

Infrared (IR) thermography is an optical measuring method, where a radiometrically calibrated matrix of detectors senses the IR radiation coming from the target. The IR radiation is partially emitted by the surface, according to its temperature and partially reflected, according to its optical characteristic, described by the emissivity and absorptivity parameters [3]. Therefore, the readings integrate thermal effects due to heat and mass transfer and the surface optical effects. In many cases, the thermographic inspection is not difficult, because the emissivity of most building materials is quite high and diffusing. Sometimes, special care must be paid in order to unscramble the two parts.

The main limit of thermography is that the object must be seen. Luckily, nowadays the cost of the equipment is a limiting factor of much less importance. In fact, the cost of a thermographic device spans between 1 up to 100 thousands of euros or more.

## 2. THE ENERGY SAVING IN BUILDINGS

The development of energy monitoring systems for buildings is currently receiving considerable research attention all around the world. The directive 2002/91/EC of the European Parliament of 16 December 2002 started a new age about the energy performance of buildings.

It requires member countries to:

- Develop a comprehensive methodology for calculation of integrated energy performance of buildings and heating, ventilation and air-conditioning (HVAC) systems including lighting;

- Set minimum requirement of energy performance for new buildings;

- Apply energy requirements in existing buildings;

- Develop energy certification systems for both new and existing buildings;

- Regularly inspect HVAC devices.

As a matter of fact, the most used practice for the energy audit of a building is through a standard, dealing with the calculation method for measurement of thermal resistance and thermal transmittance of buildings, *in situ* [4]. More recently, the standard [5] annual energy use for heating and cooling of the building, upgrade the computation procedure, taking into account the transient thermal regime.

On the contrary, the standard describing the Heat Flow-Meter (HFM) measurement method of the thermal transmission properties is much less used [6]. This is obvious, because the whole envelope must be taken into account for the energy performance of buildings, including parts as ceiling or basement very difficult or impossible to be reached. On the contrary, for practical and methodological reasons the standard refers to a measurement of a reference area. Hence, a faster and reliable experimental energy audit is highly desirable.

Buildings energy consumption is mainly due to the heat losses through the envelope. For this reason a thermal insulation layer is added to reduce the conduction trough the walls. Nevertheless, the air infiltration or exfiltration could be sometimes more important than the heat flux due to conduction. Therefore, it is important to characterize the building envelope with a global parameter accounting for the ability to reduce the heat flow through it. This means the thermal transmittance that is the inverse of the thermal resistance of the wall.

Thermal resistance ($R$) and thermal transmittance from environment to environment ($U$) ($Wm^{-2}K^{-1}$) or conductance from surface to surface are evaluated for plane building components, primarily consisting of opaque layers perpendicular to the heat flow and having no significant lateral heat flow, that is essentially one-dimensional (1D). Actually, the main drawback of the standard method is the long time (4-6 days) needed to have a local transmittance measurement. Hence, a single value has to be considered as representative of the whole building. In details, the testing method depends on the apparatus to be used, the calibration procedure for the apparatus, the installation and the measurement procedures, the analysis of the data, including the correction of systematic errors and the reporting format. In practice, accuracy given by the measurement of thermal resistance and thermal transmittance of the thermal insulation of building elements, *in situ* is pretty poor.

The two main sources of errors are the transient behaviour of temperature gradient across the building and the used probes requiring a contact with surfaces. About the second issue, it is well known the important role played by thermal contact resistance, but it is less considered the different time constant of the HFM device, compared with the building element during transient thermal state. Furthermore, the nature of HFM in practice does not allow to be used during summertime, using for example the air conditioning systems to produce the thermal gradient through the wall. This con is due to the sharp changes in the air temperature and to the unavoidable averaging between air and outer surface temperature made by the probe. Otherwise, IR thermography gets rid of all these problems, due to its optical nature.

Generally speaking, the building classification like electrical appliances is a goal still far away!

## 2.1. How Thermography can be Useful

Up to now, the use of IR thermography for detection of thermal irregularities in building is barely qualitative, according to ISO 6781 or EN 13187 [7, 8]. Recently, thermography has become supportive to the HFM measurement in order to identify where to apply the flux and temperature probes. Some attempts to take advantage of the impressive improvements in thermography show promising results [9]. Nevertheless, there is the need for a more systematic approach.

### 2.1.1. Tips for in Field Measurements

The first step is related to the correct temperature measuring, starting from the thermal image given by the IR camera. The emissivity ($\varepsilon$) is a parameter that accounts for the difference between the examined surface and a black body, which is a perfect absorber and emitter of radiation in the working spectral band [3]. Furthermore, in spite of opaque and relatively rough surfaces, semi-transparent bodies make the measurement much more challenging. For instance, the glass panels are opaque for most of the IR bands within which the camera works, but they are mirror like reflecting. Specular reflectors require paying great attention to the local temperature of the surrounding objects and the taking direction, because the amount of reflected radiation collected by the camera depends very much from the viewing angle.

Temperature is achieved through the thermographic software by input additional data, that is: emissivity of the surface, reflected radiation, distance from the surface, air temperature and relative humidity (RH). The first two issues are definitely the most important. Fortunately, building materials are generally suitable for thermography, because $\varepsilon$ is high, as in case of:

- Very oxidized metallic surfaces,

- Plastic components,

- Painted surfaces,

- Cavities of the structure.

Furthermore, most surfaces act as diffusing reflectors. For an opaque body, the reflection coefficient ($\rho$) of the surface is the complement to unity of the emissivity [10]. The amount of IR radiation coming from the environment to the surface could be measured looking at a diffusing mirror attached just to the surface. The reflected IR component that must be subtracted from the total one detected by the IR camera is the product $\rho$ by the environmental IR radiation [11].

Evaluating temperature patterns in space and in time domains as temperature difference is very informative and reliable. Actually, a relative approach allows getting rid of most of the common sources of noise, due to the environment.

IR Thermography may be applied from indoor or outdoor. The former way is more accurate, the latter more productive. It is generally suggested to avoid windy places and possibly, measure on shadow or after sunset.

For the sake of simplicity, the thermal model of buildings is developed assuming a quasi-stationary regime. Basic concepts of Energy Saving are related to one-dimension (1D) and constant heat flux both in space and in time. The use of more sophisticated thermal models is becoming a common practice, more and more, thanks to powerful computers available.

A good and old idea is to exploit the temperature map given by thermography in order to evaluate soundness of thermal insulation [10]. The index ($F$) is a relative parameter useful for this purpose. Equation 1 gives a definition of $F$, as the difference between *normal* ($T^{*}_{i}$) and *defective* ($T_{i}$) zones, normalized by the temperature difference between indoor and outdoor wall temperature. The outdoor surface temperature ($T^{*}_{o}$) could be measured by means of a separate thermographic scanning or with another contact probe.

$$F = \frac{(T_i - T_0^*)}{(T_i^* - T_0^*)} \tag{1}$$

Normally, $F$ is measured from the inside of the building, where in case of thermal bridges the local temperature will be lower than the normal one, during wintertime and *vice versa* in summer.

A weak point of this very simple approach is how to define the *normal* condition. A possibility is to assume it as the average of the whole wall temperature. More important, the *normal* temperature will not be the same, all around the building.

A further step is to try to standardize the $F$ parameter, using a threshold accounting for:

- the minimum $F$-value as the weighted average ($F$) of local measurements, as quantified by Equation 2;

- the maximum allowed defective area, according to a fixed $F$-value:

$$\overline{F} = \frac{\Sigma(F_{x,y} \cdot a_{x,y})}{\Sigma a_{x,y}} = \frac{\Sigma(F_{x,y} \cdot a_{x,y})}{A} \cong \frac{a}{A}\Sigma F_{x,y} \tag{2}$$

where, $a_{x,y}$ is the area of a pixel view at position $(x, y)$, which an average value $(a)$ within the field of view; $A$ is the total inspected area.

Starting from a thermal scanning, the following procedure allows a fast quality control of the existing thermal insulation of the building:

1.  Select a minimum $\overline{F}$-value $(F_t)$, according to the case (*e.g.*, $F_t = 0.75$);

2.  Compute the low thermal threshold for applying a segmentation of the defective area, as the *Critical* Temperature $(T_c)$, given by equation 3; where the *normal* surfaces temperature for inside $(T^*_i)$ and outside $(T^*_o)$ are taken as the corresponding mean values $\overline{T}_i$ and $\overline{T}_o$;

$$T_c = F_t(T^*_i - T^*_0) + T^*_0$$
$$T^*_i \approx \overline{T}_i; \, T^*_0 \approx \overline{T}_0 \tag{3}$$

3.  It is now possible to count the $N_d$ pixels below $T_c$; assuming to know the elementary area $A_p$ of a pixel, the defective area $(A_d = N_d \, A_p)$ and finally the percentage of $A_d$ *vs.* the total surface $(A)$ is a simple informative parameter of a defective insulation problem.

Of course, this very simplified approach is useful just as a rough estimation of the thermal insulation quality, because many points are fuzzy. In particular, it is difficult to define a single threshold temperature independently from the geographic location and particular climatic conditions. In addition, the area estimation is very inaccurate.

Anyway, the largest error source is assuming constant the temperature gradient across the building structure. This is not generally true, due to the heat capacity of the structure. As a consequence, inspections carried out at different time, probably gives different results.

## 2.2. Transmittance Measured *in situ* (*U-value*)

If the building envelope was in steady thermal state the physical model described by the Fourier equation of conduction becomes the same as *Ohm* law for the electric circuit (electrical analogy). Now, the thermal resistance $(R)$ of a wall is given as the sum of the individual resistance $(R_i)$ of any layer constituting the structure, adding the effects of heat exchange on both sides, as boundary thermal resistances. Equation 4 accounts for the relationship between heat flux $(Q)$, thermal gradient $(\Delta T)$ and thermal resistance.

$$\frac{Q}{A} = q = \frac{\sum R_i}{T_i - T_0} = \frac{R}{\Delta T^a_{i,0}} \tag{4}$$

This could be done in the workshop, but *in situ* temperature gradients vary continuously in time and the heat gain, too. Hence, the tested structure is working in transient regime. In addition, the heat flux varies also in space. Fig. **1** shows clearly how the temperature pattern is dependent on the building technology. Actually, any thermal bridges is due to different material thermal properties, hence any discontinuity is a small or large defect. The usage of the building is another reason for a peculiar thermal pattern. In the case of Fig. **1**, the thermal images taken from outdoor of a brick building, plastered with cement mortar show the wet wall of the bathroom.

In practice, it is very important to consider also the effects of defects, like insulation deficiency of structural components with reduced thermal resistance. The only way to do this is to evaluate locally the temperature using many temperature points, as thermography does.

### 2.2.1. Local Thermal Performances by Quantitative Thermography

Assuming that temperature differences are important and heat exchanges vary locally, the transmittance $(U$-value $[Wm^{-2}K^{-1}])$ could be obtained on local basis from the local heat flux $(q^{x,y})$. In addition, at thermal equilibrium the local heat flux crossing the surface is the sum of the heat exchanged with the environment, both by convection ($q_c^{x,y}$) and radiation ($q_r^{x,y}$). Equations 5 gives a reliable way to measure these quantities starting from local temperature readings [12].

$$q = \sum q^{x,y} = \sum q_c^{x,y} + q_r^{x,y} \tag{5}$$

with

$$q^{x,y} = \alpha_i^{x,y}\left(Ta_i - Tw_i^{x,y}\right) + \sigma\left[\left(T_e^{x,y}\right)^4 - \varepsilon\left(T_w^{x,y}\right)^4\right]$$

where: $\alpha_i$ is the local convective heat exchange coefficient for the indoor surface; $T_{wi}$ and $T_{ai}$ are respectively the temperature (K) of each internal wall point with coordinate $(x,y)$ and the indoor air, while $T_e$ is the radiant temperature from the environment; $\sigma$ is the Stefan-Boltzmann constant ($\sigma = 5.6704 \times 10^{-8}$ W m$^{-2}$ K$^{-4}$) and $\varepsilon$ is the surface emissivity.

Quantitative IR thermography, cooperating with other temperature measurement devices is able to evaluate the local thermal resistance of a building component, as a mean of local data set [13]. This technique, called *RIR* (patent pending) [14] allows a more accurate and useful measurement, because it allows weighting the effect of thermal bridges or defective areas of the building surface. This result is exactly what people need. *RIR* has the ability to evaluate separately the convective and radiative heat fluxes exchange between the indoor surface and the environment. In fact, heat exchange is a local phenomenon. In this context, facing so many problems, a global approach is generally applied giving a single exchange coefficient for a surface. Otherwise, IR thermography is able to detect a map of temperature difference between air and surfaces and in this way allows evaluating heat fluxes on a local basis.

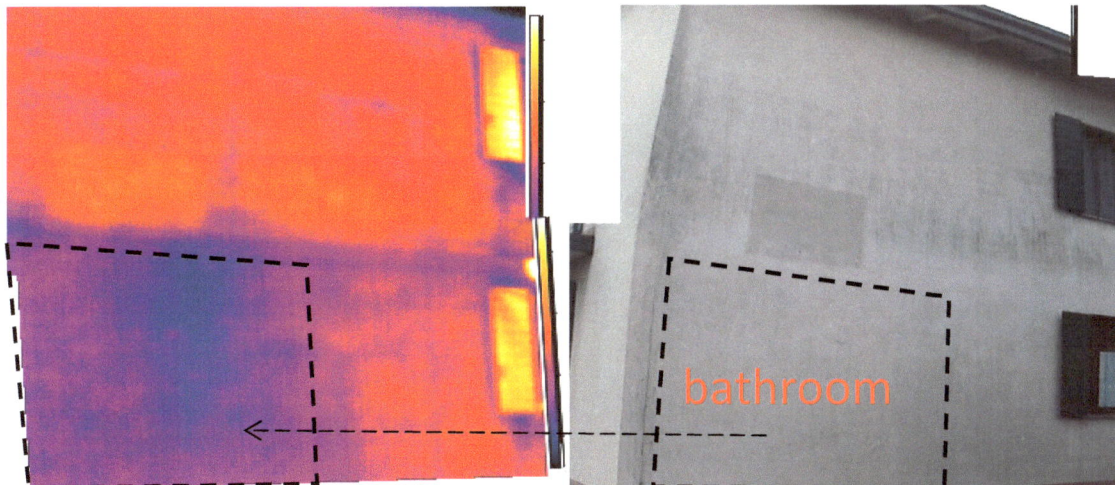

**Figure 1:** Thermal image from outdoor of a brick building, plastered with cement mortar (left) and the corresponding visual image (right); the wet wall of the bathroom shows a lower temperature.

In order to account for the transient thermal conditions the average of instantaneous values is performed for a time long enough, up to a constant value is reached. Equation 6 shows the double integration in space and time, which allows a correct measurement. For this purpose, an automatic device for long lasting measurements has been set up and applied both in laboratory and *in situ* [15].

$$Q = \frac{1}{(t_1 - t_0)} \int_{t_0}^{t_1} dt \int_s q(x, y, t) dx dy \tag{6}$$

Measurements are taken from inside and outside the building. Fig. **2** summarizes the basic steps of the measuring technique. At first, a thermal scanning from outdoor is carried out giving the detailed surface temperature map. Individual thermograms are calibrated into temperature and corrected for perspective distortion. At this stage, a common request is the geometric composition of different thermograms, with the needed correction for the perspective distortion, in order to have a total view of the building. This process is not trivial, because the quantitative temperature information must be preserved during the images processing [16]. A mosaic composition using proprietary software gives the total view, where average values are computed on suitable areas. The same scanning from outside and the processing procedure is repeated also after the internal scanning (see Fig. **2d**). Fig. **2a** shows the surface temperature distribution over the tested wall.

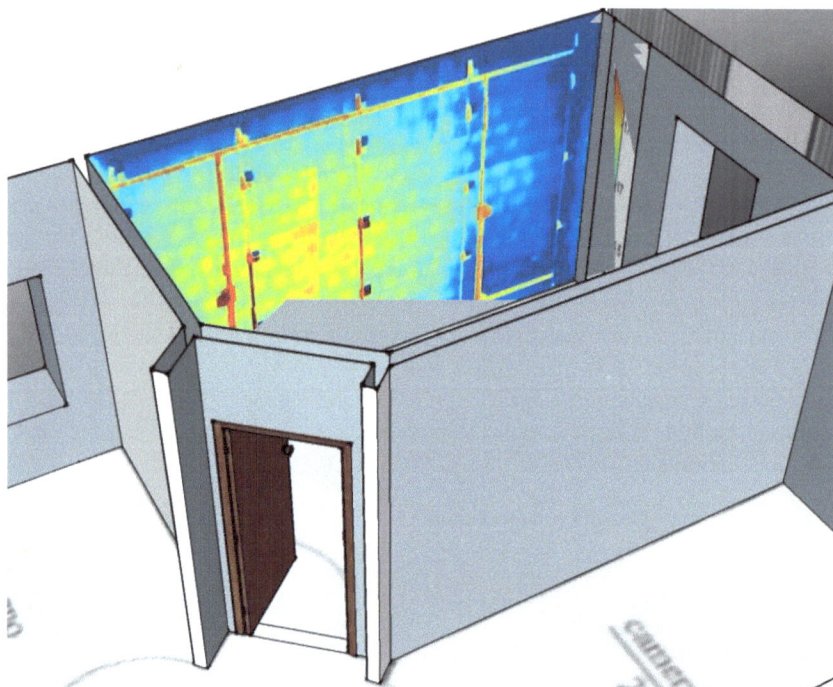

**Figure 2a:** Local values of surface temperature (°C) of the inner wall projected on the 3D building drawing.

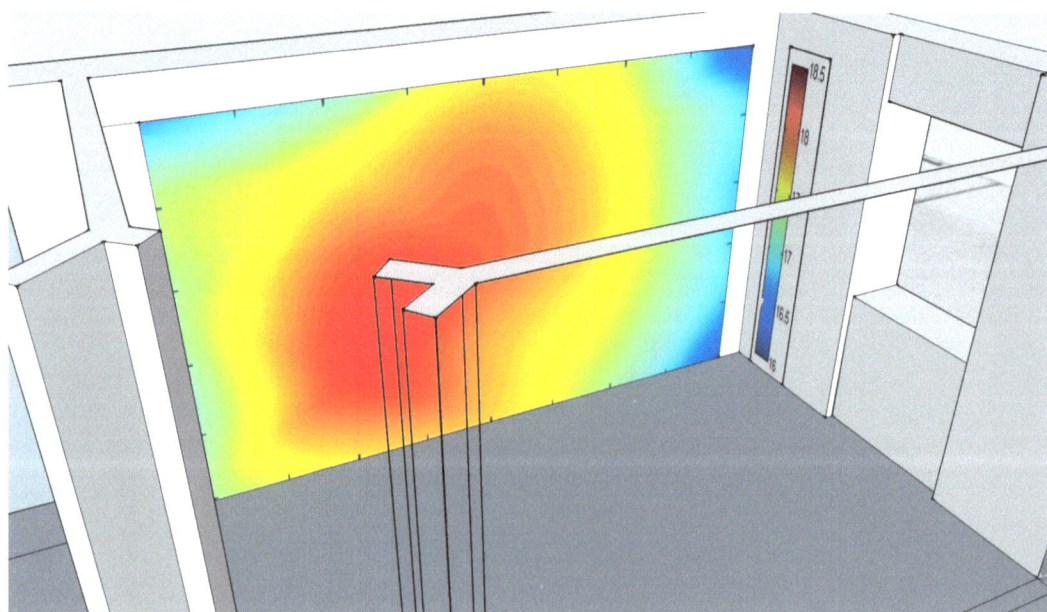

**Figure 2b:** Local values of air temperature (°C) 10cm far from the inner wall over the real building model.

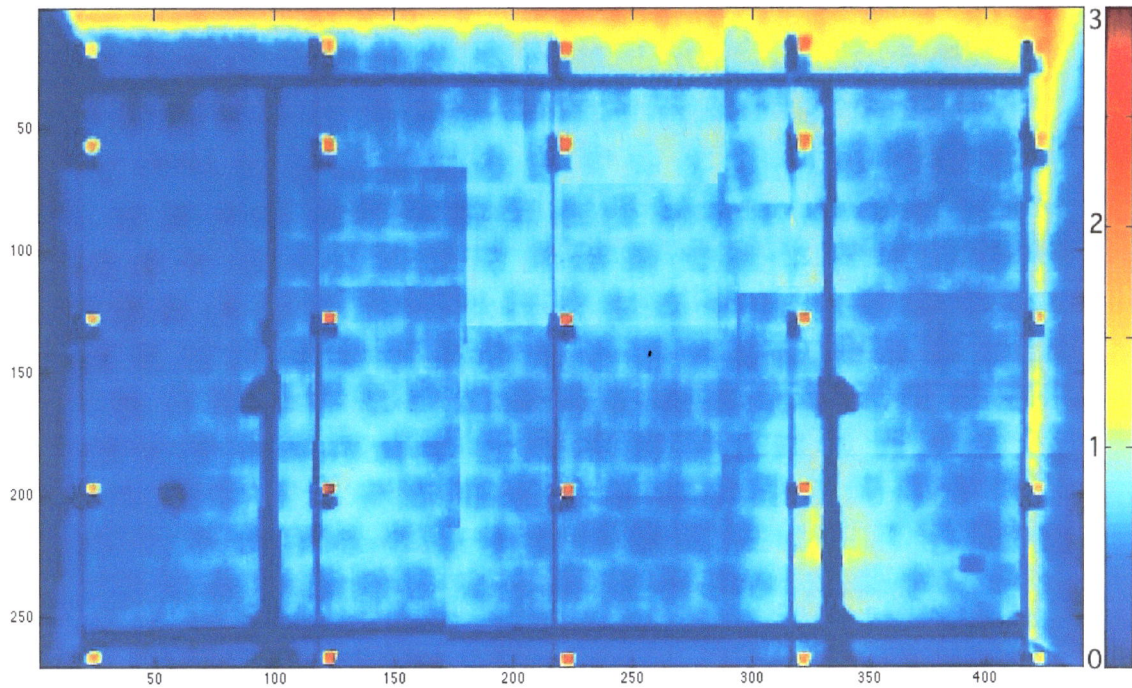

**Figure 2c:** Local values of air-surface temperature difference (K); distances are in cm; the reference structure holding several targets is clearly seen.

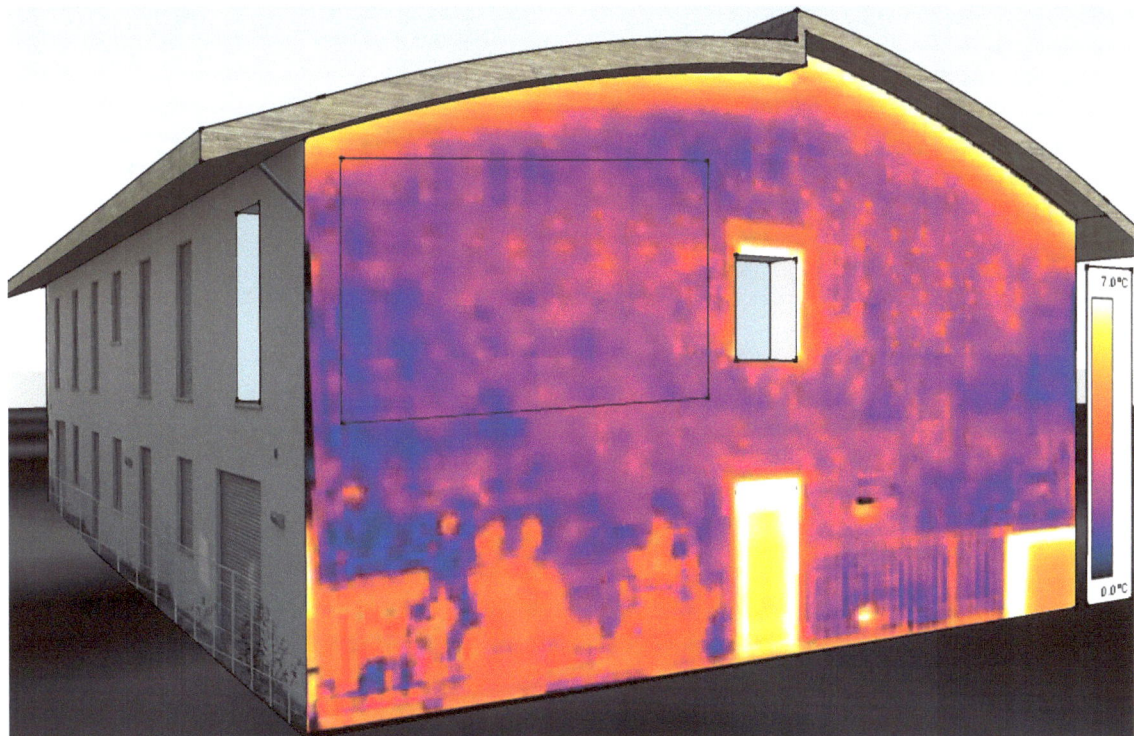

**Figure 2d:** Surface temperature (°C) of the external wall over the model of the building.

The 3D geometry of the real building is necessary to evaluate correctly the area of the measured temperature. The calibration of both temperature [17] and space is performed using a light frame, which hold several dedicated targets. The air temperature close to the wall is measured using the targets, as well

[18]. Fig. **2c** shows the temperature difference between air and surface, this is the driving gradient of the heat flux, as measured from inside. Once again, the needed high accuracy is accomplished by means of the reference grid and a calibrated temperature probe. The scale units are cm, demonstrating the effectiveness of the calibration. Tenth of thermograms are fused in order to set up the map of Fig. **2c** that is projected over the 3D drawing of the real building.

After the automatic computation of local heat flux value, the external thermal scanning is carried out, as shown in Fig. **2d**. The room tested from inside is marked with a rectangle. In the lower left corner an insulation deficiency is clearly seen, as well. A mini data logger has been recording the outdoors air temperature, relative humidity and air speed for the entire testing period.

Finally, Fig. **3** shows the map of local values of heat flux (Wm$^{-2}$) crossing the inner wall. Results are projected over the 3D model of the real building. The testing device is shown as well. The automatic pan-tilt unit and the thermal camera are placed inside the room on a tripod. The control unit has put outside avoiding any disturbance to measurements.

The new method is more sophisticated and a bit more time consuming than the usual thermographic inspection, but much more accurate and rich of useful data. We have to bear in mind that, the direct measure is the only possible way, when data about the building are not available for *U*-value computation.

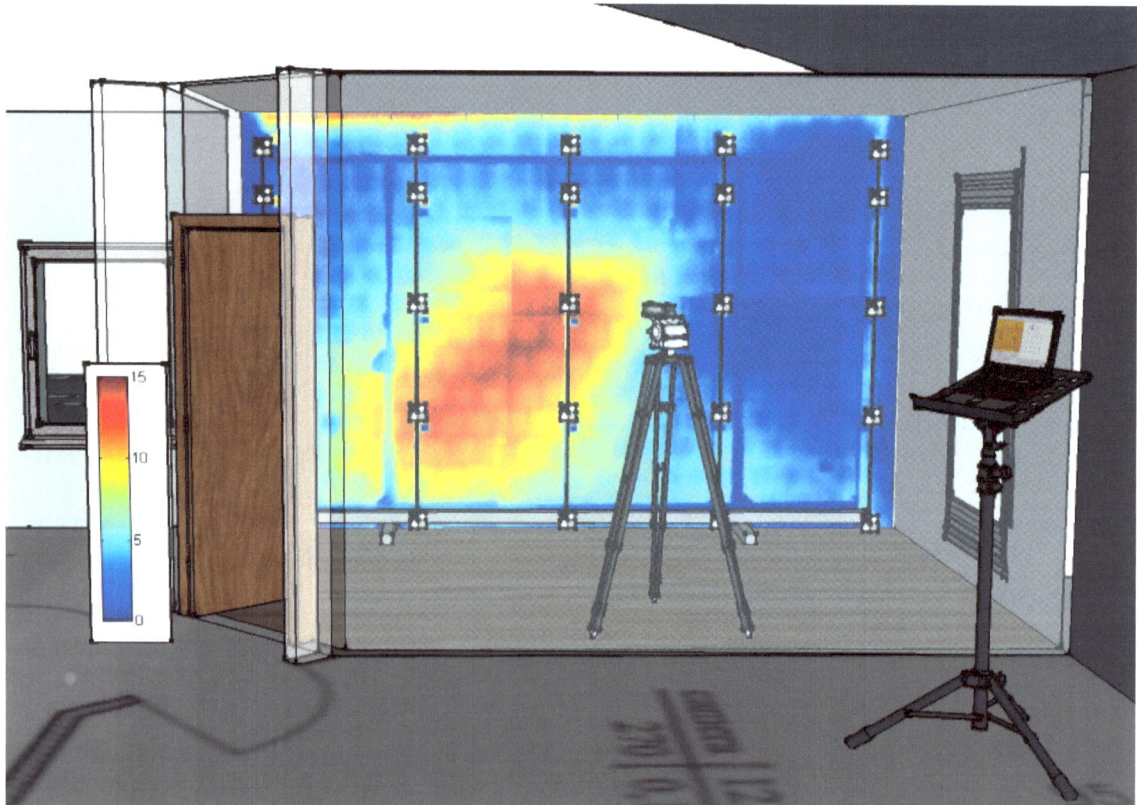

**Figure 3:** Map of local values of heat flux (Wm$^{-2}$) crossing the inner wall projected over the 3D model of the real building; the testing device is shown; the control unit is outside of the tested room.

## 3. NON-DESTRUCTIVE EVALUATION OF BUILDINGS BY THERMOGRAPHY

The previous Fig. **2d** shows an example of detection of a defective thermal insulation in the warmer area located at the bottom-left corner of the building. The defect coincides with a lower local thermal resistance, where a higher amount of heat is transferred from inside to the exterior surface. In such a case, the detection is performed in a quasi-steady thermal regime. Otherwise, when the defect is close to the surface it is more convenient to impose a heat flux and observe the associated temperature changes. For instance, in the case of some detachment of the finishing layer, an air layer is trapped in between. Hence, an additional thermal resistance is added, making the heat penetration more difficult. This condition will be associated with a local warmer pattern. The analysis of a temperature sequence in time allows characterizing width, depth and thickness of the discontinuity [19, 20].

In theory, the surface temperature history provides evidence of any step or even progressive changes in thermophysical properties, related to the presence of materials discontinuity or ageing. Discontinuities may be evaluated by comparison of sound zones with anomalous ones. In practice, the signal due to the discontinuity is generally very low (in the order of tenth of kelvin). To achieve such a high level of detectivity a special set of hardware and software tools have been developed and implemented [2, 3, 10]. It is worth saying, that Thermal Non-Destructive Testing or Evaluation (NDT, NDE) is highly sensitive to a wide variety of noise sources, as most of inverse problems. The electronic noise is not the only one, but also the variation of the surface in emissivity, shape, colours, composition *etc.* or the evenness of the heating generate much serious disturbs [21].

## 3.1. Strengthening of Reinforced Concrete

Non-destructive control by thermographic analyses has been adopted on reinforced concrete beams strengthened with Carbon Fibre Reinforced Polymer (CFRP) laminates to evaluate the quality of the interface among reinforcement and substrate before and under loading [22] during experimental laboratory bending tests (see Fig. **4a**).

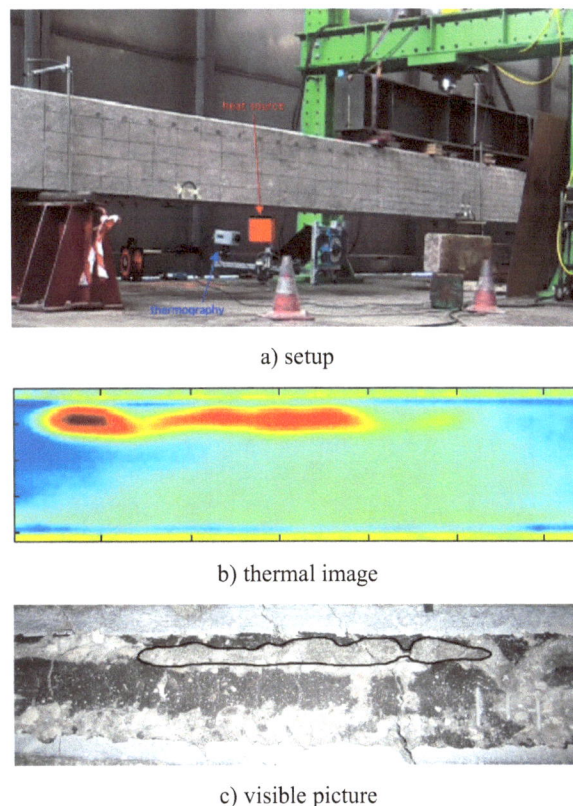

a) setup

b) thermal image

c) visible picture

**Figure 4:** Results of active thermography on reinforced concrete beam.

The proper algorithm for the processing of data acquired from the thermographic system has been preliminarily selected by testing reduced–scale CFRP strengthened concrete slabs including artificial defects and anomalies, specifically provided at the interface [23]. A movable heating equipment has been on purpose setup to allow continuous scanning along the beams under loading (Fig. **4a**) [24]. The results showed that IR thermography can supply significant qualitative and quantitative information on bond of CFRP materials applied to structural substrate, both for preliminary and *in situ* investigations [25], by using a reliable procedure low-time consuming.

A cross evaluation of the defects identified by thermography and the picture of the same point after removing the CFRP plate are shown in Fig. **4b** and **4c**. The defective pattern during the bending test and thermographic results are also usefully explaining how the failure mechanic works. The measure of the defect area has been performed during the test. In fact, IR thermography is the only way to follow the evolution in time of the defect, due to its non-destructive nature.

### 3.2. Structural Analysis

Structural analysis of masonry buildings requires an adequate level of knowledge based on the importance of structural geometry, which may include construction damage details and properties of materials. For identification and classification of masonry it is necessary to find shape, type and size of the elements, texture, size of mortar joints, assemblage. The recognition can be done through a visual inspection of the surface of walls, which can be examined, where is not visible, removing a layer of plaster [26].

Thermography is an excellent tool for fast survey and collection of vital information for this purpose, but it is extremely important to define a precise procedure developing more efficient monitoring tools. There are many examples of fruitful application of thermography for this purpose [27, 28]. Less usual is the use of thermography to help the crack analysis [29]. As far as, the appropriate physical process is addressed, the defects are reliably detected and characterized from the surface temperature. Fissures are an easy way for water intrusion, this could be observed as a local variation of thermal inertia [30].

Cracks very often are coupled to detachments of the plaster from the wall, because of the local strength applied and to the mechanical properties of the masonry and the plaster. Fig. **5** depicts an example of that, with a *first aid service* given to a frescoed church, located close to L'Aquila (Italy) a few days after an earthquake (April, the 6th 2009). The testing procedure required the heating of the interior air and processing a bunch of thermographic images, acquired regularly both in time and space. In such a way, it was possible to evaluate the risk of collapse of the whole frescoed wall in a few hours. Fig. **5b** shows precisely the weak points to bear. Fortunately, the fast intervention saved the precious work of art. By the way, the masonry bonding structure of the wall, made of rounded natural stones, is visible on Fig. **5b**.

**Figure 5a:** The apse of S. Lucia church after an earthquake, (L'Aquila, April 2009).

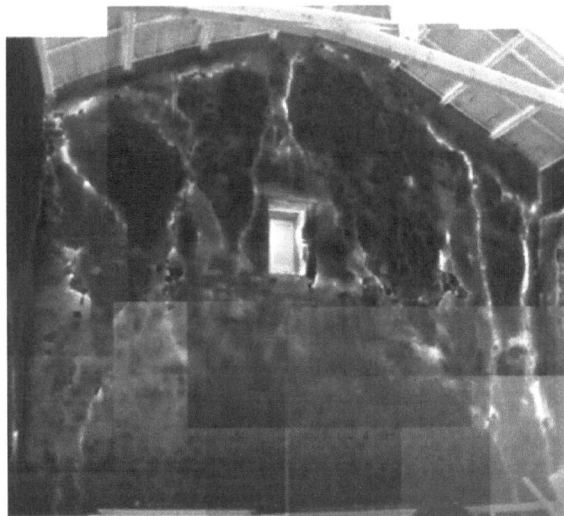

**Figure 5b:** Cracks due to the earthquake on the frescoed wall indicated by lighter lines.

## 4. MOISTURE DETECTION ON BUILDINGS

Up to date, no fully satisfying moisture detectors for buildings are on the shelf. The most common moisture detectors are based on electrical resistance measurements, but hygroscopic salts make readings not reliable. Other methods, requires to sample a small amount of building material, with aesthetic and others drawbacks. Visual inspection could be misleading as well, due to water diffusion and unevenness of materials. Quite often the dark colour, induced by the biological activity, triggered by moisture generates artefacts because it absorbs a higher amount of solar energy. This process is not reversible and sometime dark areas are recovered to normal conditions.

IR thermography is very effective for the Moisture monitoring, but a quantitative approach is difficult, due to the coupled heat and mass transfer [31, 32]. Generally, people are motivated to search for a 3D distribution of the water concentration within the building. This is not possible using thermography, because thermograms are not NMR-images. IR cameras do not see moisture, air or defects nor does the IR camera see "inside the wall", it sees barely the radiant energy.

There are different ways to use IR radiometry for the moisture detection. The most common is looking for areas of the wall, which appear colder than the normal surface. This qualitative signature is caused by the very strong latent heat of vaporisation of water. Whether or not a temperature-related signature can even be seen and, if so, when it may be seen as well as what it might look like, all depend on the conditions and the timing. Trying to measure the moisture content by surface temperature is an excellent example of ill-posed inverse problem. Actually, the variables of conduction, capacitance and transient heat flow cycles combined with the amount of interstitial water are the main determinants for water inside a structure. Expect that accurately locating interstitial water by its thermal signature will always be challenging. Temperature and humidity, and air velocity, determine if we can see evaporative signatures of surface water. Simply setting a "dew-point indicator" or isotherm alarm in the camera is insufficient. Only by continuously monitoring changes in temperature and relative humidity throughout the job and understanding the thermal relationships will you avoid making errors of judgment [33].

### 4.1. Moisture Footprints Mapping

IR thermography is not able to quantify the water content, but it is an easy way to find the source of water. The imaging feature of thermography is paramount indicating characteristic patterns of the most frequent pathologies due to moisture. The key-point is the ability to segment the surface by qualitative or quantitative techniques. For instance, capillary rising is characterized in such a way that an almost horizontal line delimits wet zones. Fig. **6** reports the temperature on the internal wall of a church; thermograms are overlapped and registered to the inspected surface. This picture shows clearly a colder band, parallel to the floor. Unfortunately, when air speed is very low, a colder lower surface is also due to natural temperature distribution of the surrounding air [34]. For this reason, boundary conditions must be taken into account. On the same picture there are seen reference targets, which indicate the reflected ambient temperature in the square plate [17].

Generally, the location and shape of the most evaporating zones, correlated with the meteorological and microclimatic data are practical tools for understanding the cause of the moisture defects. Water infiltrations are documented by cold spots, for this reason inspection before a rainfall is recommended. In order to simulate such a case a few grams of water have been intentionally sprayed on the wall of Fig. **6**, where a colder area is visible in the centre.

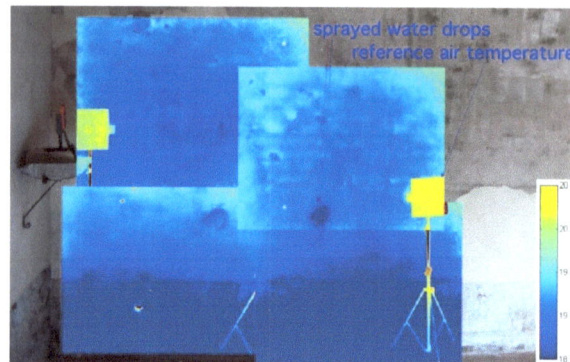

**Figure 6:** Signature of capillary rising on the interior surface of a Church (Cornaredo-Italy) by IR thermography; a passive reference target is also seen.

After a reliable identification of a defective zone, a few micro-destructive measurements are acceptable, linking thermograms to quantitative measures of thermo-hygrometric conditions. Supportive tests by means of alternative methods are regularly planned to confirm the thermographic output.

### 4.2. The Condensation Risk

Another important and common moisture problem is due to the water vapour condensation. The condensation risk is analyzed recording the history of thermo-hygrometrical parameters, *viz.*: air temperature and relative humidity, complemented by surface temperature. Then, the dew point has to be calculated and compared to surface temperature. Fig. **7** shows an example of plot *vs.* time of the air and surface temperatures, where a surface point at the time indicated by the circle is below the dew point. Some thermographic cameras offer a dedicated tool for the detection of the surface condensation conditions. Of course, this tool is operative at particular conditions of the survey time. Condensation may happen later or sooner.

### 5. MICROCLIMATIC MONITORING BY THERMOGRAPHY

The indoor microclimate is very often the major hazard issue for the conservation of buildings. For an efficient monitoring, the continuous thermographic measurement in space is crossed with the almost continuous data logging in time of conventional sensors [35]. Unfortunately, it is very expensive to cover a building with a suitable network of sensors.

Existing instruments to record environmental conditions are not appropriate to map local values such as temperature, light, humidity, and air velocity inside buildings. Actually, it is totally unpractical to move a measurement head comprising a temperature sensor, anemometer, hygrometer every 10cm in a given room! Moreover, sometimes variations of interest are below the range of common instrumentation. For instance, low limit for ordinary anemometers is about $0.5ms^{-1}$ while natural convection occurring in buildings could be even smaller than 0.1ms⁻¹. Tight building envelopes are another rising problem since, in case of poor air change, the high level of humidity might conduct to health risks related to developed fungus or others poisonous products.

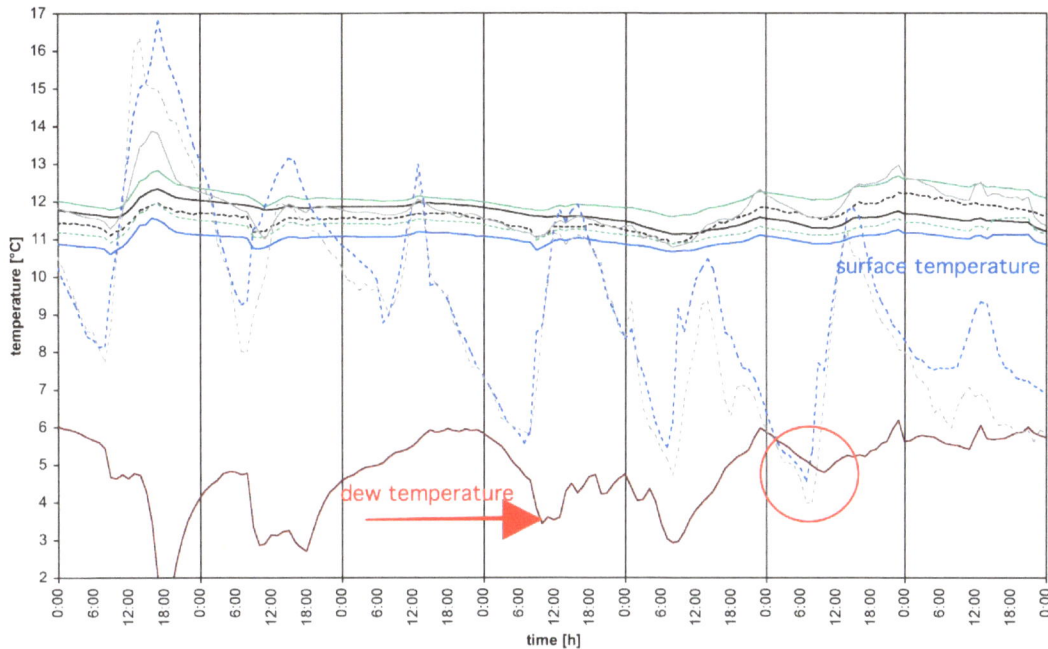

**Figure 7:** Temperature plot *vs.* time measured inside a building, where a surface point at the time marked by the circle is lower than dew point.

It is a matter of fact that some measures are not easy to achieve. RH sensors are notoriously unreliable. They must be calibrated on installation. The traditional hair, of horse or human, is still used, but high quality solid-state RH sensors are spreading. RH sensors are delicates; water soluble salts, which are nearly universal in walls, easily contaminate them.

Sometime, IR thermography is applied, but not yet on regular basis, due to lack of precise procedure and standards. It is applied only to some case study or for forensic purposes, ought to the cost. A new approach for making cost effective and reliable the analysis is to make short term monitoring of the air temperature and relative humidity using thermography, but extending results to characterize the building. Quantitative IR thermography is used to measure the humid air conditions in equilibrium with a building surface. The method is based on accurate temperature measuring as the driving parameter of the physical process. At the same time, the thermographic system images the wall surface behind the reference grid and so measures the wall's temperature, as well. Fig. **8** shows the change in the measurement scenario from past to nowadays. On the top, there is a sketch of traditional monitoring of just a few points. Below, there is a continuous map of the air main ambient parameters given by thermography with a space resolution of 1cm.

This is a part of a project dealing with the full monitoring of the indoor environment on thermodynamic basis called *aIRview* (patent pending) [36].

The use of a special ancillary device and a dedicated procedure allows the measurement of the air temperature, relative humidity and speed using only IR thermography. This understanding is fundamental to identify and quantify the moisture exchange process between the porous material and the boundary. As a consequence thereof the evaporative process can be studied on quantitative basis [37-40].

Essentially, *aIRview* enables the wet and dry bulb temperature measurements needed by the psycrometric method [41]. Equations to achieve indirectly the hygrometric quantities of air are well known and implemented into the controlling device [42]. Results obtained in laboratory conditions are compared with standard measurements of temperature and relative humidity [43].

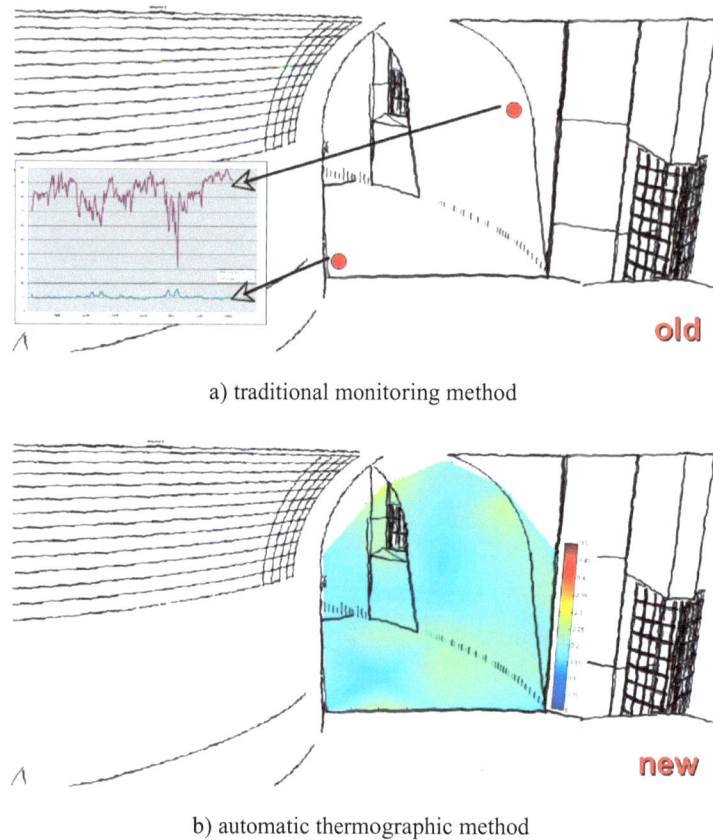

a) traditional monitoring method

b) automatic thermographic method

**Figure 8:** On the top there is a traditional monitoring method applied to a few points; below it is shown a much clear air temperature map using the automatic thermographic method.

An automatic device, shown in Fig. **9** allows continuing the measurement *vs.* time, as long as necessary [15].

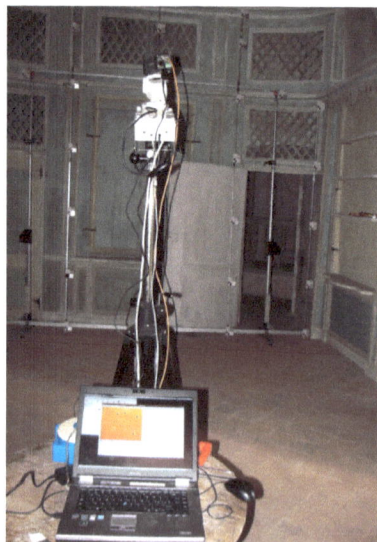

**Figure 9:** Automatic device for the air temperature, relative humidity and air velocity measurement by thermography (*aIRview*).

Such a new method allows facing a crucial phase for the building life, in a fruitful, cost effective and reliable way. The technique is accurate, because even excellent, but different detectors add their individual errors, when used together. Using only one calibrated thermographic equipment, it exploits the impressive resolution of modern IR cameras. Above all, a grid of reference targets makes such a technique very reliable.

One of the most important features of the procedure is the accurate positioning of results on the building geometry. Thermograms regarding the envelope or the boundary conditions are composed of images corrected for optical, geometric and radiometric distortions. Fig. **10** shows the air temperature map at 10cm far from the wall. A very high space resolution is achieved scanning the whole surface. This procedure has been especially developed for indoor building study, where very fast response is not so important. Actually, the surface of about 8 x 4m² is covered with 12 million of temperature readings taken in a pair of minutes. It is also worth noting the very high resolution of the resulting map. In fact, the effect of warm air infiltration is clearly visible around the right window. The natural air vertical gradient it is also reported, especially, near to the ceiling.

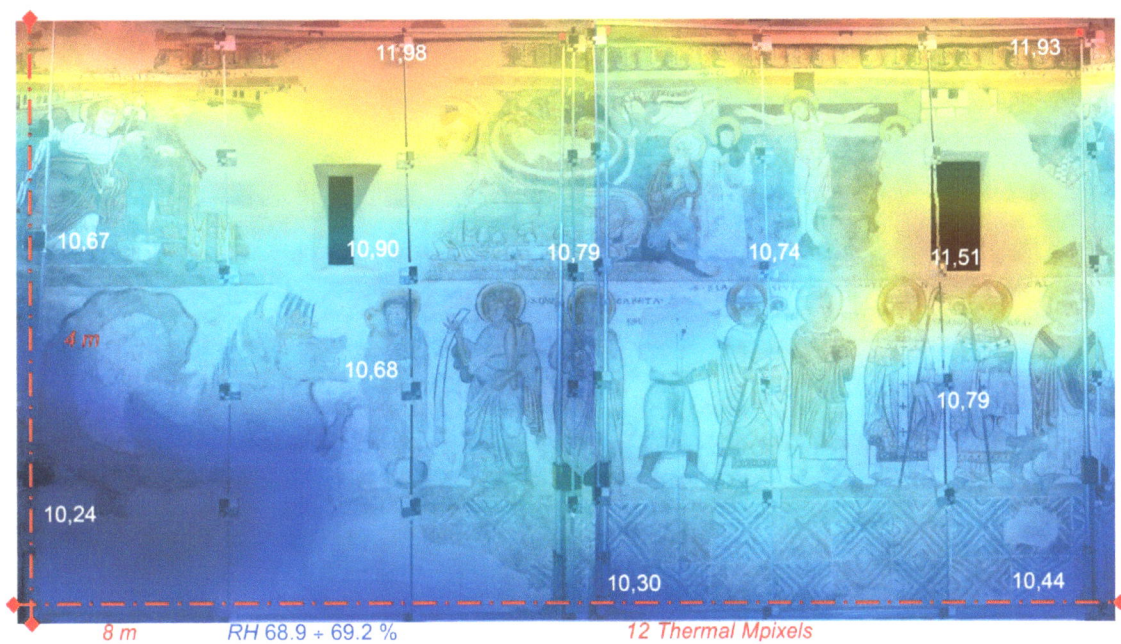

**Figure 10:** Air temperature 10cm far from a frescoed wall by *Θγ*, the write window is partially open.

### 5.1. *aIRview:* A New Diagnostic Tool

The new ongoing project for a more advanced Non-Destructive Evaluation of buildings is mainly based on two steps: at first a short run monitoring using a Quantitative IR Thermographic technique; then the modelling of micro-climatic state. Indoor heat and mass exchange are computed as a function of particular conditions, resulting from weather, different occupancy policies or Heating, Ventilating and Air Conditioning (HVAC) solutions [44]. An additional step could be added for verifying of the plant efficiency and tuning it. The first and third steps are almost accomplished, the second one is in an advanced experimental phase on some case study, including heritage, industrial and residential buildings.

The following illustrates a current application of *aIRview* in order maintaining the indoor temperature and RH values of a heritage building suitable for conservation of frescoes and usage as library [45].

The Fig. **11** shows a picture taken from the west side and the 3 dimensions (3D) sketch of the Masino castle (Italy) owned by a no-profit institution, with marked the tower where the tests have been carried out. A global view of the building is produced in 3D in order to place the building in the geographic location. In fact, it is very important to take into consideration the height above the sea level (257m) of the spot and the dominant winds direction. The climatic conditions could be supplied either by local data logging (as in the case) or by nearby data collection sites (*e.g.,* airports or weather forecasting stations).

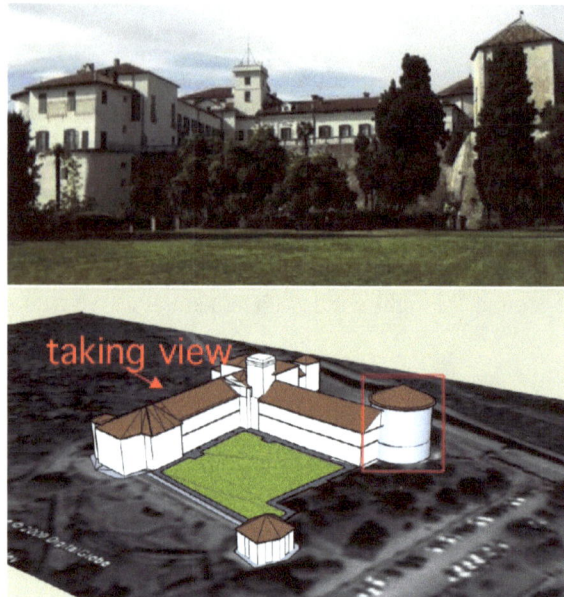

**Figure 11:** Photograph (from South) and 3D sketch of the Masino's castle (Italy), tests have been carried out within the tower surrounded by the box.

The temperature distribution inside the complex rooms of the tower at ground and first floors has been carefully mapped, considering different seasonal conditions and the effect of the opening of windows. Furthermore, adjacent rooms have been taken into account and monitored, when the air circulation could have significative effects. For instance, Fig. **12** shows the surface temperature of the wall depicted in Fig. **9**. Here, it is evident as the reference grid improves a lot the mosaic composition of 20 different thermograms considering the very small temperature difference (the side scales are expressed in centimetres).

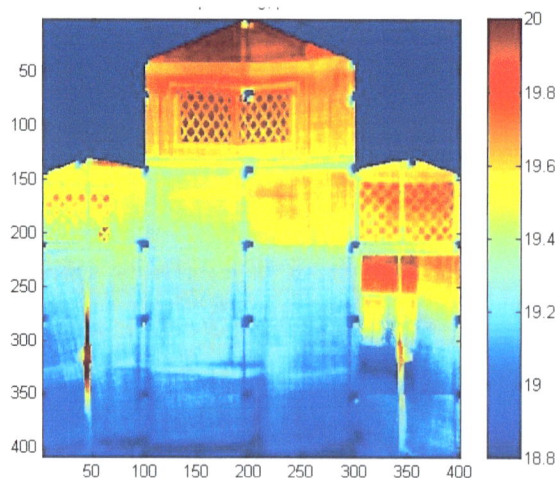

**Figure 12:** Temperature map measured on the wall seen on Fig. (**9**) within the Masino's castle.

The effects on surface temperature of the air circulating from the nearby window in the case of either opening or closure are deeply investigated. The Fig. **13** illustrates on the top the temperature of the wall in face to that reported in Fig. **12**, when the nearby window is open. The comparison of this picture with Fig. **12** demonstrates the warming effect of hot air entering from the window.

The bottom picture of Fig. **13** shows the temperature difference recorded on the same wall after the closing of the window. Almost the surface cools down, except for the left side of a fireplace, which is affected by a new hot air stream. It is important to notice the high thermal resolution that is achieved on this picture.

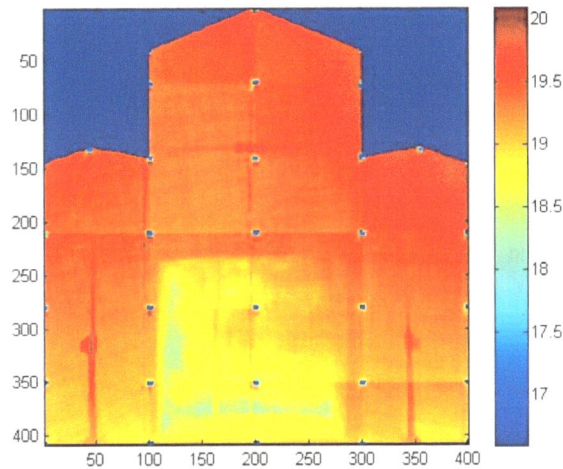

a) open window

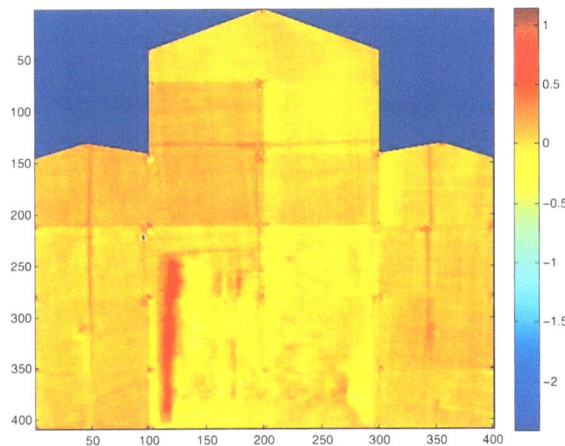

b) closed window

**Figure 13:** Effects on surface temperature measured on the wall opposite to that seen in Fig. (**9**) if the nearby window is open (a) or closed (b).

Fig. **14** reports the mean temperature history for 16 hours on four areas of the same wall, taken during the winter 2010. The fairly steady temperature *vs.* time confirms the very high thermal mass of the wall.

Finally, in order to study the air circulation inside the building a set of map of the air speed have been measured by means of *aIRview*. Fig. **15** gives the speed of the air (ranging from 0 up to $0.3 \text{ms}^{-1}$) parallel to the wall of Fig. **13**. On the left, there is the case of the opened window, on the right the closed window case.

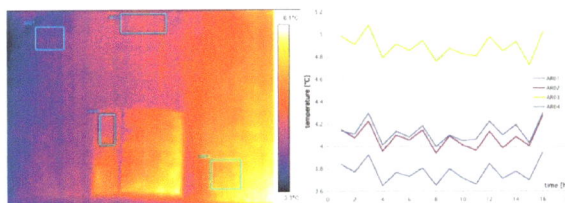

**Figure 14:** Temperature history measured on the wall of Fig. (**13**) for 16 hours during the winter 2010.

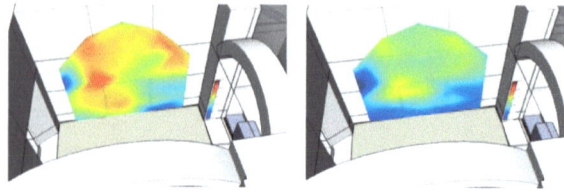

**Figure 15:** Mapping of air speed (ranging from 0 up to 0.3ms$^{-1}$) parallel to the wall of figure 13; left side: open window, right side: closed window.

## 5.2. Computational Fluid Dynamic

A complete scanning of the tower during different seasons is very informative about the internal microclimate. Nevertheless, the data acquired in a particular condition are not enough for understanding how to manage the building for a particular use.

For instance in the studied case, the need to place a library inside the tower put tight limits to the air temperature and humidity values. A powerful tool for analysis of indoor microclimate is the Computational Fluid Dynamic (CFD) [45]. Of course, the results are strictly related to the input data in terms of surface temperature. Furthermore, some initial choice for the so-called wall functions are controversial. The followed approach is to use real experimental data, efficiently gathered by thermography as input and the maps supplied by *aIRview* for the validation and tuning of CFD.

Fig. **16** shows an example of this procedure, where the air speed has been mapped for different cross sections inside the tower. The 3D representation of microclimatic data is very useful to quantify the effects of the openings. Actually, air leakages through the old window frames have been documented.

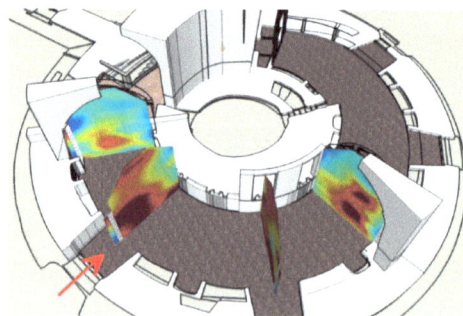

**Figure 16:** Speed of the air (ranging from 0 up to 0.3ms$^{-1}$) on different cross sections inside the tower.

The following Fig. **17** represents the mesh used to describe the geometry of the studied building section by means of CFD. The volume has been divided into 100000 elements, implemented within the ANSYS code (http://www.ansys.com). The measurement of the radiative heat exchange by *aIRview* simplifies the duty, because otherwise it must be computed by the code in parallel to CFD.

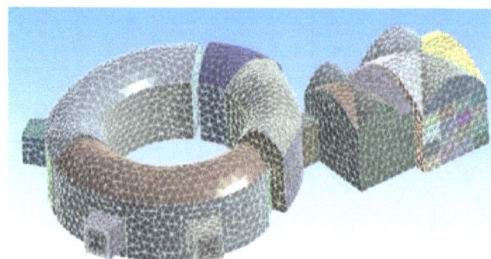

**Figure 17:** Computational Fluid dynamics of the investigated rooms for accurate modeling and simulation of indoor microclimate.

Now, it is possible to evaluate the indoor conditions in any case of interest. For instance, CFD evaluates effects of the number of people inside the library in critical winter conditions. Again, the effect of air infiltration trough the windows frame has been simulated in Fig. **18**. The upper picture shows the variation in temperature due to air infiltration corresponding to the cross section #2 (marked by the arrow in Fig. **16**. The intermediate picture shows the air velocity map on the median cross section and the bottom picture the air speed at 1 m from the floor in the whole tower.

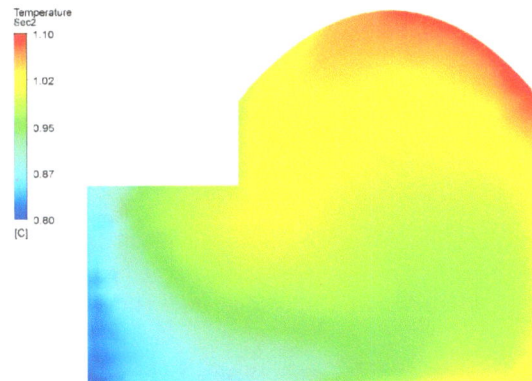

a) temperature.

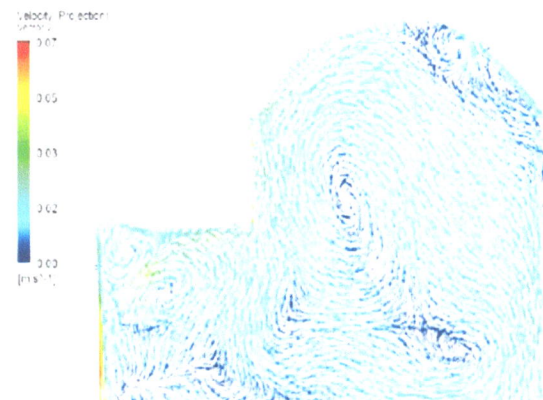

b) air velocity on the same cross section.

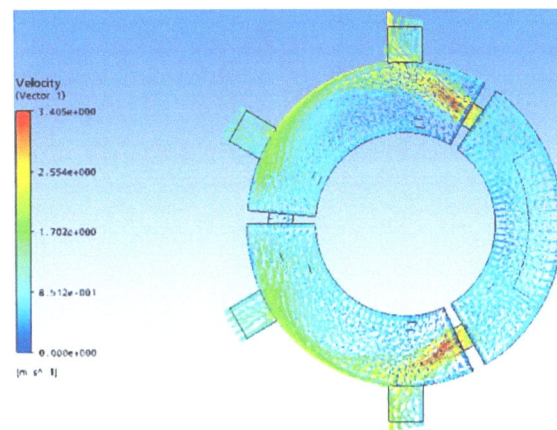

c) air velocity on a plane parallel to the tower floor, at 1m height.

**Figure 18:** Results of CFD in case of not tight windows; a) temperature increase of the air due to air infiltration in the cross section #2; b) air velocity on the same cross section; c) air velocity on a plane parallel to the tower floor, at 1m height.

## 6. MONITORING OF HVAC PERFORMANCES

The most common application of thermography is the detection of loose electrical contacts that generate heat for Joule effect [2]. Thermography early visualizes failures of components *e.g.,* misaligned pumps or revolving appliances, as electrical motors ought to the friction. Generally speaking, it is possible to check and evaluate the mass flow rate evenness starting from the surface temperature pattern. In case of liquids, as hot water of refrigerating fluids circulating inside a heat exchanger, thermography is very effective, due to the high heat exchange coefficient of the internal flow toward the surface. A signature of the reduction in the flow rate is indicated as a temperature difference along the piping. In many cases a comparative analysis among similar components is very helpful. Fortunately, the fluid is normally distributed in parallel coils. The main drawback comes from the optical nature of thermography, *viz.* the need to see the surface, which is sometimes difficult or impossible.

Fig. **19** illustrates the case of two cooled turbines of identical design used in a commercial building. A temperature increase of more than 10K demonstrates that the hottest turbine has oil cooler problem. Sometimes, insufficient thermal evidence exists to determine whether the problem is related to oil flow, water flow, or both given that both machines were operating at identical loads [11].

When the exchange with the air is involved, the assessment is more difficult, but still possible if a more quantitative analysis is performed [46]. The airflow velocity can be indirectly computed using sophisticated apparatus in the lab in order to characterize the geometric configuration [47]. This application has a long tradition for high-speed fluid, especially in the aerospace industry. More recently, thermography is starting to be applied also for low speed applications [48]. Another fruitful way to use thermography for the checking of HVAC appliances is related to the additional thermal resistances due to limited or lack of adhesion between two solids layers. This could be the case for instance of fined heat exchangers [49].

For thick piping systems, as in power plant stations, thermography may be very useful for detecting the hidden corrosion and evaluating the material loss [50, 51].

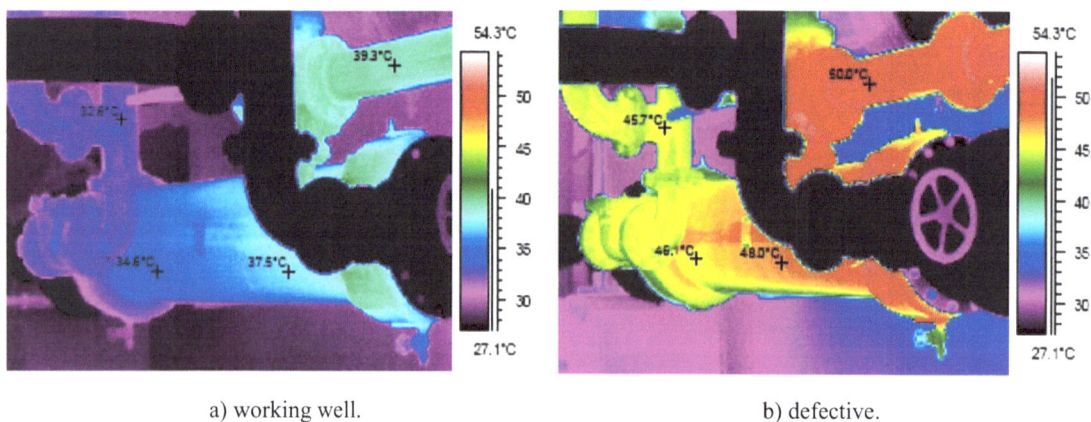

a) working well.                    b) defective.

**Figure 19:** Two identical turbines, the right one is defective, due to cooling problems.

HVAC plants used inside buildings are an excellent application of thermography. In this case, an additional issue is the integration between the plant and the envelope. In fact, phasing of load and plant is done by the plant control, but it needs the knowledge of the building *thermal mass*. Higher thermal mass reduces the temperature swing. Higher thermal mass increases the inside-outside temperature phase lag. This implies the measurement of thermal inertia, as will be described in section 7.

Thermography gives surface temperature for the evaluation of both the efficiency of appliances and the resulting performances. Also in this application, there are two main advantages comparing with others methods: the local evaluation and the visualization of defects or malfunctioning. The former feature gives the capability to evaluate quantitatively how can be optimized a complex system. The latter is extremely important when large buildings are inspected and the intervention must be limited to a defective particular point, may be. An example of this will be supplied with the analysis of a heating floor plant.

## 6.1. Radiant Heating Systems

The two major goals assessed by thermography are: the location of a water leakage from the piping system and the measurement of the energy flux delivered into the ambient.

The first issue is quite a common practice, because the pattern of the warm piping is clearly visible on the surface of the floor, as Fig. **20** shows. The detection of water leakage can be performed both in steady or transient regimes. Nevertheless, the blob given by the warm water diffusion within the screed may be confused with the reduced depth of the pipe buried within the screed. The best way to unscramble the problem is to perform a transient test. The following rules states: the closer the pipe is to the surface, the faster the temperature increasing.

**Figure 20:** Surface temperature indicating a water leak from the piping of a heating floor.

More sophisticated is the capability to quantify the thermal power delivered by the radiant heating to the room. Fig. **21** shows the surface temperature generated by a floor heating of reduced thickness (30mm). Starting from this temperature map, calibrated by radiometric targets and using the surface heat exchange coefficient ($h$) ($Wm^{-2}K^{-1}$) given by the standard [52], the delivered power ($q$) ($Wm^{-2}$) is computed on local basis using the equation 7:

$$q = \frac{1}{A}\Sigma_n h \left(T^s_{x,y} - T^a\right) \tag{7}$$

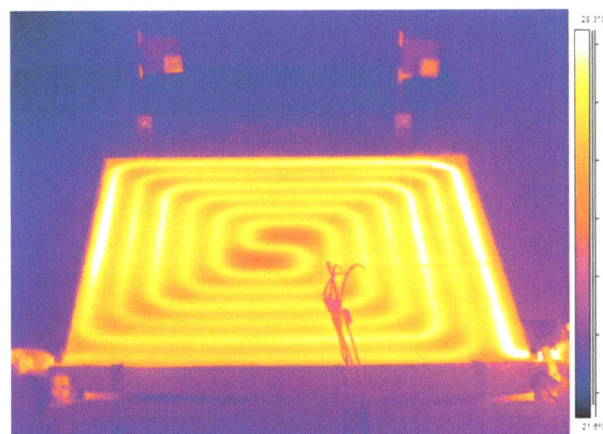

**Figure 21:** Surface temperature over a heating floor; the two targets on the top are used for air and radiant temperature measurements.

where $A$ is the area of the heating floor, $T^s_{x,y}$ is the temperature of any surface element in the location $x,y$ and $T^a$ is the so called operative temperature. The operative temperature is a blend of radiant and air temperature, but in practice is very close to the air temperature at 1 m from the floor, in the middle of the room. The radiant temperature and the air temperature may be measured by thermography as well, by means of some special targets, visualized on the top of Fig. **21**.

Using thermography it is also possible to detect voids between the screed and the finishing layer. These adhesion defects may have a significant contribution in decreasing the system efficiency.

## 7. EVALUATION OF THERMAL PROPERTIES OF BUILDING MATERIALS

When dealing with building materials it is common to encounter a lack of information about the thermal properties. On the other hand, being these materials generally multi-layered, the knowledge of each layer is crucial for the solution of the inverse problem. Another issue is the HVAC efficiency, which must be harmonized with natural heat flux, not only with the thermal load. Mathematical modelling is used more and more in order to set up an efficient control strategy. Nevertheless, it is paramount to known thermo-physical properties of materials.

A reliable database it is not yet available, because of weatherisation, *viz.* the absorption of moisture, deterioration and ageing. Hence, change of thermal properties with depth from the surface is expected. The needed tests are normally performed at lab, but the use of thermography is conceivable to work on site, even without the sampling. A set of non standard techniques, using thermography, as the main measuring device have been set up for evaluating thermal properties, by direct or indirect ways.

The final goal is accessing the building envelope global and local performances in steady and transient regime. Thermal conductivity is the first issue, but also thermal diffusivity and inertia (effusivity) are very important.

### 7.1. Thermal Conductivity

The standard methods for thermal conductivity measurements are not always satisfying, due to the needed kind of the sample and the time involved in any test.

A variant of the standard guarded hot plate method for the measurement of the thermal conductance of solid plane building materials has been developed and continuously tested. In steady state conditions the heat flux ($Q$ [W]), flowing through a slab sample, is equal to the temperature difference between the two sides of the slab, times a constant factor typical of the material, that is the effective thermal conductivity ($\lambda$) ($Wm^{-1}K^{-1}$), times the ratio between the surface of the slab $A$ and the thickness $z$. Equation 8 allows to determine $\lambda$ if the geometry of the sample, the heat flux and the temperature gradient across it are known [53].

$$\lambda = \frac{z\,Q}{A(T_{sh}-T_{sc})} \qquad\qquad (8)$$

The underlying idea is to use a thermographic camera, equipped with a close up lens for measuring the temperature gradient across the slab of the tested material. The heat flux is generated by means of a couple of thermoelectric devices applied to both sides of the sample. The same devices allow measuring the heat flux, if the characteristic constants and the imposed electrical current are known. The sides non in contact with the sample of the thermoelectric devices are maintained at controlled temperature. The temperature of the hot and cold surfaces of the sample are equally spaced from the room temperature, in order to reduce as much as possible the heat exchange with the environment. In such a way, the net heat losses are almost zero and the heat flux is nearly perpendicular to the sample surface [54].

Fig. **22a** illustrates the experimental apparatus, with the thermal camera and the two thermoelectric devices sandwiching the sample (black painted). Fig. **22b** shows a thermogram recorded during test using a known sample of AISI 304 for calibration purposes. Averaging hundreds of columns of the thermogram we have a better signal to noise ratio. The resulting temperature profile across the sample is presented in Fig. **22c**. Here, it is also visible the contact thermal resistance between the thermoelectric devices and the sample.

One of the great advantages of this test is the small size of the samples (volume of $40x40x10mm^3$). This allows running a test in a few minutes. Therefore, several samples of the same material and several tests with different heat fluxes are normally performed in order to obtain a conductivity value. In addition, thermography allows a continuous gradient evaluation throughout the sample cross section, which implies a continuous assessment of the local value of conductivity. This means that the contact thermal resistance it is not anymore a problem, because this segment of the thermal profile is skipped.

**Figure 22a:** Experimental set up for the thermal conductivity measurement; the sample is sandwiched between two thermoelectric devices.

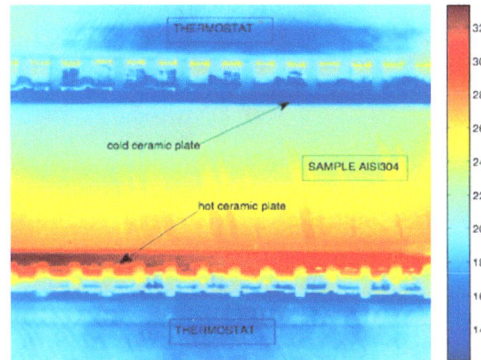

**Figure 22b:** Temperature gradient obtained with a close-up lens of a sample of AISI 304.

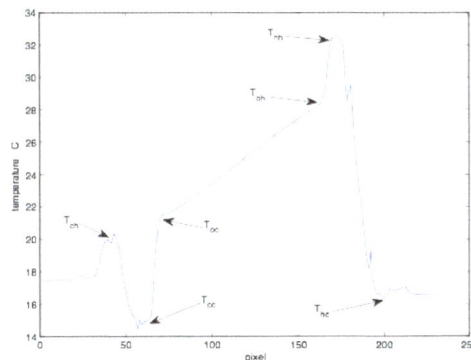

**Figure 22c:** Average temperature profile, from top (left) to bottom (right).

## 7.2. Thermal Diffusivity

Recovering quantitative information about the internal structure of an object from the surface temperature evolution i explained in section 3. The solution of the inverse thermal problem requires the knowledge of some material properties. In dynamic thermal tests one of the most important parameters is the thermal diffusivity ($\alpha$), expressed in $m^2s^{-1}$, which is a measure of the rate at which the heat is diffusing through the material. The key role of thermal diffusivity in solving the direct and inverse dynamic thermal problems makes its knowledge necessary in most of the cases. Unfortunately, the literature data do not help enough especially for ancient buildings, due to the large range of reported values. The best is the direct measuring.

In 1863 Ångström proposed a method for measuring the thermal diffusivity of long bars with a small cross-section applying at one end periodic variations of temperature, which caused thermal waves to travel along the bar. The measure was carried out once the periodic state was reached. Several works based on this method have been proposed [55]. The most used heating sources are chopped laser beam and lamps coupled with a moving mask. The major drawback of these sources is the difficulty to obtain a uniform heating and the long time required to reach the periodic state.

The use of a thermoelectric device, that works alternatively as heat source and sink, shortens the time needed to reach the periodic state (being the average temperature close to the room temperature). Moreover, a good thermal contact with the object under test ensures a homogeneous stimulus. The only important disadvantage of this technique is due to the high inertia of the device that prevents the use of high stimulation frequencies [56].

The only standard method for the thermal diffusivity measurement is the so-call *Flash method* proposed by Parker in 1961 [57] that become an ASTM standard in 1992 [58]. It is based on the analytical solution of Fourier equation with adiabatic boundary conditions for an infinite slab subjected to a thermal Dirac pulse [53]. The test is performed according to the transmission scheme: the sample is located between the heater and the sensor. An energy pulse is deposited on a sample surface (called front surface) and the temperature evolution is measured on the other surface (rear surface). Thermal diffusivity is linked to the sample thickness $L$ and to the half-maximum rise time through the equation 9:

$$\alpha = \frac{0.139\, L^2}{\tau_{1/2}} \qquad (9)$$

$t_{1/2}$ is the time in seconds needed to reach half of the maximum temperature value [57]. Fig. **23** shows in red the temperature history on the rear side of a plaster sample, indicating $t_{1/2}$. It is clear that in case of non-adiabatic testing conditions (green curve) the maximum temperature will be lower than in the adiabatic case (blue curve), as far as the $t_{1/2}$. This implies a systematic overestimation of diffusivity.

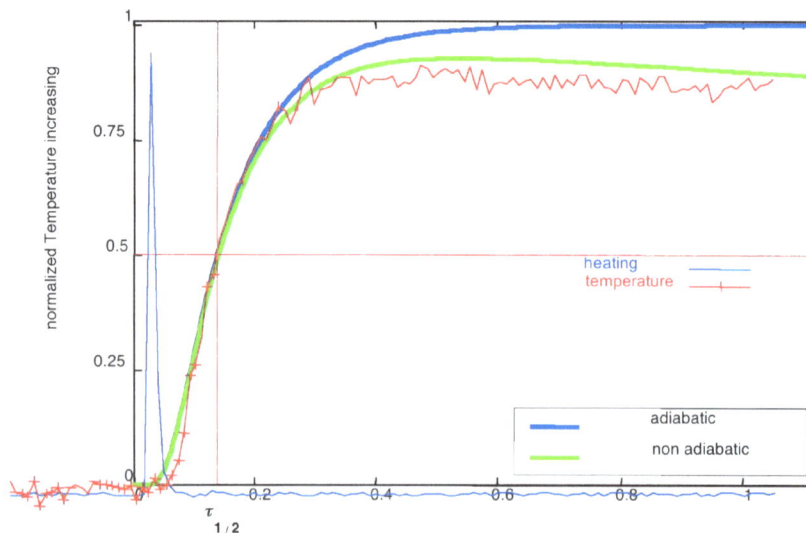

**Figure 23:** Temperature history on the rear face of a sample during the diffusivity measurement by the flash method in adiabatic and non-adiabatic cases.

The standard test requires a complex apparatus to ensure that the theoretical assumptions are met. To bring this method into field applications of IR thermography, the most restrictive requirements are the need to stimulate the sample with a short duration pulse and the necessity to carry out the test in adiabatic conditions. Furthermore, it is required machining the sample in a way that is not conceivable with the common building materials. Indeed, testing a low diffusivity object generally requires a large amount of energy and therefore a long duration pulse. Moreover, if the test is carried out in a non-controlled environment, the surface heat exchange must be taken into account. In such conditions equation 9 does not provide correct results and a more sophisticated mathematical model is needed. For these reasons it is now popular a variant of the standard method, that uses as detector thermography and works in non-adiabatic conditions. The diffusivity value is obtained as a result of an optimisation procedure (for instance least squares fitting) based on mathematical expression. In addition to the estimate of $\alpha$, the procedure provides estimates of $(q/h)$ and the Biot number $(Bi=hz/\lambda)$ [59], where h is the surface thermal exchange coefficient.

This method has been successfully applied to a wide variety of building materials, including stones, bricks, plasters, composites, metals, plastics, *etc.* As an example, Table **1** reports the diffusivity measurement of a brick sample, coming from the historical arsenal in Venice [60]. The diffusivity measurement of a 5.5mm thick sample has been repeated six times. The sample was heated with a pulse of duration ranging from 1.5s to 2.5s and the temperature evolution on the rear surface was observed through an infrared camera sensitive in the 8–13μm wavelength band. The sampling interval was set to 0.5s and the observation interval to 60s. Hence, for each measurement a sequence of 120 thermograms was acquired. It is worth noting that the test is repeatable with a precision about 2%.

**Table 1:** Diffusivity measurements of a 5.5 mm thick historical brick sample, repeated six times.

| *Test number* | *RMS* | $[10^7 m^2 s^{-1}]$ | *q/h* [K m$^{-1}$] | *Bi* |
|---|---|---|---|---|
| 1 | 0,0004 | 4,999 | 1653 | 0,172 |
| 2 | 0,0002 | 4,920 | 2131 | 0,123 |
| 3 | 0,0032 | 4,998 | 2379 | 0,121 |
| 4 | 0,0007 | 5,011 | 2444 | 0,224 |
| 5 | 0,0002 | 4,994 | 3439 | 0,231 |
| 6 | 0,0024 | 4,971 | 2420 | 0,191 |

## 7.3. Thermal Effusivity

Thermal inertia, a.k.a. effusivity $(\beta)$, is the property of the material that measures the reaction to the delivery of a certain amount of energy with a temperature variation. By definition, the thermal inertia is $\beta = (\lambda \rho c_p)^{0.5}$; in SI units $(Jm^{-2}K^{-1} s^{-0.5})$.

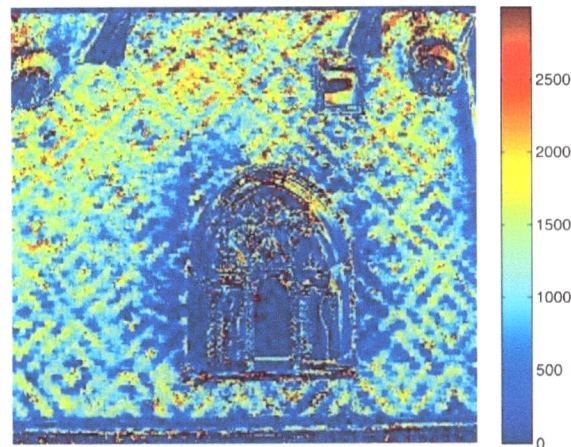

**Figure 24:** Map of thermal diffusivity measured on the facade of Palazzo Ducale in Venice.

An important feature of the thermal effusivity is to be correlated to the materials decay or the moisture content [61]. The effusivity evaluation *in situ* for any point $(x,y)$ is given through an energy balance between the solar radiation absorbed by the surface $(q_a)$ and the temperature evolution $(T_t\text{-}T_0)$ at time $t$, starting from the initial temperature $T_0$, described by equation 10:

$$\beta_{x,y} = 2\frac{q_a}{(T_t-T_0)}\sqrt{\frac{t}{\pi}} \tag{10}$$

Fig. **24** shows the map of thermal diffusivity measured on the facade of Palazzo Ducale in Venice.

Because, thermal diffusivity $\alpha = \lambda/(\rho c_p)$, the knowledge of thermal inertia and thermal diffusivity gives the thermal conductivity and $(\rho\,c_p)$.

## 8. CONCLUSIONS

IR Thermography is growing importance because the temperature is a global monitoring parameter. Thermography maps the IR radiative energy flux and can determine heat flux and surface temperature. The first step is to transform a thermal image from the energy to the surface temperature.

Modern and historical buildings can be regularly inspected by IR thermography for energy saving and maintenance purposes. Thermography is often applied qualitatively to see something more than visually, mainly because of fascinating imaging, but a deep knowledge is required to achieve useful results. A well defined procedure is needed either exploiting operator's expertise, simply looking at thermal images or using sophisticated data reduction algorithms and extremely sensitive equipment. In any case, only trained personnel could achieve good results.

Quantitative Thermography is expanding its applications to new fields including comfort evaluation and structural analysis. In practice, a new approach allows a fast diagnosis of the cause of deterioration, when the indoor environment may become armful. Different remedies are studied and validated by the combined use of tests and numerical modelling achieved by quantitative thermography and Computational Fluid Dynamic.

The thermal resistance of the building envelope can be evaluated *in situ*. IR thermography could be developed to estimate the *R*-value of buildings.

Thermographic Non-Destructive Evaluation, especially using transient techniques, may efficiently characterize discontinuities within the building envelope.

Established and new thermographic techniques for moisture evaluation have been set up. The cooling effect of surface evaporation is quite strong, and a distribution of the high moisture concentration is shown by thermography using a passive approach. The same technique is useful for the environmental conditions monitoring.

The combined thermal performance of plants integrated on the envelope, can be studied by IR thermography. The heat flux is a critical factor that thermography helps to monitor.

IR thermography is very effective, quickly locating defective areas on large surfaces. Others defects, as missed insulation, leakages or pipes blockage are detectable.

Efficiency of HVAC *e.g.,* radiant floor heating is studied by Thermography.

Thermal properties are very informative and thermography is a suitable measuring tool. Thermal conductivity, effusivity and diffusivity of building materials are measured at laboratory using thermoelectric and photo-thermal heat sources.

## ACKNOWLEDGEMENTS

Author is highly in debt with many friends and colleagues working together on the presented research projects. Among others I would like to acknowledge the precious help and their important contribution given by Paolo Bison, Chiara Bressan, Gianluca Cadelano, Mara Gava, Mario Girotto, Davide Maistro, Sergio Marinetti, Fabio Peron, Elisabetta Rosina, Ayse Tavukcuoğlu, Vladimir Vavilov, Monica Volinia.

The author wish to acknowledge the Fondo Ambiente Italiano, the Soprintendenza Beni artistici, storici e ambientali di Torino and of Trento for the kind cooperation. We are also grateful to Yousave S.p.a. and the Fondazione CRT of Torino for the support.

## REFERENCES

[1]     Ljungberg SA. Infrared Techniques in Buildings and Structures. Operation and Maintenance, Infrared Methodology and Technology, Non-destructive Testing Monographs and Tracts, Vol.7, G. Maldague Ed., Gordon & Breach Science Publishers USA 1994; 211-52.

[2]     ASNT, Nondestructive Testing Handbook, 3rd edition: Volume 3, 'Infrared and Thermal Testing'. Technical Editor: X. Maldague. Editor: Patrick O. Moore, cap.18, 2001.

[3]     Maldague X. Theory and Practice of Infrared Technology for NonDestructive Testing. John Wiley-Interscience, 2001; 684.

[4]     ISO 6946: *In situ* measurement of thermal resistance and thermal transmittance-calculation method.

[5]     ISO 13790: 2008 Calculation methods for assessment of the annual energy use for space heating and cooling.

[6]     ISO 9869: Building elements-*In-situ* measurement of thermal resistance and thermal transmittance.

[7]     ISO 6781:1983 Thermal insulation-Qualitative detection of thermal irregularities in building envelopes-Infrared method.

[8]     EN 13187 Thermal performance of buildings. Qualitative detection of thermal irregularities in building envelopes. Infrared method.

[9]     Madding R. Finding R-Values of Stud Frame Constructed Houses with IR Thermography. Inframation proceedings ITC 126 A 2008-05-14.

[10]    Vavilov VP. Nondestructive testing handbook, Thermal/Infrared testing, vol. 5, book 1, VV Klyuev (Ed), Spektr publisher, Moscow, 2009.

[11]    ISO 18434 part 1, Condition monitoring and diagnostics of machines-Thermography

[12]    Grinzato E, Bison P, Cadelano G, Peron F. R-value estimation by local thermographic analysis; Proceedings of Thermosense XXXII°, SPIE vol.7661, Orlando (USA) 2010; 7661-15.

[13]    Grinzato E, Bison P, Cadelano G, Peron F. Automatic U-Value Measurement By Local Thermographic Analysis, Proceedings of Inframation 2010, Las Vegas (USA), 2010; 159.

[14]    Grinzato E. Patent pending n° PD2010C000121, 16 April, 2010, titled: thermographic system for measuring the thermal heat flux and transmittance of buildings by quantitative IR thermography (Metodo termografico di misura del flusso termico attraverso l'involucro edilizio e della trasmittanza termica), (in italian).

[15]    Pretto A, Menegatti E, Bison P, Grinzato E. Automatic indoor environmental conditions monitoring by IR thermography, in: NDT in Canada national Conference, London (Ontario), 2009 (August 27-28).

[16]    Grinzato E. Dedicated image processing for thermographic non-destructive testing, Infrared techniques for NdT, Ed. X. Maldague; cap.IV°, Gordon and Breach Science Publisher, vol.7, 1994; 103-28.

[17]    Ohman C. Practical methods for improving thermal measurement, Development Department, AGA Infrared System AB, Box 3, S-182 Danderyd, Sweden.

[18]    Bison.PG, Bressan C, Grinzato E, Marinetti S. Automatic air and surface temperature measure by IR Thermography with perspective correction, 1992 SPIE 1821, Boston (USA); 252-60.

[19]    Meola C, Carlomagno GM. Infrared Thermography in Non-Destructive Inspection: Theory and Practice, in: C Meola (Ed) Recent Advances in Non-Destructive Nova Science Publishers. 2010 ISBN: 978-1-61728-082-5,

[20]    Vavilov, V, Kauppinen T, Grinzato E. Thermal characterisation of defects in building envelopes using long square pulse, and slow thermal wave techniques, Research in Nondestructive Evaluation, Springer-Verlag (1997) vol.9 No 4; 181-200.

[21]    Grinzato E, Vavilov V, Bison PG, Marinetti S, Bressan C. Methodology of processing experimental data in Transient Thermal NDT, SPIE 2473, Orlando (USA), 1995; 167-78.

[22] Neubauer U, Rostasy FS. Bond failure of concrete fiber reinforced polymer plates at inclined cracks – experiments and fracture mechanics model, in: Proceedings of 4th International Symposium on Fiber Reinforced Polymer Reinforcement for Reinforced Concrete Structures, SP-188, American Concrete Institute, Farmington Hills (MI), USA, 1999, 369-82.

[23] Grinzato E, Trentin R, Bison PG, Marinetti S. Control of CFRP strengthening applied to civil structures by IR thermography, Thermosense XXIX°, SPIE 6541, Orlando (USA), 2007.

[24] Valluzzi MR, Grinzato E, Pellegrino C, Modena C. IR thermography for interface analysis of FRP laminates externally bonded to RC beams, Materials and Structures RILEM DOI 10.1617/s11527-008-9364-z, 2008.

[25] Brown JR, Hamilton HR. NDE of Fiber-Reinforced Polymer Composites bonded to concrete using IR thermography, Spie 5405, Thermosense XXVI, 2004; 414-24.

[26] On-site investigation techniques for the structural evaluation of historic masonry buildings, www.onsiteformasonry.bam.de, European Research Contract No. EVK4-CT-2001-00060.

[27] Binda L, Saisi A, Tiraboschi A. Investigation procedures for the diagnosis of historic masonries, Construction and Building Materials 2000; 14(4): 199-233.

[28] Grinzato E, Cadelano G, Bison P, Petracca A. Seismic risk evaluation aided by IR thermography Thermosense XXXI°, (2009) SPIE 7299.

[29] Sakagami T, Izumi Y, Kubo S. Application of infrared thermography to structural integrity evaluation of steel bridges. J Mod Opt 2010; 57 (18): 1738–46.

[30] Tavukçuoğlu A, Akevren S, Grinzato E. *In situ* examination of structural cracks at historic masonry structures by quantitative infrared thermography and ultrasonic testing. J Mod Opt 2010; 57 (18): 1779–89.

[31] Eckert EG, Drake RM. Analysis of Heat and Mass Transfer, McGraw Book Company, NY, USA, 1972.

[32] Grinzato E, Vavilov V, Kauppinen T. Quantitative infrared thermography in buildings. Energy and Buildings Elsevier Science 29;1-9

[33] Snell J. The thermal behavior and signatures of water in buildings, Thermosense XXIX SPIE, 6541:654109.1–654109.9, Orlando (USA) April 9-13, 2007.

[34] Grinzato E, Ludwig N, Cadelano G, Bertucci M, Gargano M Bison P. Infrared thermography for moisture detection; laboratory study and on site test. Material Evaluation 2011; 69 (1):97-110.

[35] Grinzato E, Peron F, Strada M. Moisture monitoring of historical buildings by long period temperature measurements, Thermosense XXI° (1999) SPIE 3700; 471-82.

[36] Grinzato E. A method and apparatus for thermal-hygrometric monitoring of wide surfaces by IR Thermography (2006 PCT/EP2007/054713 and PD2006A000191).

[37] Massari G, Massari M, "Damp buildings, old and new, Rome: ICCROM 7–12, 1993.

[38] Grinzato E, Bison PG, Marinetti S, Vavilov V. Thermal/Infrared nondestructive evaluation of moisture content in building: theory and experiment, in Proceedings of Dealing With Defects In Building, Varenna (Italy), Eds. M. Moroni and P. Sartori, September 1994, Part 1, 1994; 345-56.

[39] Ludwig N, Rosina E. Restoration mortars at IRT: optical and hygroscopic properties of surfaces, Proceeding of 8th QIRT, 2006, http://qirt.gel.ulaval.ca

[40] Grinzato E, Cadelano G, Bison P. Moisture map by IR thermography. Journal of Modern Optics, Taylor & Francis Volume 57, Issue 18 October 2010; 1770–78.

[41] Grinzato E. Humidity and air temperature measurement by quantitative IR thermography. Quantitative InfraRed Thermography J (QIRT) 7/1 2010; 55-72.

[42] Camuffo D. Microclimate for cultural heritage. Elsevier, 1998.

[43] Tavukçuoğlu, Grinzato E. Determination of critical moisture content in porous materials by IR thermography. Quantitative InfraRed Thermography Journal (QIRT) 3, 2006; 231-45.

[44] Grinzato E, Peron F, Gava M. The environmental and anthropogenic hazards of monuments Visualization procedure by IR thermography and numerical modelling, 8th International Symposium on the Conservation of Monuments in the Mediterranean Basin, Patras (Grece) 2010.

[45] Grinzato E, Bressan C, Peron F, Romagnoni P, Stevan AG. Indoor climatic conditions of ancient buildings by numerical simulation and thermographic measurements. Thermosense XXII° (2000) SPIE 4020; 314-23

[46] Astarita T, Cardone G, Carlomagno GM. Infrared thermography: An optical method in heat transfer and fluid flow visualization. Optics and Lasers in Engineering 2006, 44(3-4): 261-81.

[47] ASHRAE, handbook of fundamentals, http://www.ashrae.org/publications.

[48] Altmayer EF, Gadgil AJ, Bauman FS, Kammerud RC. 1983, Correlations for convective heat transfer from room surfaces. ASHRAE Transactions 1983, 89: 61-77

[49]    Grinzato E, Fornasieri E, Turra R. Contact thermal resistance measurement on finned tubes heat exchangers, by IR thermography., 2$^{nd}$ European Thermal Science and 14$^{th}$ UIT Conference, Rome (Italy), May 1996; 1051-5.

[50]    Cramer KE, Jacobstein R, Reilly T. Boiler tube corrosion characterization with a scanning thermal line, SPIE 4360, Thermosense XXIII (2001); 594-605.

[51]    Grinzato E, Vavilov V, Bison PG, Marinetti S. Hidden corrosion detection in thick metallic components by transient IR thermography. Infrared Physics & Technology Elsevier vol. 49, 2007; 237-48.

[52]    EN 1264: 2008, Water based surface embedded heating and cooling systems.

[53]    Carslaw HS, Jaeger JC, Conduction of heat in solids, Oxford University Press, 1959.

[54]    Bison PG, Grinzato E. Fast estimate of solid materials thermal properties by IR thermography. Quantitative InfraRed Thermography J (QIRT) 7/1-2010;17-34

[55]    Almond, DP, Patel PM. Photothermal Science and Techniques. Chapman and Hall (1996).

[56]    Cernuschi F, Bison PG, Figari A, Marinetti S, Grinzato E. Thermal diffusivity measurements by photothermal and thermographic techniques. International Journal of Thermophysics 25/2; 439-57.

[57]    Parker WJ, Jenkins RJ, Butler CP, Abbott GL. Flash Method of Determining Thermal Diffusivity, Heat Capacity and Thermal Conductivity. J Appl Phys 1961; 32(9): 1679-84.

[58]    ASTM E1461-01, Standard Test Method for Thermal Diffusivity of Solids by the Flash Method.

[59]    Bison PG, Marinetti S, Mazzoldi A, Grinzato E, Bressan C. Cross comparison of thermal diffusivity measurements by thermal methods. J Infrared Phy Technol 2002; 43: 127-32.

[60]    Grinzato E, Bison PG, Marinetti S. Monitoring of ancient buildings by the thermal method. J Cult Herit 2002; 3: 21-9.

[61]    Baggio P, Bison PG, Bonacina C, Bressan C, Grinzato E. Thermal inertia evaluation in porous material by multispectral optical analysis, Proc. of XVIII International Congress of Refrigeration. August 10-17 (1992) Montreal, Quebec, Canada.

# Appendix A

## RELEVANT STANDARDS

ISO 13790: annual energy use for heating and cooling of the building

ISO 9869: Building elements *in situ* measurement of thermal resistance and thermal transmittance

ISO 6946: *In situ* measurement of thermal resistance and thermal transmittance-calculation method

ISO 13790: 2008 Calculation methods for assessment of the annual energy use for space heating and cooling

ISO 6781:1983 Thermal insulation-Qualitative detection of thermal irregularities in building envelopes-Infrared method

EN 13187 Thermal performance of buildings. Qualitative detection of thermal irregularities in building envelopes. Infrared method.

ASTM C1046-95 (2001) Standard Practice for *in situ* Measurement of Heat Flux and Temperature on Building Envelope Components

ANSI ASHRAE 105-1984. Standard Methods of Measuring and Expressing Building Energy Performance

ASNT ASTM (American Society for Testing and Materials) C 1153-97 (2003): Location of Wet Insulation in Roofing Systems,

ASTM C1060-90 (2003): Thermographic Inspection of Insulation Installations in Envelope Cavities of Frame Buildings.

ISO 18434, Condition monitoring and diagnostics of machines-Thermography

# Concluding Remarks to Part II – Section 4 and Final Closure

Section 4 of Part II, or better Chapter 9, was devoted to the use of infrared thermography in civil engineering and architecture. Indeed, this is a field for which the use of an infrared imaging device is particularly useful. In fact, it is possible to take, in a click, a temperature map of the entire façade of a building; and then, it is possible to derive performance indicators from the temperature pattern.

This chapter went through both classical and very innovative techniques, which are now available for dealing with usual or more complex problems. In particular, it is shown as using a novel method, it is possible to "see" the environmental main quantities as air temperature, relative humidity and velocity, obtained from thermographic readings.

It is worth noting that, from the temperature map, one can:

- Ascertain the envelope tightness for energy saving purposes;

- Discover water infiltration;

- Detect presence of anomalies (cracks, voids, *etc.*).

- An infrared camera can be used to control the conservation state of civil edifices and antique structures as well of works of art.

As a final remark, several examples of application of infrared thermography in many fields spanning from architecture to fluid-dynamics to foodstuff conservation to medicine were presented and discussed through the different chapters of this eBook. However, this eBook is far away the presumption of being exhaustive since there are many other application fields not included herein. Amongst them, it is worth mention the aerial thermography which allows for a broad scene on the territory and helps dealing with specific questions like discovering pollution sources in rivers, lakes and streams, finding steam leaks and water leaks in district heating and cooling systems, counting wild animals, controlling high voltage transmission lines, defining the extent of landfill fires etc. Many other applications are possible, it is important to take into account that a great result derives from the combination of two factors. One is a correct choice of the most adequate instrument since today there is availability of a multitude of cameras for any needs. The other and more important factor is linked to the operator skill; but now, training is a common practice and there are training centers and schools for preparation and qualification of the personnel around the world which grow together with the dissemination of infrared devices.

# Subject Index

## A

Absolute zero, 3
Absorptance, 6-7
Absorption, 7-10, 15-21
absorption coefficient, 7, 9
accuracy, 39, 42, 48, 50, 62, 64, 68, 105, 121, 131, 135, 202, 208
acquisition frequency, 139
active heating, 137
adhesion, 220, 222
adiabatic effectiveness, 136
adiabatic wall temperature, 135,136, 149, 151, 158, 159
aerodynamic, 135, 153
aerospace, 139, 199, 220
air, iii, 17, 25, 51, 64, 89-90, 92-93, 103-106, 109, 112-127, 136, 149, 154, 156, 158, 161, 167, 169-170, 175, 178, 200-202, 205, 207-222, 231
aircraft, iii, 29-30, 149, 187-188, 190-191
air flow, 103, 106, 109, 113-114, 117-126, 169, 175
airfoil, 153, 154
algorithm, 120, 139, 209, 226
alloy, 23
amorphous, 20, 57
angular speed, 158-160
animal welfare, ii, 85, 92, 93, 102
arsenide, 20-21, 57, 166
atmosphere, 6, 17-18-19, 51, 165, 168
atmospheric pressure, 170
atmospheric temperature, 17
atmospheric transmittance, 17-18
atoms, 3, 8
azimuthally, 158

## B

Background radiation, 22, 71, 91
Band (infrared), 4, 8, 10-11, 16-19, 23, 24, 49, 52, 63, 134, 136, 166, 168, 202, 212, 225
Bandgap, 8, 20-21, 24, 166
Bandwidth, 19, 21
Barrier, 20, 24
Battery, 35, 38, 44, 46, 48
Beams, 209-210
Beam splitter, 167
Bending, 20, 209-210
Bias, 20, 192, 195
Biological, 77, 92, 105, 113, 211
Biot number, 137-138, 140, 156, 225
Blackbody, 4, 11-13, 108, 110-111, 168, 181
Blood, 62, 66-67, 69, 71-74, 85, 87-88, 90, 92, 96
Body temperature, ii, 63, 73-74, 87-88, 92, 104
Bolometer, 20, 22, 32, 44, 46, 48, 50, 52-54, 56-57, 107, 182, 194, 196-197
Boltzmann's constant, 4, 205

www.ingramcontent.com/pod-product-compliance
Lightning Source LLC
Chambersburg PA
CBHW050827220326

41598CB00006B/326